THE SCIENCE OF HEALTH & FITNESS

Kendall Hunt
publishing company

KIRSTIN BREKKEN SHEA | GAYDEN DARNELL | DOTTIEDEE AGNOR | BETH NETHERLAND

Cover image © Shutterstock, Inc.

www.kendallhunt.com
Send all inquiries to:
4050 Westmark Drive
Dubuque, IA 52004-1840

Copyright © 2014 by Kendall Hunt Publishing Company

ISBN 978-1-4652-4939-5

All rights reserved. No part of this publication may be reproduced, stored in a retrieval system, or transmitted, in any form or by any means, electronic, mechanical, photocopying, recording, or otherwise, without the prior written permission of the copyright owner.

Printed in the United States of America

Brief Contents

CHAPTER 1
Wellness and Longevity

CHAPTER 2
Stress and Performance

CHAPTER 3
Aging and Disease

CHAPTER 4
Complementary and Alternative Medicine

CHAPTER 5
Exercise Science and Sports Medicine Topics

CHAPTER 6
Nutrition and Metabolism

CHAPTER 7
Scientific Principles of Weight Management

CHAPTER 8
The Basics of Neuroscience and Drug Addiction

CHAPTER 9
Psychoactive Drug Types

CHAPTER 10
The Reproductive System

CHAPTER 11
Fetal Development

Contents

CHAPTER 1
Wellness and Longevity   1

    Distracted Driving   *8*
    Drowsy Driving   *9*

    The Facts   *10*
    The Cause   *10*
    Safety Precautions   *10*

    Precontemplation   *12*
    Contemplation   *12*
    Preparation   *12*
    Action   *12*
    Maintenance   *13*

Summary   17
References   17
Wellness Goal   18
Activities   21
    Homework Experience   *21*
    Journal—Goal Setting   *21*

CHAPTER 2
Stress and Performance   23

    Neurotransmitters control   *29*
    How Stress Affects Neurotransmitters   *29*
    Main Causes of Neurotransmitter Deficiencies   *30*
    How to Treat Neurotransmitter Imbalance   *30*
    Alternatives to Medication   *30*

    Results of Endorphin Secretion   *31*

    Simple Sugars   *32*

Stress Awareness Log   37

CHAPTER 3
Aging and Disease 41

Who Is At Risk for CVD? 51
CVD Prevention 52
Arteriosclerosis 53

References 75
Contacts 77
Recommended Reading 77
Activities 77
Notebook Activities 77
NOTEBOOK ACTIVITY 79
Healthy Back Test 79
NOTEBOOK ACTIVITY 81
Is Osteoporosis in Your Future? 81

CHAPTER 4
Complementary and Alternative Medicine 83

References 102
Contacts 103
Recommended Books For Further Reading 104

CHAPTER 5
Exercise Science and Sports Medicine Topics 105

Measuring Health Risk 109
Fitness or Fatness 110

References   139
Contacts   140
Activities   141
   Notebook Activities   141
NOTEBOOK ACTIVITY   143
   Calculating Your Activity Index   143
   Karvonen Formula   145
   Developing an Exercise Program for Cardiorespiratory Endurance   147
   Assessing Your Current Level of Muscular Endurance   149
   Check Your Physical Activity and Heart Disease I.Q.   151
   Answers to the Check Your Physical Activity and Heart Disease I.Q. Quiz   152
   Assessing Cardiovascular Fitness: Cooper's 1.5-Mile Run   153

CHAPTER 6
Nutrition and Metabolism   155

Summary   172
KEY TERMS   173
TEST YOURSELF   174
THOUGHT QUESTIONS   175
MEDICAL QUESTIONS   175

CHAPTER 7
Scientific Principles of Weight Management   177

References   212
Recommended Reading   213
Activities   213
   Activities   213

Homework Experience 215
   Body Mass Index Calculator 215
   Facts about My Favorite Fast-Food Meal 217
WRITING PROMPTS 219

CHAPTER 8
The Basics of Neuroscience and Drug
   Addicition 221

   Intravenous injection (IV) 223
   Other methods of injection 223

   Transdermal Patch 224
   Intranasal 224

References 231
IN-CLASS QUIZ DRUGS: AND YOUR BODY
   CHAPTER 8 233

CHAPTER 9
Psychoactive Drug Types 235

   Drinking Problems 240
   Alcoholism 240
   Chronic Effects 242

   Drinking and Driving 244
   Alcohol Use in College 245
   Binge Drinking 246
   Alcohol Poisoning 246

   Hydrocodone 277
   Codeine 277
   Morphine 277
   Oxycodone 277

References   283
IN-CLASS ACTIVITY   287
    The Physical Effects of Smoking   287
NOTEBOOK ACTIVITY   289
    "Why Do You Smoke?" Test   289
    Scoring Your Test   290
    "Do You Want to Quit?" Test   291
    Scoring Your Test   291
    Alcohol Screening Self-Assessment   293
    Making Changes   295

CHAPTER 10

The Reproductive System   297

    Duct System   301
    Epididymis   302
    Ductus Deferens   304
    Ejaculatory Duct   305
    Urethra   305

    Seminal Vesicles   306
    Prostate Gland   306
    Bulbourethral Glands   307

    Testicular Descent   309

    Males   309
    Secondary Sex Characteristics   311
    Menopause   311
    Meiosis and Mitosis   311
    Spermatogenesis   313
    Other Testosterone Effects   314

    Initiation   315
    Erection   315
    Emission   315
    Ejaculation   316

    Prostate Cancer   317
    Hormonal Imbalances   317
    Causes of Male Infertility   318

    Common Uterine Disorders   322

    Embryonic Origin of Reproductive Tissues   325
    Puberty   325
    Secondary Sex Characteristics   326
    Menopause (Climacteric)   326

    Female Sex Hormones   330

    Summary of Hormone Interactions and the
        Female Reproductive Cycle   333

    Disorders of the Female Reproductive
        System   334
    Breast Cancer   335
    Causes of Female Infertility   336

    Bacterial Diseases   337
    Parasites   339
    Viral Diseases   339

    Contraceptive Methods   340
    Behavioral Methods   340

chapter review   344

Answers to end-of-chapter questions 347
Thought Questions 348

CHAPTER 11
Fetal Development 349

Stage 1, Fertilization, Day 0 351
Polyspermy 352
Multiple Births 354
Stage 2, Zygote Formation, Days 1 to 4 355
Stage 3, Blastocyst Formation, Days 4 to 5 356
Stage 4, Implantation, Human Chorionic
    Gonadotropin Secretion, Days 5 to 6 356
Stage 5, Implantation Complete, Placental
    Circulation Begins, Days 7 to 12 357
Stages 6a and 6b, Gastrulation,
    Chorionic Villi Formation, Day 13 359

Neurulation and Notochord Formation,
    Days 14 to 16 359
Formation of the Notochord, Days 17 to 19 360
Appearance of Somites, Days 19 to 21 361
Fusion of the Neural Fold, Days 21 to 23 361
Days 23 to 25 361
Days 25 to 27 361
Days 27 to 29 361
Days 29 to 33 363
Days 33 to 38 363
Days 38 to 41 363
Days 41 to 43 364
Days 44 to 46 364
Days 47 to 48 364
Days 49 to 51 365
Days 52 to 53 365
Days 53 to 55 365
Days 56 to 57 365
In Review… 365

Fetal Monitoring 367

Week 12 to 13 368
Week 14 to 15 369
Week 16 to 17 369
Week 18 to 19 369
Week 20 to 21 370
Week 22 to 23 370
Week 24 to 25 370
Week 26 to 27 370
Week 28 to 29 370
Week 30 to 31 370
Week 32 to 33 370
Week 34 to 35 371
Week 36 to 37 371
Week 38 to 39 371
Week 40 371

Amniotic Sac 374
Placenta 374
Yolk Sac 375
Allantois 375
Gut and Body Cavities 375
The Face 376
Skin 376
Skeleton 376
Muscle 376
Nervous System 376
Respiratory System 377
Urinary System 377
Cardiovascular System 378
Reproductive System 379
In Review… 380

In Review… 384

Post Parturition 387

Initiation of Breathing 387
Cardiopulmonary Changes 387
Thermoregulatory Changes 388
Digestive Changes 388
Integumentary Changes 389
Immunological Changes 389
Apgar Scores 389

In Review… 391

Life Stages 392
Aging 392
In Review… 395

Linkage Analysis 397
Dominant and Recessive Genes 397
Sex-Linked Traits 399
Sex Limited Traits 399
Sex Controlled Traits 400
Sex Imprinting 400
Polygenic Traits 400
Incomplete Dominance 400
Codominance 400
Multiple Allele Series 401
Modifying and Regulating Alleles 401
Incomplete Penetrance 402
Late Onset 402
Epigenetics 402
Pleiotropy 402
Environmental Influences 403

Genetic Counseling 404

In Review… 405

Pregnancy Problems 405
Ectopic Pregnancy 405
Placenta Previa 406
Preeclampsia 406
Dystocia and Cesarean Section 406
Spontaneous Abortions 406
Premature Infants 407

Clinical Terms 407
Chapter Review 407

Answers to end-of-chapter questions 411
Thought Questions 412
Index 413

Chapter 1
Wellness and Longevity

OBJECTIVES

Students will be able to:
- Differentiate the definitions of health and wellness.
- Identify the seven dimensions of wellness.
- Explain the link between preventative behaviors and wellness.
- Discuss health behaviors that increase the quality and longevity of life.
- Identify the significance of *Healthy People 2020*.
- Identify the five stages of change.
- Identify key elements for a successful behavior change.

Adaptation of *Health and Fitness: A Guide to a Healthly Lifestyle, 5/e* by Laura Bounds, Gayden Darnell, Kirstin Brekken Shea and Dottiede Agnor. Copyright © 2012 by Kendall Hunt Publishing Company.

> "Take care of your body with steadfast fidelity. The soul must see through these eyes alone, and if they are dim, the whole world is clouded."
>
> —Johann Wolfgang Von Goethe

Health is a universal trait. The World Health Organization defines **health** as a "state of complete physical, mental, and social well-being and not merely the absence of disease or infirmity." Webster's Dictionary offers "the condition of being sound in body, mind, or spirit; especially: freedom from physical disease or pain . . . the general condition of the body" as a definition of health. However, health also has an individual quality; it is very personal, and unique.

Early on, definitions of health revolved around issues of sanitation and personal hygiene. Today, the definition of health has evolved from a basis of physical health or absence of disease, to a term that encompasses the emotional, mental, social, spiritual, and physical dimensions of an individual. This current, positive approach to health is referred to as wellness. **Wellness** is a process of making informed choices that will lead one, over a period of time, to a healthy lifestyle that should result in a sense of well-being.

Dimensions of Wellness

Wellness is a holistic approach to life that integrates mind, body, & spirit. Wellness emphasizes an individual's potential and responsibility for his or her own health. It is a process in which a person is constantly moving either away from or toward a most favorable level of health. Wellness results from the adoption of low-risk, health-enhancing behaviors. The adoption of a wellness lifestyle requires focusing on choices that will enhance the individual's potential to lead a productive, meaningful, and satisfying life.

It is the complex interaction of each of the seven dimensions of wellness that will lead an individual, over time, to a higher quality of life and better overall health and well-being. Constant, ongoing assessment of one's behaviors in the following dimensions is key to living a balanced life. In addition to the seven dimensions shown in Figure 1.1, there is discussion of other factors that influence wellness.

Emotional Wellness

Are you engaged in the process of Emotional Wellness?

Evaluate your own emotional wellness with this brief quiz.

- Am I able to maintain a balance of work, family, friends, and other obligations?
- Do I have ways to reduce stress in my life?
- Am I able to make decisions with a minimum of stress and worry?
- Am I able to set priorities?

If you answered "No" to any of the questions, it may indicate an area where you need to improve the state of your emotional wellness.

FIGURE 1.1

Seven Dimensions of Wellness

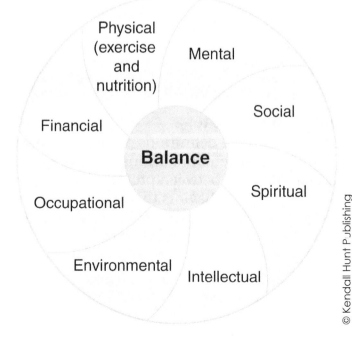

Emotional

Emotional wellness is being aware and accepting of not only your feelings and emotions but also the feelings and emotions of others. An individual who is emotionally healthy is able to enjoy life despite unexpected challenges and potential problems. Effectively coping with life's difficulties and unexpected events is essential to maintaining good health. Equally important to good personal wellness is the ability to understand your feelings and express those feelings or emotions outwardly in a positive and constructive manner. "Bottled-up" negative emotions can affect the immune system and result in chronic stress, which in turn can lead to serious illnesses such as high blood pressure and can potentially lead to a premature death.

The result of being emotionally well adjusted and being able to express emotions appropriately are healthy, mutually rewarding relationships with others and an overall enjoyment of life.

Intellectual

The mind can have substantial influence over the body. To be intellectually healthy, it is essential to continue to explore new avenues and interests and to regularly engage in new and ongoing learning opportunities and experiences. An individual must be open to new and different ideas & concepts. The more "unknowns" an individual faces or explores, the more opportunities he or she has to learn and grow intellectually.

Reading for pleasure, taking a dance class, learning a foreign language, traveling and learning about other cultures are all excellent ways to improve or maintain intellectual wellness. Truly well individuals are able to combine what they have learned through their formal education with what they experience outside of a traditional classroom.

Intellectual Wellness

Are you engaged in the process of Intellectual Wellness?

Evaluate your own intellectual wellness with this brief quiz.

- Am I open to new ideas?
- Do I seek personal growth by learning new skills?
- Do I search for lifelong learning opportunities and stimulating mental activities?
- Do I look for ways to use creativity?

If you answered "No" to any of the questions, it may indicate an area where you need to improve the state of your intellectual wellness.

Social

Social health is an individual's ability to relate to and interact with others. Socially healthy people are able to communicate and interact with the other people they come in contact with each day. They are respectful and caring of their family, friends, neighbors, and associates. A key component to being socially well is thinking about how one's actions impact, either positively or negatively, other people. For example, choosing to text while driving could impact the person texting by being killed in a car accident. Obviously their decision impacts them but, did they ever stop to think how their decision to text and drive would forever impact their family and friends or the police and paramedics who respond to the scene? Being socially well as an adult is more than being able to do well on a presentation, going out, or having friends. Although reaching out and communicating with others may be difficult or uncomfortable initially, it is extremely important to a person's social health and their overall sense of well-being.

Spiritual

Spiritual health helps a person achieve a sense of inner peace, satisfaction, and confidence it comes from within and can help give the sense that all is right with the world. A person's ethics, values, beliefs, and morals can contribute to their spiritual health. It can teach one to appreciate their own ethics, beliefs, morals and values and live a life true to them but at the same time it encourages a person to be tolerant of differences in other's ethics, values, morals and beliefs. Good spiritual health can help give life meaning and purpose.

The ability to relate and interact with others is important to a person's overall sense of well-being.

> **Social Wellness**
> Evaluate your own social wellness with this brief quiz.
>
> - Do I plan time to be with my family and friends?
> - Do I enjoy the time I spend with others?
> - Are my relationships with others positive and rewarding?
> - Do I explore diversity by interacting with people of other cultures, backgrounds, and beliefs?
>
> If you answered "No" to any of the questions, it may indicate an area where you need to improve the state of your social wellness.

Physical

Physical wellness is maintaining a healthy body through regular exercise, good nutrition, and the avoidance of harmful habits. Ensuring good physical health begins with devoting attention and time to attaining healthy levels of cardiovascular fitness, muscular strength and endurance, flexibility, and body composition. When coupled with good nutritional practices such as consuming foods and beverages that are known to enhance good health rather than those that impair it, good sleep habits, and the avoidance of risky social behaviors such as drinking and driving or unprotected sexual intercourse, a physically healthy body results. This is the component that is most often associated, at first glance, with a person's health.

Occupational

An occupationally well individual is able to enjoy the career they have chosen and the way they contribute to society. They have chosen a path that is consistent with their own values, interests, and beliefs. Attaining occupational wellness begins with determining what roles, activities, and commitments take up a majority of an individual's time. These roles, activities, or commitments could include but are not limited to being a student, parenting, volunteering in an organization, or working at a part-time job while pursuing one's degree. It is when each of these areas are integrated and balanced in a personally and professionally fulfilling way that occupational wellness occurs.

Environmental

An individual's health and wellness can be substantially affected by the quality of their environment. An environmentally well individual recognizes that they are dependent on their natural environment just as the environment is dependent on them. They are aware that they have a responsibility toward the upkeep of environmental quality and have taken the time to identify ways in which they can be more environmentally "friendly." Access to clean air, nutritious food, sanitary water, and adequate clothing and shelter are essential components to being well. An individual's environment should, at the very least, be clean and safe.

Through wellness, an individual manages a wide range of lifestyle choices. How a person chooses to behave and the decisions he or she makes in each of the seven dimensions of wellness will determine their overall quality of life. Making an active effort to combining and constantly trying to balance each of the seven dimensions is key to a long and fulfilling life.

Financial Wellness

There are many different wellness models and most include multidimensional elements. One element which has typically not been included is financial wellness. Financial wellness has an impact on an individual and society as a whole. The first step to gaining financial wellness is financial responsibility. There are numerous ways to be financially responsible, some of which include:

- Have a monthly budget and do not overspend
- Wait for items to go on sale
- Use coupons
- Avoid credit card debt (pay the balance every month)
- Use credit cards only in emergency situations

Financial wellness has an impact on an individual and society as a whole.

ELEMENTS OF WELLNESS

ELEMENT	SYMBOL	SYMBOL EXPLAINED	EXAMPLES
EMOTIONAL WELLNESS • The awareness and acceptance of feelings and emotions.		• The heart represents the general source of emotions (vs. the mind/brain as a symbol for rational thought).	• Fitness and exercise • Relationships with friends/family • Balancing work and family • Laughing and crying • Adequate sleeping patterns • Personal contact, ie. hugging
ENVIRONMENTAL WELLNESS • The recognition of interdependence with nature.		• The tree is a general, easy-to-identify representative of the natural realm.	• Reduce, reuse, recycle • Reusing materiels • Adopting Leave-No-Trace • Conserving water and fuels • Spend time in a state/national park • Finding value in surroundings • Positive workplace & attitude • Air quality
INTELLECTUAL WELLNESS • The openness to new concepts and ideas.		• A light bulb is a general representative for fresh thinking, innovation, and creativity.	• Reading & learning for fun • Participation in class, organization • Adopting a new hobby • Traveling • Adequate sleeping patterns • Self-help information
OCCUPATIONAL WELLNESS • The ability to enjoy a chosen career and/or contribute to society through volunteer activities.		• A gear with cogs or teeth represents a unit of labor or effort.	• Continuing education • Satisfying career/profession • Volunteering • Workplace safety • Exercise
PHYSICAL WELLNESS • The maintenance of a healthy body through good nutrition, regular exercise, and avoidance of harmful habits.		• An individual person represents the human body.	• Fitness and exercise; stretching • Personal hygiene • Walk/cycle to work • Know your numbers: cholesterol, blood sugar, blood pressure • Good nutrition • Adequate sleeping patterns • Regular medical/dental exams
SOCIAL WELLNESS • The ability to perform social roles effectively, comfortably, and without harming others.		• Three persons with connected hands represent harmony, networking, and friendship.	• Establish and maintain personal friendships • Community involvement • Attending social settings, ie. festivals, neighborhood events • Group fitness classes • Hobby/activity organizations
SPIRITUAL WELLNESS • The meaning and purpose of human existence.		• The sun represents the beginning and end of a day; a reflection of the source of growth and vitality in the universe. • Solar symbols can have meaning in astrology, religion, mythology, mysticism, and divination.	• Meditation; prayer • Religious affiliation • Explore and enjoy the flora & fauna of a wilderness area. • Watch a sunrise or sunset • Exercise • Freedom • Outdoor activities

Copyright © University of Nebraska-Lincoln. Reprinted by permission.

> **Spiritual Wellness**
>
> Are you engaged in the process of Spiritual Wellness? Evaluate your own spiritual wellness with this brief quiz.
>
> - Do I make time for relaxation in my day?
> - Do I make time for meditation and/or prayer?
> - Do my values guide my decisions and actions?
> - Am I accepting of the views of others?
>
> If you answered "No" to any of the questions, it may indicate an area where you need to improve the state of your spiritual wellness.

> **Physical Wellness**
>
> Are you engaged in the process of Physical Wellness? Evaluate your own physical wellness with this brief quiz.
>
> - Do I know important health numbers, like my cholesterol, weight, blood pressure, and blood sugar levels?
> - Do I get annual physical exams?
> - Do I avoid using tobacco products?
> - Do I get sufficient amount of sleep?
> - Do I have an established exercise routine?
>
> If you answered "No" to any of the questions, it may indicate an area where you need to improve the state of your physical wellness.

- Save at least 5 percent of net income in case of an emergency
- Do not get a car loan for more than five years (three- or four-year notes are even better)
- Always pay your bills on time
- Shop and trade at resale stores
- Pay off a mortgage early
- Check out books from a library instead of purchasing them from a bookstore
- Carpool
- Ride your bike
- Eat at home
- Go for a hike instead of going to a movie

Factors That Influence Health and Wellness

In addition to the dimensions of wellness, the factors shown in Figure 1.2 also influence health and wellness, as well as physical fitness. You will see that lifestyle is only one component that works in tandem with other factors to make up good health and wellness.

Leading Causes of Death

Heart disease: 597, 689
Cancer: 574,743
Chronic lower respiratory disease: 138,080
Stroke (cerebrovascular disease): 129,479
Accidents (unintentional injuries): 120,859
Alzheimer's disease: 83,494
Diabetes: 69,071
Nephritis, nephritic syndrome, and nephrosis: 50,476
Influenza and pneumonia: 50,097
Intentional self-harm (suicide): 38,364

Addressing each dimension of wellness contributes to a well-rounded individual, as well as one's longevity. However, when reviewing the leading causes of death, we must acknowledge that unintentional injuries are high on the list. If we look specifically at people from 1 to 44 years of age, unintentional injury and violence account for over 50% of deaths each year, more than non-communicable and infectious diseases combined (National Center for Injury Prevention and Control, 2014). Therefore, in order to improve our chances to live a health and long life we must consider ways to reduce our risk of unintentional injuries.

Injury: The Leading Cause of Death Among Persons 1–44

In 2010 in the United States, injuries, including all causes of unintentional and violence-related injuries combined, accounted for 50.6% of all deaths among persons ages 1–44 years of age—that is more deaths than non-communicable diseases and infectious diseases combined.

Injury Facts

- More than 180,000 deaths from injury each year—1 person every 3 minutes
- Leading cause of death for people ages 1–44 in the U.S.
- An estimated 2.8 million people hospitalized with injury each year

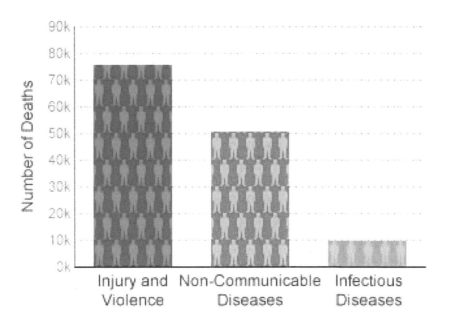

Injury Deaths Compared to Other Leading Causes of Death for Persons Ages 1–44, United States, 2010*

- An estimated 31.7 million people treated in Emergency Department for injury each year
- Violence and injuries cost more than $406 billion in medical care and lost productivity each year

In general, there are a few common characteristics of an accident victim: young, male, prone to experimentation, more likely to take unnecessary risks and with poor impulse control (Thygerson, 1992). Please consider that all of these characteristics need not be met in order to be at risk of accidents and just because you are not in any of these groups does not mean you are not at risk. According to the research team of Frances E. Jensen and David K. Urion, of Children's Hospital Boston and Harvard Medical School, not only do adolescents have a slower to develop prefrontal cortex (responsible for controlling impulses and emotion and helping us understand cause and effect, right and wrong), but they are also often sleep deprived, have little exercise/activity on a regular basis and are constantly trying to multitask (Ruder, 2008). Does any of this sound familiar? Each of these has a negative effect on decision-making when it comes to accidents and violence. Consider a study conducted at the McLean Hospital Brain Imaging Center by Debroah Yurgelun-Todd and colleagues. They used functional magnetic resonance imaging to compare the activity of teen and adult brains. Because the prefrontal cortex is the last region of the brain to fully develop, they found that while adults could often use rational decision-making skills, teens lacked the ability to do the same. Similarly, Jay Giedd and colleagues at the National Institutes of Mental Health used brain imaging to measure the structure, as opposed to activity, of adolescent brains when compared to adults, finding that development of the frontal lobe continues well into one's early twenties (Talukder, 2013).

*Note: Injury includes unintentional injury, homicide, suicide, legal intervention, and those of undetermined intent. Non-communicable diseases include cancer, cardiovascular, kidney, respiratory, liver, diabetes, and other diseases. Infectious diseases include HIV, influenza, pneumonia, tuberculosis, and other infectious diseases. Data Source: National Vital Statistics System using CDC Wonder (http://wonder.cdc.gov).

Occupational Wellness
Are you engaged in the process of Occupational Wellness? Evaluate your own occupational wellness with this brief quiz.

- Do I enjoy going to work most days?
- Do I have a manageable workload at work?
- Do I feel that I can talk to my boss and co-workers with problems arise?

If you answered "No" to any of the questions, it may indicate an area where you need to improve the state of your occupational wellness.

Environmental Wellness
Are you engaged in the process of Environmental Wellness? Evaluate your own environmental wellness with this brief quiz.

- Do I recycle?
- If I see a safety hazard, do I take the steps to fix the problem?
- Do I volunteer time to worthy causes?
- Am I aware of my surroundings at all times?

If you answered "No" to any of the questions, it may indicate an area where you need to improve the state of your environmental wellness.

The leading cause of unintentional death is motor vehicle crashes. In 2009, over 35,900 individuals died in motor vehicle crashes in the United States. The 15–24 year-old age group had nearly 8,000 deaths for that year (NSC, 2011).

One of the leading factors in motor vehicle crashes is driver inattention. According to the National Highway Traffic Safety Administration (NHTSA), nearly 80 percent of crashes involve some form of driver inattention. This signifies the importance of the driving task and that the need for attention is critical (NHTSA, 2010).

Distracted Driving

Operating a motor vehicle is the single most dangerous activity that we do on a daily basis and yet we feel confident that we can drive and do others things at the same time.

We pride ourselves in our ability to multitask. We have been conditioned to think that we are more productive and successful if able to focus on more than one thing at a time. But can we truly multitask?

John Medina, author of *"Brain Rules"* says, "research shows that we can't multitask. We are biologically incapable of processing attention-rich inputs simultaneously." The brain focuses on ideas and concepts one after another instead of both at the same time. The brain must let go of one activity to go to another, taking several seconds. According to Professor Clifford Nass at Stanford University, the more that you multi-task, the less productive you become.

When we operate a motor vehicle, many things can be considered distractions. Any secondary activity like texting, talking on a cell phone, putting on make-up, eating and drinking, adjusting your music and even your GPS can all cause problems while driving. Taking your eyes off the road for as little as two seconds can be dangerous. We have added more distractions by using our smart phones for Facebook, Twitter and other social media.

There are 3 main types of distraction:

- Visual—taking your eyes off the road. For example, glancing down at your phone or changing songs on your CD player or iPod
- Manual—taking your hands off the wheel. Examples range from hand held cell phone use, texting, eating or changing clothes
- Cognitive—taking your mind off the task. Daydreaming or your current emotional state could play a role in your attention on the driving task.

All three types can be a serious, life-threatening practice.

Statistically, distracted driving has passed drunk driving as the number one safety concern for the driving public. Drivers on cell phones are more impaired than drivers with a .08 BAC, which is considered legally intoxicated in all states.

This is not to say that drinking and driving is not a serious problem, it brings to light, the impact of distracted driving. What about an intoxicated driver on a cell phone or texting? and the odds of being involved in an accident increases greatly.

According to NHTSA research, distraction-related fatalities represented 16 percent of overall traffic fatalities in 2009. Nearly 5,500 people were killed and 448,000 were injured in crashes involving a distracted or inattentive driver (NHTSA, 2009).

We know what distractions are out there, but what can we do to solve this problem? With nearly 5,500 fatalities in 2009 that number will only continue to grow unless we, as the driving public, make some positive changes. Legislation may work to some degree but changing our behavior is crucial in solving this problem.

For more information on Distracted Driving visit there websites: enddd.org—the official U.S. Government Website for Distracted Driving distraction.gov—sponsored by the Department of Transportation (DOT) "Faces of Distracted Driving" caseyfeldmanfoundation.org.

U.S. Department of Transportation Secretary, Ray LaHood, cautions that researchers believe the epidemic of distracted driving is likely far greater than currently known. Police reports in many states still do not document routinely whether distraction was a factor in vehicle crashes, making it more difficult to know the full extent of the problem.

With that in mind, The Department of Transportation has hosted two Distracted Driving Summits in Washington, D.C., the most recent in September of 2010. Many states have considered passing laws prohibiting cell phone use and texting while driving. There are 30 states, and the District of Columbia that have banned text messaging for all drivers. Twelve of these laws were enacted in 2010 alone.

Changing your behavior and encouraging your friends to change may help bring an end to the senseless tragedy of distracted driving.

Drowsy Driving

Fatigue on the road can be a killer. It happens frequently on long trips, especially long night drives. There is no test to determine sleepiness and no laws regarding drowsy driving; therefore, it is difficult to attribute crashes to sleepiness. According to the NHTSA, drowsy driving accounts for approximately 100,000 accidents each year, injuring 71,000 and producing 1,550 fatalities (NHTSA, 2011). In a 2006 poll conducted by the National Sleep Foundation (NSF, 2007) reported driving a vehicle while feeling drowsy during the prior year, with 37 percent reporting that they actually dozed off while driving. It is equally as dangerous if not more dangerous to drive when you are drowsy than intoxicated. Some drivers abstain from alcohol but no one can resist the need to sleep. People are less likely to admit that they are feeling fatigued and therefore continue to drive when drowsy, leaving it up to self-regulation. Results from a recent study by the Stanford Sleep Disorders Clinic, performed by Dr. Nelson Powell, concluded that the sleepy drivers performed the same as the drunk drivers on basically all skills tested.

The NSF has created the "Drive Alert . . . Arrive Alive" campaign to help people become aware of the dangers of drowsy driving. One very important detail pointed out by this campaign is that people fall asleep more often on high-speed, long, boring, and rural highways. The more monotonous the drive, the more likely the driver will suffer some fatigue. According to the NSF, drivers who pose a greater risk for drowsy driving are those who are sleep deprived, drive long distances without breaks, drive through the night, drive alone, or those drivers with undiagnosed sleep disorders. Shift workers also pose a greater threat because they typically have non-traditional work schedules. Young people are more prone to sleep-related crashes because they typically do not get enough sleep, stay up late, and drive at night. NSF has a few warning signs to indicate that a driver may be experiencing fatigue. These include not remembering the last few miles driven, drifting from their lane, hitting rumble strips, yawning repeatedly, having difficulty focusing, and having trouble keeping the head up. The NSF also offers these tips for staying awake while driving:

- Get a good night's sleep.
- Schedule regular stops.
- Drive with a companion.
- Avoid alcohol.
- Avoid medications that may cause drowsiness.

If anti-fatigue measures do not work, of course the best solution is sleep. If no motels are in sight and you are within one to two hours of your destination, pull off the road in a safe area and take a short twenty to thirty-minute nap.

The leading cause of accidental death is motor vehicle accidents.

> Traumatic Brain Injury (TBI) can be caused by a bump, blow, or jolt to the head that disrupts normal brain function. An estimated 1.7 million people sustain a TBI annually in the U.S., killing 52,000. The leading cause of TBI deaths is motor vehicle traffic injuries, with the highest death rate among adults 20–24 years old.

Most drowsy driving crashes involve males between the ages of 16–25 (NSF, 2008). Because of this growing problem, many colleges and universities, are providing awareness programs to try to prevent this tragedy from occurring.

In any kind of motor vehicle crash, a seat belt may save your life! The lap/shoulder safety belts reduce the risk of fatalities to front seat passengers of cars by 45 percent and for trucks by 60 percent. They also reduce the severity of injuries by 50 percent for cars and 65 percent for trucks (NSC, 2008).

Air bags combined with safety belts offer the best protection. There has been an overall 14 percent reduction in fatalities since adding air bags to vehicles (NSC, 2008). Buckle up!

Current Issues Impacting College Campuses:
High risk drinking
Illegal drug use
Prescription drug use
Sexual assault
Stalking
Relationship violence
Hazing
Hate crimes
Fire safety

Source: The Clery Center for Security on Campus http://clerycenter.org/national-campus-safety-awareness-month.

Fire Safety 101
A Factsheet for Colleges & Universities

Every year college and university students experience a growing number of fire-related emergencies. There are several causes for these fires; however most are due to a general lack of knowledge about fire safety and prevention.

The U.S. Fire Administration (USFA) offers these tips to help reduce and prevent the loss of life and property in dormitory and university housing fires.

The Facts

In cases where fire fatalities occurred on college campuses, alcohol was a factor. There is a strong link between alcohol and fire deaths. In more than 50% of adult fire fatalities, victims were under the influence at the time of the fire. Alcohol abuse often impairs judgment and hampers evacuation efforts. Cooking is the leading cause of fire injuries on college campuses, closely followed by careless smoking and arson.

The Cause

Many factors contribute to the problem of dormitory housing fires.

- Improper use of 911 notification systems delays emergency response.
- Student apathy is prevalent. Many are unaware that fire is a risk or threat in the environment.
- Evacuation efforts are hindered since fire alarms are often ignored.
- Building evacuations are delayed due to lack of preparation and preplanning.
- Vandalized and improperly maintained smoke alarms and fire alarm systems inhibit early detection of fires.
- Misuse of cooking appliances, overloaded electrical circuits, and extension cords increase the risk of fires.

Safety Precautions

- Provide students with a program for fire safety and prevention.
- Teach students how to properly notify the fire department using the 911 system.

- Install smoke alarms in every dormitory room and every level of housing facilities.
- Maintain and regularly test smoke alarms and fire alarm systems. Replace smoke alarm batteries every semester.
- Regularly inspect rooms and buildings for fire hazards. Ask your local fire department for assistance.
- Inspect exit doors and windows and make sure they are working properly.
- Create and update detailed floor plans of buildings, and make them available to emergency personnel, resident advisors, and students.
- Conduct fire drills and practice escape routes and evacuation plans. Urge students to take each alarm seriously.
- Do not overload electrical outlets and make sure extension cords are used properly.
- Learn to properly use and maintain heating and cooking appliances.

www.usfa.fema.gov

Crime Awareness and Campus Security Act of 1990 The Clery Act was named for Jeanne Clery, a 19-year-old freshman who was raped and murdered in her dorm room at Lehigh University in 1986. Her parents were later informed that there had been thirty-eight violent crimes on this campus and the students were unaware of this problem. As a result, her parents, Connie and Howard Clery, along with other campus crime victims, convinced Congress to enact the law known as the "Crime Awareness and Campus Security Act of 1990." The law was amended in 1992 and 1998 to include rights to victims of campus sexual assault and to expand the reporting requirements of the colleges and universities. In 1998, the law was officially named the "Clery Act."

The Clery Act requires all colleges and universities to accurately report the number of campus crimes per category to the campus community and prospective students. Follow up with your campus Police Department to find out how this report is dispersed to current & prospective students, staff, & faculty. College campuses have often been the site for criminal activity. These offenses include sex offenses, robbery, aggravated assault, burglary, arson, and motor vehicle theft. Hate crimes as well as hazing issues can be included in the reports as well as alcohol and weapons violations. Approximately 80 percent of the crimes that take place on college campuses are student-on-student, with nine out of ten felonies involving alcohol or other drugs.

As of 2002, the Clery Act also requires all states to register sex offenders, under Megan's Law, if they are students or employees of the college or university. This information is available to the campus police as well as students who request such information.

Under this law, colleges and universities can be fined for failure to report campus crimes. Omission of this information is not only illegal but it poses a threat to students' safety. The fines send a strong message for schools to take the obligation of reporting crimes and protecting students seriously.

A Wellness Profile

Living well requires constant evaluation and effort on an individual's part. The following list includes important behaviors and habits to include in your daily life:

- Be responsible for your own health and wellness. Take an active role in your life and well-being.
- Learn how to recognize and manage stress effectively.

- Eat nutritious meals, exercise regularly, and maintain a healthy weight.
- Work towards healthy relationships with friends, family, and significant others.
- Avoid tobacco and other drugs; use alcohol responsibly, if at all.
- Know the facts about cardiovascular disease, cancer, infections, sexually transmitted infections, and injuries. Utilize this knowledge to protect yourself.
- Understand how the environment affects your health and take appropriate measures to improve it.
(adapted from Insel & Roth, 2009)

Changing Behavior and Setting Goals

The Stages of Change

Living well requires constant evaluation and effort.

The Stages of Change Model (SCM) was originally developed in the late 1970s and early 1980s by James Prochaska and Carlo DiClemente when they were studying how smokers were able to quit smoking. The SCM model has been applied to many different behavior changes including weight loss, injury prevention, alcohol use, drug abuse, and others. The SCM consists of five stages of change precontemplation, contemplation, preparation, action, and maintenance. The idea behind the SCM is that behavior change does not usually happen all at one time. People tend to progress through the stages until they achieve a successful behavior change or relapse. The progression through each of these stages is different depending upon the individual and the particular behavior being changed. Each person must decide when a stage is complete and when it is time to move on to the next stage.

Precontemplation

The stage at which there is no intention to change a specific behavior in the foreseeable future. Many individuals in this stage are unaware of their unhealthy behavior. They are not thinking about change and are not interested in any help. People in this stage tend to defend their current behavior and do not feel it is a problem. They may resent efforts to help them change.

Contemplation

The stage at which people are more aware of the consequences of their unhealthy behavior and have spent time thinking about the behavior but have not yet made a commitment to take action. They consider the possibility of changing, but tend to be ambivalent about change. In this stage, people straddle the fence, weighing the pros and cons of changing or modifying their behavior.

Preparation

A stage that combines intention and behavioral criteria. In this stage, people have made a commitment to make a change. This can be a research phase where people are taking small steps toward change. They gather information about what they will need to do to change their behavior. Sometimes, people skip this stage and try to move directly from contemplation to action. Many times, this can result in failure because they did not research or accept what it was going to take to make a major lifestyle change.

Action

The stage at which individuals actually modify their behavior. This stage requires a considerable commitment of time and energy. The amount of time people spend in the action stage varies. On average, it generally lasts about six months. In this stage, unhealthy people depend on their own willpower. They are making efforts to change the unhealthy behavior and are at greatest risk for relapse. During this stage, support from friends and family can be very helpful.

Along the way to a permanent behavior change, most people experience a relapse. In fact, it is much more common to have at least one setback than not. Relapse is often accompanied by feelings of discouragement. While relapse can be frustrating, the majority of people who successfully change their behavior do not follow a straight path to a lifetime free of unwanted behaviors. Rather, they cycle through the five stages several times before achieving a consistent behavior change. Therefore, the SCM considers relapse to be normal. Relapses can be important opportunities for learning and becoming stronger. This is where a behavior change journal and weekly reflections can help an individual see how much progress has been made, as well as what may trigger relapses. The main thing to remember is that the goal is getting closer. Do not get upset by life or setbacks, but keep moving forward and get closer to the end goal.

Maintenance

The stage in which people work to prevent relapse and focus on the gains attained during the action stage. Maintenance involves being able to successfully avoid temptations to return to the previous behavior. The goal of the maintenance stage is to continue the new behavior or lack there of without relapse. People are more able to successfully anticipate situations in which a relapse could occur and prepare coping or avoidance strategies in advance.

Behavior Change and Goal Setting

Listed below are some tips for successful behavior change.

Choose a behavior that an individual is really invested in. Utilize the Lifestyle Assessment Inventory at the end of this chapter to see what behavioral areas need the most attention. Ideas for behavior change include: better communication, working on a particular relationship (parent, friend, significant other), increasing exercise, quitting smoking, decreasing procrastination, decreasing or eliminating sodas, eating more fruits and/or vegetables, flossing teeth every day, stretching, and so on.

Only change one behavior at a time. After reviewing lifestyle behaviors, people tend to get excited and want to change several different behaviors. Even if the behaviors are related, it is best to choose only one to focus on at a time. After a specific behavior has become a habit (at least six months in the maintenance stage) the individual can consider working on another behavior.

The goal should be specific and measurable. The more specific the goal and the plan to achieve this goal are, the more likely the behavior change will be successful. If an individual wants to increase fitness, it would be best to be very specific about the short- and long-term goals. For example, the individual should consider their baseline (where they are right now). If someone is not exercising at all, they should not begin working out five times per week the following week. During the first week the individual may want to exercise two times for fifteen minutes each exercise session. The following week the goal could be three times at twenty minutes each exercise session. The final goal may be five days per week for thirty minutes each time. This particular goal should take at least a month or two to achieve. The Behavior Change and Goal Setting notebook activity at the end of the chapter can help outline a plan of change.

Any behavior change target should be realistic. Often, behavior change goals include weight loss. To increase the long-term success rate, the most a person should lose is two pounds per week. One pound is equal to 3,500 calories. In order to lose two pounds per week the caloric deficit would need to be 7,000 calories. This translates to a deficit of 1,000 calories per day, which is not easy to achieve. The best way to achieve this caloric deficit is to include both exercise and limit caloric consumption. For example, an individual could expend part of the needed caloric deficit with exercise (approximately 500 calories per day) as well as consume fewer (approximately 500) calories per day for a total daily caloric deficit of 1,000. Remember, this is the most an individual should lose per week.

Have a reward system. It is nice to have short- and long-term goals that have a small reward when a goal is reached. These rewards should never be counterproductive. For example, if an individual is trying to lose weight, the worst type of reward would be to have a dessert. Some constructive reward ideas could be to go to a movie, go for a specific hike, buy a new pair of shorts, purchase a book or magazine.

Keep a journal. Recording notes on a regular basis is a great way to keep a behavior change project on an individual's mind. It also creates a method to track progress and setbacks. A lot can be learned from looking at what worked and what did not in previous weeks. It is best to journal a minimum of three days per week and include a weekly reflection statement summarizing how the week progressed. This can give the individual critical insight that they may not have had without the journaling process.

Have a support group. Tell friends and family members who will be supportive about a particular behavior change. The more people who know about the behavior change, the more likely the change will be successful.

By regularly evaluating your lifestyle and making small changes, you can maintain a healthy lifestyle. There are significant benefits to choosing healthy behaviors early on. The earlier these healthy behaviors are achieved, the more graceful aging will be. In Figure 1.2 the life expectancy is differentiated between healthy life expectancy and unhealthy years. In Figure 1.5 the number one

FIGURE 1.2

The Stages of Change Model consists of precontemplation, contemplation, preparation, action, and maintenance

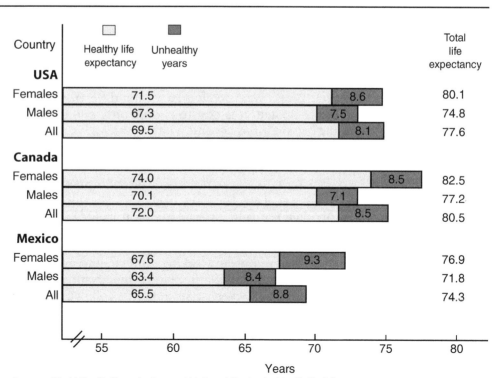

Country	Healthy life expectancy	Unhealthy years	Total life expectancy
USA			
Females	71.5	8.6	80.1
Males	67.3	7.5	74.8
All	69.5	8.1	77.6
Canada			
Females	74.0	8.5	82.5
Males	70.1	7.1	77.2
All	72.0	8.5	80.5
Mexico			
Females	67.6	9.3	76.9
Males	63.4	8.4	71.8
All	65.5	8.8	74.3

Sources: World Health Organization and National Center for Health Statistics

cause of death is unintentional injury until the age of 44. After age 44, the leading causes of death are cancer and heart disease. These two figures demonstrate how critical it is that healthy behavior choices are made now rather than waiting until an injury or disease has occurred.

The best way to avoid injuries and disease is through prevention. There are three types of prevention: primary, secondary and tertiary. **Primary prevention** utilizes behaviors to avoid the development of disease. This can include getting immunizations, exercising regularly, eating healthy meals, limiting exposure to sunlight, using sunscreen, having safe drinking water, and guarding against accidents. The focus of this textbook will be primary prevention to help individuals choose behaviors that will prevent disease and premature death.

Secondary prevention is aimed at early detection of disease. This can include blood pressure screenings, mammograms, and annual pap tests to identify and detect disease in its earliest stages. This is before noticeable symptoms develop, when the disease is most likely to be treated successfully. With early detection and diagnosis, it may be possible to cure a disease, slow its progression, prevent or minimize complications, and limit disability. Another goal of secondary prevention is to prevent the spread of communicable diseases. In the community, early identification and treatment of people with communicable diseases, such as sexually transmitted infections, not only provides secondary prevention for those who are infected but also primary prevention for people who come in contact with infected individuals.

The Stages of Change Model consists of precontemplation, contemplation, preparation, action, and maintenance.

Tertiary prevention works to improve the quality of life for individuals with various diseases by limiting complications and disabilities, restoring function, and slowing or stopping the progression of a disease. Tertiary prevention plays a key role for individuals with arthritis, asthma, heart disease, and diabetes.

Healthy People 2020: Improving the Health of Americans

There are twelve major public health areas, which include leading health indicators. They are:

- Access to health services
- Clinical preventive services
- Environmental quality
- Injury and violence
- Maternal, infant, and child health
- Mental health
- Nutrition, physical activity, and obesity
- Oral health
- Reproductive and sexual health
- Social determinants
- Substance abuse
- Tobacco

www.healthypeople.gov/2020/LHI.default.aspx

There are four major factors that influence personal health:

1. personal behavior
2. heredity
3. environment
4. access to professional health care personnel

The importance of prevention is made clear in *Healthy People 2020. Healthy People* was first developed in 1979 as a *Surgeon General's Report.* It has been

> Researchers at the Human Population Laboratory of the California Department of Health published the following list of healthrelated behaviors that have been associated with good health and a long life. These behaviors include:
>
> - 1. Regular exercise
> - 2. Adequate sleep
> - 3. A good breakfast
> - 4. Regular meals
> - 5. Weight control
> - 6. Abstinence from smoking and drugs
> - 7. Moderate use of (or abstinence from) alcohol
>
> It was shown that by following six of the seven listed behaviors, not only is an individual's quality of life greatly improved, but also, men could add eleven years to their lives and women could add seven years to their lives.

reformulated since 1979 as *Healthy People 1990: Promoting Health/Preventing Disease, Healthy People 2000: National Health Promotion and Disease Prevention and Healthy People 2010; Objectives for Improving Health*. The original efforts of these programs were to establish national health objectives and to serve as a base of knowledge for the development of both state-level and community-level plans and programs to improve the nation's overall health. Much like the programs *Healthy People 2020* is based on, it was developed through broad consultation programs and the best and most current scientific knowledge in the public and private sectors. It is also designed in a way that will allow communities to measure the success rates, over time, of the programs they choose to implement.

Healthy People 2020 has four overarching goals. The first goal is to attain high quality, longer lives free of preventable disease, disability, injury, and premature death. The second goal is to achieve health equity, eliminate disparities, and improve the health of all groups. The third goal is to create social and physical environments that promote good health for all. One last goal is to promote quality of life, healthy development, and healthy behaviors across all life stages.

Summary

Health is "a state of complete physical, mental, and social well-being and not merely the absence of disease or infirmity" according to the World Health Organization. By definition, health is a universal trait. Due to the fact that personal behaviors are one of the four major factors that influence a person's lifespan and quality of life, health also takes on a very individual and unique quality.

The idea of wellness is an individual-based approach to health. Wellness is grounded in behavior modification strategies that result in the adoption of low-risk, health-enhancing behaviors. By balancing the seven components of wellness—emotional, intellectual, social, spiritual, physical, occupational, and environmental—a person can, to some degree, prevent disease and premature death.

Changing behaviors and setting goals to achieve healthy change are major steps to wellness. Using the SCM can be helpful in making behavior changes. The SCM consists of precontemplation, contemplation, preparation, action, and maintenance stages. These stages are very important to recognize when preparing for a behavior change project. The key elements in a successful behavior change include planning, research, and individual willpower. Healthy behaviors chosen early in life affect an individual's wellness now and in the years to come.

Prevention is a fundamental factor in promoting wellness. The three types of prevention include primary, secondary, and tertiary. Primary prevention utilizes behaviors to avoid injuries, the development of diseases, and premature death. Secondary prevention focuses on early detection of disease and tertiary prevention works to improve the quality of life for individuals with various disease processes.

Each decade, since the 1979 *Surgeon General's Report,* the nation has refined its health agenda—first through *Healthy People* 1990 through *Healthy People 2000: National Health Promotion and Disease Prevention* through *Healthy People 2010: Objectives for Improving Health,* and currently through *Healthy People 2020.* When an attempt is made to understand the four goals of *Healthy People 2020:*

- Goal 1—Attaining high-quality, longer lives free of preventable disease, disability, injury, and premature death
- Goal 2—Achieving health equity, eliminating disparities, and improving the health of all groups
- Goal 3—Creating social and physical environments that promote good health for all
- Goal 4—Promoting quality of life, healthy development, and healthy behaviors across all life stages

and connect the topic areas of this program with these goals, the overwhelming importance of prevention in promoting an individual's level of wellness is made clear.

References

Centers for Disease Control and Prevention, National Center for Injury Prevention and Control (2014). The leading cause of death for persons 1-44 in U.S.

Centers for Disease Control and Prevention, National Center for Injury Prevention and Control (2014). FASTSTATS: Leading causes of death.

Corbin, C. B., Welk, G. J., Corbin, W. R. and Welk, K. A. *Concepts of Physical Fitness* (16th ed). McBrown. 2010.

Finkelstein, E. A., Corso, P. S., and Miller, T. R., Associates. Incidence and Economic Burden of Injuries in the United States. New York, NY: Oxford University Press. 2006.

Floyd, P., Mims, S., and Yelding-Howard, C. *Personal Health: Perspectives and Lifestyles* (4th ed). Morton Publishing Co. 2007.

http://ahha.org http://wellness.ndsu.nodak.edu/education/dimensions.shtml

http://who.int/aboutwho/en/definition.html

http://www.cdc.gov/nchs/data/hp2k99.pdf

http://www.healthypeople.gov/2020/default.aspx

http://www.m-w.com/dictionary.htm

http://www.wellnesswise.com/dimensions.htm

Hyman, B., Oden, G., Bacharach, D., and Collins, R. *Fitness for Living* (3rd ed). Kendall-Hunt Publishing Co. 2006.

Insel, P. M. and Roth, W. T. *Core Concepts in Health* (12th ed). McGraw Hill Publishing. 2009.

NCIPC: Web-based Injury Statistics Query and Reporting System (WISQARS) http://www.cdc.gov/injury/wisqars

Payne, W. A., Hahn, D. B., and Lucas, E. B. *Understanding Your Health* (10th ed). McGraw Hill Publishing. 2008.

Pruitt, B.E. and Stein, J. *Health Styles.* Allyn & Bacon. 1999.

Ruder, Debra Bradley. (2008). The teen brain: a work in progress. *Harvard Magazine*, 8–10.

The Clery Center for Security on Campus. http://clerycenter.org/national-campus-safety-awareness-month

Talukder, Gargi. (2013). Decision-making is still a work in progress for teenagers. http://brainconnection.brainhq.com/decision-making-is-still-a-work-in-progress-for-teenagers/

Thygerson, Alton L. (1992). Safety. Boston, MA: Jones and Bartlett Publishers.

Wellness Initiative/University of Nebraska-Lincoln

Wellness The Total Package, 2nd Edition.

Wellness Goal

Reflect on your own wellness. Identify a behavior you would like to change during the semester and establish one **Wellness Goal** to work towards this semester.

Describe your wellness goal using the SMART approach. Then ask yourself. What behavioral goals support your SMART goal? Include a detailed action plan for each behavioral goal. Make sure the action plan incorporates specific and measurable targets.

Describe your goal using a SMART approach. Specific, Measurable, Attainable, Realistic, & Time Sensitive
Goal: *My goal is to exercise more.*

Ask yourself. Can my goal be more specific? If yes, rewrite.
Ask yourself. Can I measure my goal? If no, rewrite.

Revised Goal: *My goal is to exercise daily for at least 10 minutes. Depending on my mood, energy, wants, & needs, I'll increase the duration of my exercise practice.*

Ask yourself. Is this is a realistic goal? Is it attainable? If no, rewrite.
Ask yourself. Is there a specified time period for my goal? If no, rewrite.

Revised Goal: *My goal is to exercise daily for at least 10 minutes. Depending on my mood, energy, wants, & needs, I'll increase the duration of my exercise practice. I will continue this exercise practice for a lifetime.*

You've identified your SMART goal! *My goal is to exercise daily for at least 10 minutes. Depending on my mood, energy, wants, & needs, I'll increase the duration of my exercise practice. I will continue this exercise practice for a lifetime.*

Ask yourself. What behavioral goals support your SMART goal? Describe each behavioral goal in detail.

1) *I'll focus on my exercise practice daily.*
   - I'll keep my yoga mat in the corner of my bedroom next to my nightstand.
   - I'll practice a minimum of 10 minutes each night before I brush my teeth and take my shower.
   - On some nights, I'll practice longer based on my mood, energy, wants, & needs.

2) *I'll practice ultimate frisbee once a week with my roommates.*
   - Every Thursday night before dinner, I'll join my roommates for a game of ultimate frisbee.
   - I'll pack water and a change of clothes after lunch.

3) I'll attend cross-fit on the weekend with my buddy Jane.
   - Every Saturday morning at 8:00am, we will meet for cross-fit practice at 8:00am.
   - I'll set out my workout clothes the night before and will go to bed by 12:00pm. I'll wake up by 7:00am for a light breakfast.

Name _____ Section _____ Date _____

Homework Experience

Journal—Goal Setting

Locate the Stages of Change in Chapter 1 of your health textbook. Choose one health behavior in your life that you would like to change. You may choose to eliminate a behavior (ex. Smoking or eating fried food) or choose to begin a new behavior (like exercising or mediating).

Health Behavior:

What dimension of wellness encompasses this behavior?

What stage of change are you currently in?

What support, tools, or resources do you need to progress through the stages of change?

Set three specific short-term goals that will help you change this behavior

1. _____
   _____

2. _____
   _____

3. _____
   _____

How will you measure your success?

Chapter 2
Stress and Performance

OBJECTIVES

- Identify the body's response to acute and chronic stress.
- Differentiate stress theories and types of stress.
- Identify cognitive behavioral techniques and mindfulness techniques for managing stress.

From *Critical Issues in Health, 2nd Edition* by Lana Zinger. Copyright © 2011 by Lana Zinger. Reprinted by permission.

What Do You Know About Stress?

True or False

1. Anxiety and stress are the same thing.
2. High stress levels and chronic stress can contribute to heart disease, high blood pressure and strokes, and can depress the immune system.
3. It is impossible to treat stress without seeking professional help.
4. Good stress helps keep us alert, motivates us to face challenges and drives us to solve problems.
5. Any challenge, whether a physical or a psychological one, is a source of stress, and challenging situations should be avoided in order to avoid stress.
6. Smoking a cigarette helps to relieve stress.
7. Drinking coffee reduces stress.
8. Exercise helps to build the body but it depletes the body of energy, causing stress.
9. Studies have indicated that people who "vent" their emotions by talking about their problems, writing about their problems, etc., are less likely to experience physical and psychological illness.
10. Stress management includes getting enough sleep, drinking enough water, eating healthy foods, getting enough exercise, and learning how to say "NO!".

Definition of Stress

To your body, *stress* is synonymous with change. Stress is the emotional and physical symptoms that individuals experience as the result of change. Stress may be considered as any physical, chemical, or emotional factor that causes bodily or mental tension and that may be a factor in disease causation. The majority of visits to physicians are because of stress-related complaints.

Stress can be positive or negative (eustress vs. distress). A mild degree of stress and tension can sometimes be beneficial.

Symptoms of Stress

- Rapid heart rate
- Headaches, backaches
- Muscular aches
- Sweating
- Tics
- Insomnia
- Fatigue
- High blood pressure
- Impotence and other sexual problems
- Dizziness
- Depression, anxiety
- Irritation, anger, hostility
- Fear, panic attacks
- Poor concentration
- More infections, illnesses

Types of Stress

- Emotional _____
- Physical _____
- Environmental _____

Episodes of Stress

- Acute _____
- Episodic _____
- Chronic _____

What Happens to Our Bodies Under Stress?

- *Brain*—headaches, anxiety, depression, insomnia, memory loss
- *Digestive system*—slows down; mouth ulcers or cold sores; upset stomach
- *Heart*—Increased heart rate, increased blood pressure
- *Skin*—breakouts, rashes, itching, eczema
- *Muscle*—tension, tics
- *Reproductive system*—menstrual disorders, infertility in females; impotence, premature ejaculation in males.

Stress and the Immune System

- Powerful chemicals triggered by stress suppress the immune system, making the body more susceptible to illness.
- Stress interferes with the body's ability to heal.
- Increased adrenaline production causes the body to increase metabolism of proteins, fats, and carbohydrates to quickly produce energy for the body to use.
- The pituitary gland increases production of andrenocorticotropic hormone (ACTH), which in turn stimulates the release of cortisone and cortisol hormones. These hormonal releases may inhibit the functioning of disease-fighting white blood cells and suppress the immune system's response.

Health Disorders Associated with Chronic Stress

- Coronary heart disease
- Hypertension
- Diabetes
- Progression of breast cancer
- Ulcers
- Eating disorders
- Asthma
- Depression
- Migraines
- Sleep disorders
- Chronic fatigue
- Physical aches, pains

Selye's Model—General Adaptation Syndrome

Stressor = Alarm → Resistance → Exhaustion

Our bodies try to keep in balance (homeostasis), but stress may upset that balance. Hans Selye explained stress through GAS (general adaptation syndrome), a way that the body tries to keep in balance.

A stressor is anything that affects you emotionally or physically.

Examples of Stressors

- School
- Work
- Parents
- Relationships
- Road rage
- Illness and disability
- Death of a loved one
- Discrimination
- Violence
- War
- Other

The *alarm stage* is when your body is preparing to defend itself against the stressor. Your body will go into "fight or flight" response by releasing hormones such as adrenaline and insulin to allow you to flee or attack. Heart rate and blood pressure will be elevated. In ancient times, a stressor such as being chased by a tiger would be beneficial by allowing the release of adrenaline hormones to allow for running away or killing the tiger.

The *resistance stage* allows arousal to be elevated while the body is trying to defend itself against the stressor.

The *exhaustion stage* occurs when resources are limited or depleted and the ability to resist the stressor is impaired. This stage leads to increased vulnerability of health problems and an impaired immune system.

Personality Types and Stress

- *Type A*—aggressive, hard-driven, impatient; high levels of distress. Prone to stress-related diseases such as ulcers, heart disease, hypertension, and certain cancers.
- *Type B*—easygoing, laid-back, patient; low levels of distress.
- *Type C*—passive, apologetic, overly sensitive; moderate levels of distress. Prone to stress-related mental disorders.
- *Type D*—tendency towards negativity; may experience a lot of stress, anger, worry, hostility, tension, and other negative and distressing emotions. Prone to depression, heart disease, hypertension.

Stress May Trigger Anger

Researchers believe that prolonged stress and anger, result in the breakdown of the cardiovascular system. It can also increase your risk for developing **mental health concerns** such as:

- Depression
- Eating disorders
- Drug, alcohol or other addictions
- Suicidal thoughts
- Relationship problems

You might be holding in anger and not even be aware of it. Do you find yourself flying off the handle on a regular basis or having road rage or screaming or exploding at the littlest things that aggravate you?

Anger is caused by an irrational perception of reality and a low frustration point. Angry people almost never admit responsibility—they blame something or someone else for their anger. When you are in the angry "rage"—its hard to think clearly because your emotions take control of your actions. The fight or flight response takes over and increases blood pressure and heart rate and releases adrenaline into your bloodstream, which tells your body to either defend yourself, or attack someone/something.

How to Control Your Anger

- Breathe deeply, from your diaphragm.
- Slowly repeat a calming word or phrase such as "relax" or "calm."
- Use imagery; visualize a relaxing experience.
- Exercise or yoga to release the adrenaline.
- Take a "time out"—leave the scene and cool off.
- Talk with a psychologist to improve your problem-solving skills.

How to Deal With an Angry Person

1. Become an impartial observer—act as if you were watching someone else.
2. Stay calm—do not add fuel to the fire.
3. Refuse to engage.
4. Defuse them by ignoring them, looking away, or starting another conversation with a totally different topic, or find something you can agree with or praise them.
5. Walk away if the person is getting out of control.
6. Practice your deep breathing.

How to Change Your Personality to Become More Stress-Resilient

- Build greater social support networks.
- Participate in and contribute to your community in productive ways.
- Set clear boundaries and expectations for yourself.
- Develop decision-making skills.
- Practice effective communication techniques.
- Learn conflict management techniques.
- Do not try to control the outcome of every situation.

Hardiness—health authorities have now identified the concept of hardiness as a characteristic that has helped people negate self-imposed stress.

One researcher in the stress hardiness field is the clinical psychologist at City University, New York, Susan Kobasa, PhD. In the late 1970s she carried out a study on a group of executives who were under a lot of stress while their company, the Bell Telephone Company, was undergoing radical restructuring. On completion of the study, when the data was analyzed, she found that certain personality traits protected some of the executives and managers from the health ravages of stress.

These stress-hardy personality traits included:

1. **Commitment**—having a purpose to life and involvement in family, work, community, social friends, religious faith, ourselves, etc., giving us a meaning to our lives.
2. **Control**—studies have shown that how much control we perceive we have over any stressor will influence how difficult the stressor will be for us to cope with.
3. **Challenge**—how we perceive the events that occur in our lives; seeing our difficulties as a challenge rather than as a threat, and accepting that the only thing in life that is constant, is change.

You can't always influence what others may say or do to you but you can influence how you react and respond to it.

(Unknown)

Definitions

Hormone: A chemical substance released into the body by the endocrine glands such as the thyroid, adrenal, or ovaries. The substance travels through the bloodstream and sets in motion various body functions. For example, prolactin, which is produced in the pituitary gland, begins and sustains the production of breastmilk after childbirth.[31]

Neurotransmitters: Specialized chemical messengers (e.g., acetylcholine, dopamine, norepinephrine, serotonin) that send messages from one nerve cell to another. Most neurotransmitters play different roles throughout the body, many of which are not yet known.[32]

Neurotransmitters are molecules that regulate brain function. They are chemicals that relay messages from nerve to nerve both within the brain and outside the brain. They also relay messages from nerve to muscle, lungs, and intestinal tracts. They can accentuate *emotion, thought processes, joy, elation*, and also *fear, anxiety, insomnia,* and the terrible urge to *overindulge* in food, alcohol, drugs, and so on. In short, neurotransmitters are used all over the body to transmit information and signals. They are manufactured and used by neurons (nerve cells) and are released into the synaptic clefts between the neurons.

Currently, over 50 neurotransmitters have been identified, and it is estimated that around 100 neurotransmitters exist in the biological systems.

Stress Hormones/Neurotransmitters

Epinephrine, also known as *adrenaline*, the major stress neurotransmitter, is related to blood pressure and heart rate. Adrenaline prepares the body for "fright, fight, or flight" responses and has many effects, including:

- Action of heart increased
- Rate and depth of breathing increased
- Metabolic rate increased
- Force of muscular contraction improves
- Onset of muscular fatigue delayed

Norepinephrine, also known as *noradrenaline*, is a second stress neurotransmitter. High levels of this hormone are seen in states of anxiety and insomnia. It is released in response to perceived threat. The effects of the hormone noradrenaline are similar to the effects of adrenaline, the other hormone secreted by the adrenal medulla.

The actions of noradrenaline include:

- Constriction of small blood vessels leading to increase in blood pressure
- Increased blood flow through the coronary arteries and slowing of heart rate
- Increase in rate and depth of breathing

Increased amounts of both adrenaline and noradrenaline are secreted when the body is under stress.

Cortisol is secreted in times of stress. Cortisol stimulates fat and carbohydrate metabolism for fast energy, and stimulates insulin release and maintenance of blood sugar levels. The end result of these actions is *an increase in appetite*, especially cravings for sugared foods. Cortisol increases abdominal obesity. Cortisol secretion is highest in people who sleep less than six hours per night.

Major "Happy" Neurotransmitters

- *Endorphins* (opioids): Provides mood-elevating, enhancing, and euphoric effects. The more endorphins present, the happier you are! Endorphins are like natural painkillers—your body's natural heroin.

- *Dopamine*: Runs your body's pleasure center. Creates feelings of bliss and pleasure, euphoria, and focus. Also leads to appetite control and controlled motor movements. Modulates the effect of the excitatory hormones, and is necessary for states of relaxation and mental alertness.
- *Serotonin*: Manufactured from tryptophan. It is found all over the body and is necessary to modulate the levels of the stress hormones. Promotes and improves sleep, improves self-esteem, relieves depression, diminishes craving, and prevents agitated depression and worrying. Converts to melatonin and then back to serotonin. Regulates your body clock. First to fail under stress.
- *Melatonin*: "Rest and recuperation" and "antiaging" hormone. Regulates body clock.
- *Acetylcholine*: Affects alertness, memory, and sexual performance; stimulates appetite control, and release of growth hormone.
- *Phenylethylamine (PEA)*: Provides feelings of bliss, feelings of infatuation (high levels are found in chocolate).
- *Oxytocin*: Stimulated by dopamine. Promotes sexual arousal, feelings of emotional attachment, and desire to cuddle.
- *GABA (gamma amino butyric acid)*: Found throughout central nervous system, produces antistress, antianxiety, antipanic, and antipain effects. Allows individual to feel calm, maintain control, and focus.

Neurotransmitters control

- Nicotine craving
- Premenstrual syndrome (PMS)
- Irritable bowel
- Caffeine craving
- ADHD
- Anorexia and bulimia
- Migraine headache
- Panic attacks
- Alcohol craving
- Leg cramps
- Constipation
- Carbohydrate cravings
- OCD
- Aggression
- Impulsivity

How Stress Affects Neurotransmitters

The brain uses feel-good transmitters called endorphins when managing daily stress. When the brain requires larger amounts of endorphins to handle increased stress, the ratio of many of the other transmitters, one to another, becomes upset, creating a chemical imbalance. We begin to feel stress more acutely—a sense of urgency and anxiety creates even more stress. As long as the brain has a balanced amount of happy and sad messengers, everything runs smoothly and we are in homeostasis.

It is imperative that all of the major neurotransmitters be present daily and in sufficient amounts in order for the brain to be chemically balanced. When insufficient amounts of one or more of these neurotransmitters exists, it upsets the ratio and symptoms are experienced.

Depleted supplies of feel-good transmitters means it will be impossible for you to feel happy, upbeat, motivated, or on track. You will feel just the opposite: a decrease in energy and interest, feelings of worthlessness, and a pervasive sense of helplessness to control the course of your life.

Certain transmitters, when depleted, may cause you to be easily agitated or angered, experience mild to severe anxiety, and have sleep problems. You may feel more psychological and physical pain. These are all possible symptoms of neurotransmitter deficiencies.

Main Causes of Neurotransmitter Deficiencies

- *Genetics*: A person's genetic makeup is responsible for low, high, or balanced levels of transmitters from birth.
- *Stress*: Stress depletes neurotransmitters! Any type of stress (lack of sleep, everyday mental and emotional battles, or poor health) will deplete feel-good transmitters. This results in a reduction of transmitters needed for sleep, as well as a reduction in pain-blocking transmitters.
- *Diet*: The specific amino acids that our brains manufacture transmitters from are frequently not supplied by our modern diet or in the way our brain best utilizes them. Nutrient-depleted soils, fruits and vegetables not allowed to fully ripen on the vine, and overprocessing of foods have all combined over the last century to rob our diets of many life-giving nutrients. Experts in the field of brain nutrition all agree that it is very difficult to get the necessary supply of the specific amino acids from our American diet that our brain needs to create enough of the neurotransmitters that keep us feeling balanced and happy.

How to Treat Neurotransmitter Imbalance

If we treat a neurotransmitter imbalance with pharmaceutical medication (i.e., the serotonin reuptake inhibitors like Prozac, Celexa, Paxil, and Zoloft), we tend to impose an artificial and imbalanced level of a specific neurotransmitter.

These drugs will increase the amount of serotonin at the synaptic cleft, causing the body to *think* that serotonin levels are higher. Most people will feel better temporarily. When serotonin stores fall below a certain level, the medication "stops working" and a different medication must be used. However, they do *not increase the total body stores of serotonin*, and therefore are not the best permanent solution.

Alternatives to Medication

1. Increase dietary intake of tryptophan. American diets tend to be high in carbohydrate and low in protein. Foods high in tryptophan are mostly high-protein foods:
   - Cottage cheese—450 mg per cup
   - Fish and other seafood—800–1,300 mg per pound
   - Meats—1,000–1,300 mg per pound
   - Poultry—600–1,200 mg per pound
   - Peanuts, roasted with skin—800 mg per cup
   - Sesame seeds—700 mg per cup
   - Dry, whole lentils—450 mg/cup

2. Increase amount of exercise. Exercise leads to more efficient use of insulin, thus reducing insulin resistance and decreasing the amount of food that is stored as fat. When the cells process nutrients better, they make neurotransmitters better. Exercise releases endorphins, which are natural mood elevators.

3. Reduce our intake of caffeine. Caffeine makes the body think is it under stress, which raises the cortisol level, raises the insulin level, and causes carbohydrates to be deposited as fat.

Functions of Endorphins

Endorphins, chemicals produced in the brain in response to a variety of stimuli, may be nature's cure for high levels of stress.

Endorphins are among the brain chemicals known as neurotransmitters, which, as noted, function in the transmission of signals within the nervous system. At least 20 types of endorphins have been demonstrated in humans, and they may be located in the pituitary gland, other parts of the brain, or distributed throughout the nervous system.

Stress and pain are the two most common factors leading to the release of endorphins. Endorphins interact with the opiate receptors in the brain to reduce our perception of pain, having a similar action to drugs such as morphine and codeine. Unlike drugs, however, activation of the opiate receptors by the body's endorphins does not lead to addiction or dependence.

Results of Endorphin Secretion

- Decreased feelings of pain
- Feelings of euphoria
- Modulation of appetite
- Release of sex hormones
- Enhancement of the immune response

Stress and Digestion

Stress contributes to ulcers and ailing digestion.

- Good nutrition can help.
- Complex carbohydrates increase levels of serotonin.
- Stay hydrated with water to compensate for fluid lost during sweating under stress and stress-induced dry mouth.
- Eat frequent, small meals to maintain normal blood sugar, prevent fatigue and irritability, and prevent slow metabolism.
- Avoid overeating, which can increase stress.
- Limit consumption of sugar, caffeine, nicotine, and alcohol.

Vitamin and Mineral Deficiencies

Chronic stress can deplete several vitamins necessary for energy metabolism, as well as those necessary for the stress response itself. The stress response activates several hormones responsible for mobilizing and metabolizing fats and carbohydrates for energy production. The breakdown of fats and carbohydrates requires vitamins, specifically the B vitamins and vitamin C. An inadequate supply of these may affect mental alertness, promote depression, and lead to insomnia. Stress is also associated with the depletion of calcium and the inability of bones to absorb it properly.

"Pick-Me-Ups" and Stress

When people are stressed, they use *pick-me-ups* or *put-me-downs* to combat the stressor. Several substances tend to either mimic or induce the stress response, or decrease the efficiency of the body's metabolic pathways, thus setting the stage for more pronounced physiological reactions to stress. The biggest mistake people make in handling stress is using pick-me-ups to boost happy messengers, which has the effect of riding a wild roller coaster.

Sugar: Excess sugar tends to deplete vitamin stores, especially the B vitamins that are crucial for optimal function of the central nervous system. Depletion of B vitamins may result in fatigue, anxiety, and irritability. In addition, excess simple sugars can cause major fluctuations in blood glucose levels, resulting in pronounced fatigue, headaches, and irritability.

Simple Sugars

- Glucose (honey)
- Lactose (milk)
- Fructose (fruit)
- Sucrose (cane)

Fats: You've likely heard all the bad news about how fat creates artery-clogging cholesterol and weight gain. But a high-fat diet also leaves you feeling lethargic and just not feeling as well as you would on a diet high in complex carbohydrates.

Caffeine: Caffeine stimulates the release of several stress hormones, resulting in a state of hyper-alertness, and makes a person more likely to interpret events as stressful.

Alcohol: Increases short-term energy, but then blood sugar dips. Diminishes pain, but can lead to aggression and depression.

Salt: High sodium acts to increase water retention, and as water volume increases, blood pressure increases. Habitual high sodium intake may contribute to hypertension.

Tobacco: Powerful toxin; destroys trachea, bronchi, and lung function. Damages arteries, causing insufficient blood supply to the brain, heart, and organs. Carcinogenic. Increases dopamine but levels fall shortly, requiring more nicotine.

Drugs: Increases release of dopamine (pleasure center), shutting off brain's natural supply.

Your own adrenaline: Allows body to prepare for fight or flight. For example, a workaholic who is overstressed → works longer hours → feeds off his own adrenaline.

Put-Me-Downs

"Doctor, can you give me something to calm me down?"

Medications that temporarily force the body into sleeping, producing a tranquilizing effect, are referred to as *put-me-downs*. These medications produce addiction and severe withdrawal symptoms.

- Valium, Xanax, Ativan
- Barbiturates

Stress Interventions

1. Do a minimum of 30 minutes of aerobic exercise daily, or take a walk.
2. Make your body clock regular.
3. Eat five small, balanced meals per day.
4. Avoid caffeine, drugs, and tobacco.
5. Reduce intake of refined sugars and alcohol.
6. Sleep about eight hours nightly.
7. Spend time each day with relaxation techniques—imagery, daydreaming, prayer, or meditation.
8. Take a warm bath or shower.
9. Listen to music; watch a comedy.
10. Postpone making changes in your life.
11. Hug someone, hold hands, or stroke a pet.
12. Pray.

Techniques to Handle Stress

1. *Deep breathing*: Inhale through your nose slowly, counting silently to five, and then exhale through your mouth slowly, counting silently to five.

2. *Positive affirmations*: Talk in a positive manner to yourself; turn your negative comments into positive ones: "*I will pass the test.*"
3. *Stretching*: Stretch the area where tension has built up, holding for 30 seconds; relax and repeat three to five times.
4. *Progressive muscle relaxation*: Tense a muscle and hold it for a silent, slow count of 5–10 seconds, then release the muscle for a silent, slow count 5–10 seconds.
5. *Visualization*: Find a quiet space, get comfortable, close your eyes, and try to imagine yourself in a calm, enjoyable, relaxing setting. Think about the details of your tranquil setting. Who is with you? How is the weather? What does it smell like?
6. *Meditation*: Find a quiet space, get comfortable, close your eyes, and begin breathing deeply. Focus on one image or thought. Try this for 5–10 minutes every day.
7. *Yoga*: Used since ancient times to invigorate the body and calm the mind.
8. *Massage*: Research indicates that human touch is vital for well-being.
9. *Pets*: Animals decrease stress levels through touch, companionship.
10. *Talk to someone*: A friend, parent, teacher, or a college counselor.

Free and confidential counseling:
Queensborough College Counseling Center—Library Room 428, 718-631-6370 or www.qcc.cuny.edu/counseling

How to Conquer Test Anxiety

- Prepare well in advance—rehearsal and repetition.
- Get a good night's sleep—eight hours.
- Eat a nutritious meal to boost your brain's energy.
- Have a positive attitude—go in with a positive attitude, reminding yourself that you are well prepared.
- If you stumble on a question, don't linger on it. Go on to the next question.
- Ask for clarification if you don't understand the question.
- Don't pay attention to the people around you—stay focused on the test and don't get distracted.
- Take deep breaths, bringing more oxygen to the brain.
- Pace yourself—be aware of the time.

How to Manage Your Time

- Rank tasks in order of importance.
- Schedule a time frame for each task.
- List your deadlines on a calendar or daily planner.
- Delegate and ask for help with some tasks.
- Say no if demands are unreasonable.
- Schedule personal time in your daily planner for exercise, a hobby, or meditation, and stick to that schedule.
- Watch less television (1 hour per day).

Mindful Meditation

Meditation is the practice of seeing clearly, the art of moment-to-moment awareness, a nonjudgmental quality of mind that is aware of what is happening in and around oneself in the *present moment*.[34]

How to Mindfully Meditate

- Begin by sitting in a chair or on a cushion on the floor, with your back straight. Relax into your sitting posture with a few deep breaths. Allow the body and mind to become utterly relaxed while remaining very alert and attentive to the present moment. Feel the areas of your body that are tense and the areas that are relaxing. Just let the body follow its own natural law. Do not try to force or fix anything.

- Let your mind be soft, and allow a spacious awareness to wash gently through your body. Simply feel the sensations of sitting, sidestepping with your mind the tendency to see your body—to interpret, define, or think about it. Just let such thoughts and images come and go without being bothered by them, and attune to the bare sensations of sitting.
- Feel your body with an awareness that arises from within your body, not from your head. Awareness of the body anchors your attention in the present moment. Gently sweep your awareness through your body, feeling the sensations with no agenda, no goal. Allow your body to anchor awareness in the present moment by just staying mindful of these sensations.
- After some time, shift your awareness to the field of sound vibrations. Awareness of sounds creates openness, spaciousness, and receptivity in the mind. Be aware of both the pure sound vibration as well as the space or silence between the sounds. As with body sensations, incline your awareness away from the definition of the sound, or thoughts about the sound, and simply attune to the sound just as it is.
- After some minutes of awareness of body and sounds, bring your attention to your natural breathing process. Locate the area where the breath is most clear and let awareness lightly rest there. For some it is the sensation of the rising and falling of the abdomen. For others it may be the sensations experienced at the nostrils with inhalation and exhalation.
- You can use very soft mental labels to guide and sustain attention to the breath, such as "rising/falling" for the abdomen and "in/out" for the nostrils. Let the breath breathe itself without control, direction, or force. Feel each breath from within the breath, not from the head. Feel the full breath cycle from the beginning through the middle to the end.
- The awareness is a combination of light, open spaciousness and receptivity, like listening, and alert, attentive presence, touching the actual texture, shape, and form of sensations.
- Let go of everything else, or let it be in the background. Just let the breathing breathe itself. Rest in a sense of utter relaxation, in that mindful feeling, with the sensations of the breath.
- As soon as you notice the mind wandering off, lost in thought, be aware of that with nonjudging awareness; gently connect it again to your anchor. Just feel from within the stream of sensations.
- Toward the end of your sitting, not striving or anticipating, not pouncing on sensations in the present, not bending back to what was just missed or reflecting on what just happened, keep inclining to the totality of the present moment. Keep anchoring easily, deeply, and restfully. Just one breath at a time.
- Mindfulness of breath begins to collect and concentrate the mind so that the initial distractions of thoughts, emotions, sensations, and sounds soon become objects of awareness themselves. Insight is gained into the true nature of the body and mind.
- As concentration grows, mindfulness opens to the entire "flow" of body/mind experience through all the sense doors—sights, sounds, smells, tastes, touch, and mental/emotive.[35]

Progressive Muscle Relaxation

One of the most simple and easily learned techniques for relaxation is progressive muscle relaxation (PMR), a widely used procedure that was originally developed by Jacobson in 1939.[36] The PMR procedure teaches you to relax your muscles through a two-step process. First, you deliberately apply tension to certain muscle groups, and then you stop the tension and turn your attention to noticing how the muscles relax as the tension flows away. Before practicing PMR, you should consult with your physician if you have a history of serious injuries, muscle spasms, or back problems, because the deliberate muscle tensing of the PMR procedure could exacerbate any of these preexisting conditions.

How to Get Started

Sit in a comfortable chair or bed. Get as comfortable as possible—no tight clothes, no shoes, don't cross your legs. Take a deep breath; let it out slowly. What you'll be doing is alternately tensing and relaxing specific groups of muscles. After tension, a muscle will be more relaxed than prior to the tensing. Concentrate on the feel of the muscles, specifically the contrast between tension and relaxation. In time, you will recognize tension in any specific muscle and be able to reduce that tension.

Don't tense muscles other than the specific group at each step. Don't hold your breath, grit your teeth, or squint! Breathe slowly and evenly and think only about the tension-relaxation contrast. Each tensing is for 10 seconds; each relaxing is for 10 or 15 seconds. Count "1,000, 2,000 . . ." until you have a feel for the time span. Note that each step is really two steps—one cycle of tension-relaxation for each set of opposing muscles.

Do the entire sequence once a day if you can, until you feel you are able to control your muscle tensions. Be careful: If you have problems with pulled muscles, broken bones, or any medical contraindication for physical activities, ***consult your doctor first***.

1. *Hands.* The fists are tensed; relaxed. The fingers are extended; relaxed.
2. *Biceps and triceps.* The biceps are tensed (make a muscle, but shake your hands to make sure not tensing them into a fist); relaxed (drop your arm to the chair—really drop them). The triceps are tensed (try to bend your arms the wrong way); relaxed (drop them).
3. *Shoulders.* Pull them back (careful with this one); relax them. Push the shoulders forward (hunch); relax.
4. *Neck* (lateral). With the shoulders straight and relaxed, the head is turned slowly to the right, as far as you can; relax. Turn to the left; relax.
5. *Neck* (forward). Dig your chin into your chest; relax. (**Bringing the head back is not recommended—you could break your neck**).
6. *Mouth.* The mouth is opened as far as possible; relaxed. The lips are brought together or pursed as tightly as possible; relaxed.
7. *Tongue* (extended and retracted). With mouth open, extend the tongue as far as possible; relax (let it sit in the bottom of your mouth). Bring it back in your throat as far as possible; relax.
8. *Tongue* (roof and floor). Dig your tongue into the roof of your mouth; relax. Dig it into the bottom of your mouth; relax.
9. *Eyes.* Open them as wide as possible (furrow your brow); relax. Close your eyes tightly (squint); relax. Make sure you completely relax the eyes, forehead, and nose after each of the tensings.
10. *Breathing.* Take as deep a breath as possible—and then take a little more; let it out and breathe normally for 15 seconds. Let all the breath in your lungs out—and then a little more; inhale and breathe normally for 15 seconds.
11. *Back.* With shoulders resting on the back of the chair, push your body forward so that your back is arched; relax. Be very careful with this one, or don't do it at all.
12. *Buttocks.* Tense the buttocks tightly and raise pelvis slightly off chair; relax. Dig buttocks into chair; relax.
13. *Thighs.* Extend legs and raise them about 60 off the floor or the footrest—but don't tense the stomach—relax. Dig your feet (heels) into the floor or footrest; relax.

14. *Stomach.* Pull in the stomach as far as possible; relax completely. Push out the stomach or tense it as if you were preparing for a punch in the gut; relax.
15. *Calves and feet.* Point the toes (without raising the legs); relax. Point the feet up as far as possible (beware of cramps—if you get them or feel them coming on, shake them loose); relax.
16. *Toes.* With legs relaxed, dig your toes into the floor; relax. Bend the toes up as far as possible; relax.

Now just relax for a while. As your days of practice progress, you may wish to skip the body areas that do not appear to be a problem for you. After you've become an expert on your tension areas (after a few weeks), you can concern yourself only with those. These exercises will not eliminate tension, but when it arises, you will know it immediately, and you will be able to "tense-relax" it away or even simply wish it away.[37]

Name _____ Section _____ Date _____

Stress Awareness Log

A daily record of stressors can serve as a valuable tool in learning to cope with stress. For at least 3 days, keep a log of all the situations that were stressful to you. What physical or mental symptoms did you exhibit? How did you deal with the stressor—was that an effective strategy or what else should you have done?

Date	Stressor	Symptoms	How did I deal—effective or not?

Name _____ Section _____ Date _____

1. Define stress and explain how the body responds to stress using the concepts in this chapter.

2. Describe the role of stress on hormones and neurotransmitters.

3. What healthy techniques can you use to help cope with stress?

Chapter 3
Aging and Disease

OBJECTIVES

Students will be able to:
- Identify the major hypokinetic diseases afflicting Americans.
- Identify key concepts and findings related to the aging process.
- Identify cardiac risk factors and strategies to reduce controllable risk factors.
- Integrate various strategies to combat obesity.
- Determine wellness behaviors that enhance longevity and quality of life.

Adaptation of *Health and Fitness: A Guide to a Healthly Lifestyle, 5/e* by Laura Bounds, Gayden Darnell, Kirstin Brekken Shea and Dottiede Agnor. Copyright © 2012 by Kendall Hunt Publishing Company.

> "When health is absent wisdom cannot reveal itself, art cannot become manifest, strength cannot be exerted, wealth is useless and reason is powerless"
>
> —Herophilies, 300 B.C.

The life expectancy of this baby is almost double that of a baby born one hundred years ago.

The maximum recorded lifespan for humans, reported in 2010, was 122.5 years for females and 116 years for males (Biology of Aging, 2011). The life expectancy for someone born at the beginning of the twenty-first century is almost double the life expectancy of those born at the beginning of the twentieth century. The Centers for Disease Control and Prevention (CDC) has determined that *lifestyle is the single largest factor affecting longevity of life.* "If exercise could be packed into a pill, it would be the single most widely prescribed, and most beneficial, medicine in the nation" (National Institute on Aging). Wellness across the lifespan is dependent on many factors. Meeting all the needs to sustain life is the minimum, with quality and quantity of life determined by many factors. Maintaining a sense of purpose, staying active, staying connected socially, getting adequate rest and nutrition, and being engaged in lifelong learning can help individuals-from childhood through the end of life- to live life well. Scientists are interested in what influences our **health span**. Health span is maximizing the years of our life that we live in good health and good function.

Aging

Social support, whether with family or friends, plays a significant role in wellness.

Although aging is a completely natural and inevitable process, some people age more gracefully than others. As the typical American's lifespan expands, quality of life for many is compromised due to habits and lifestyle choices made earlier in life. Ensure your independence as you age by choosing how you live your life now. Balancing work, family commitments, and leisure time can be stressful. Stress takes a toll on our bodies, our minds, and our relationships. Take time to consider how well are you managing the stress in your life.

Everyone experiences age-related decline in biological functions of the body. Chronological age is our true age in years. Biological age can be different depending on our lifestyle choices. **Biological age** can be younger than chronological age with good nutrition, adequate rest on a regular basis, stress-management techniques, and consistent exercise (see Table 3.1). Biological age can be older than chronological age when unhealthy habits are the norm: poor diet, inadequate sleep, excessive alcohol use, smoking, and obesity. You are in charge of your biological age. What will your biological age be in ten, twenty, and thirty years from now? How about fifty years from now? It is interesting to note that physiological changes with aging are similar to those changes seen with inactivity or prolonged weightlessness, such as experienced by the astronauts (Bloomfield et al., 2002). An

TABLE 3.1 ♦ Effects of Physical Activity and Inactivity on Older Men

	Exercisers	Non-Exercisers
Age (yrs)	68.0	69.8
Weight (lbs)	160.3	186.3
Resting heart rate (bpm)	55.8	66.0
Maximal heart rate (bpm)	157.0	146.0
Heart rate reserve* (bpm)	101.2	80.0
Blood pressure (mm Hg)	120/78	150/90
Maximal oxygen uptake (ml/kg/min)	38.6	20.3

*Heart rate reserve = maximal heart rate – resting heart rate.

Source: From *Principles and Labs for Fitness & Wellness*, 6th edition, Wadsworth Publishing.

integrative biology professor at Berkeley, 78-year-old Marian Diamond, lists five **essentials for staying mentally vigorous:**

1. **diet,**
2. **exercise,**
3. **challenge,**
4. **novelty,**
5. **love.**

These five essentials seem to be critical to maintain quality of life at retirement age, but perhaps they are essentials for us all, at any age (Springen and Seibert, 2005).

Lack of activity can accelerate age related muscle and bone loss. However, staying active can do the opposite. Participating in resistance training as an older adult can even impact age-related declines often seen in cognition and memory. "Our research found that resistance training could positively affect cognition, information processing, attention, memory formation and executive function" (Chang et al., 2012). *It is clear that as more people live longer, it is possible to live a long health span as well as lifespan, but this is not the norm.* Disability in later years is common due to hypokinetic (a disease characterized by sedentary living) conditions such as obesity, high blood pressure and high levels of blood sugar. This chapter addresses common health conditions that are highly correlated to our everyday lifestyle choices and our biological age.

Community

In modern American culture it is not uncommon for children to move away from home and raise families far away from the grandparents. Distance can make relationships difficult, and particularly challenging when an aging parent needs assistance. Elders may face loneliness, loss of purpose and depression. When elders stay connected and involved both physical and mental health can be positively impacted. In Hispanic and Italian cultures there is a deep emphasis on family, often with grandparents living in the house with their married children and helping to care for grandchildren. These grandparents stay integrated in the family, often teaching the grandkids life lessons and about family history. In Asian cultures elders are venerated. Respect for a person only increases with age, and it is considered an honorable duty to care for parents as they age. The 60th and 70th year birthdays are special life events celebrated by the community. In Greek, "old man" is a reverent word indicating deep respect. In countries such as China, there are very few nursing homes, and there is a negative social stigma associated with sending one's relatives to live elsewhere.

As an older adult, even if family is far away, staying engaged and active will help maintain a healthy life. Joining a club, volunteering, learning a new language and staying active are just a few of the options. College Station's Jesse Coon was a great example of how it should be done-aging well (see It's never too late inset).

It's Never Too Late

He was a world champion. Jesse Coon of College Station, Texas, passed away on July 30, 2005, after a long, full, and active life. He was 94. Coon started swimming competitively at the ripe old age of 64, and he started breaking world records in his early 80's.

The former physics professor at Texas A&M University broke five world records at his last major swim meet in Munich, Germany, in the 90 through 94 age group. His stroke? The butterfly. When competing, Coon worked out in the pool ninety minutes five times per week. Jesse Coon was an active sailor and he mowed his own lawn.

Not all mature Americans need to be world record holders to benefit from a more active lifestyle. Recent studies indicate that regular exercise and physical activity can reduce or slow down the biological process of aging. Older adults can experience increased life satisfaction, happiness, and self-esteem, along with reduced stress with a regular activity. A friend noted that it never occurred to Jesse that he was old. He celebrated life and always had a positive attitude. Coon himself said "the older you get, the more important it is to exercise." Coon is an example not only to others of the gray-haired set, but to all of us of all ages.

Epigenetics: Are your Genes Malleable?

Epigenetics is the study of functional changes in the genome that do not result in changes in the DNA. The Greek prefix "epi" implies changes in gene function that are "on top of" or "in addition to" genetics. You can think of your genes as being malleable where their expression can be turned up or down and in some case off and on by environmental conditions.

Epigenetics changes, for example, have been observed in rats given dietary supplements. These supplements alter the expression of particular genes that affect

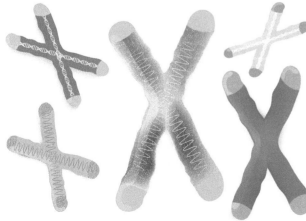

fur color, weight, and propensity to develop particular cancers. The Overkalix Study looked at physiological effects of environmental factors on the children and grandchildren of Swedish men and women exposed to famine in the late 1800s and early 1900s. Mortality risk ratios were determined for the children and grandchildren. The grandfathers' food supply (famine or normal) was linked to the mortality rate of the grandsons but not the granddaughters. However, the grandmothers' food supply was related to the mortality rate of the granddaughters. Lumley, L.H., et. al., determined that The nutrition of the grandmother during her pregnancy influences the mother's nutrition in utero which in turn influences the grandchild's birth weight.

There are also numerous studies looking at identical twins. Why do identical twins become more different as they age? Factors like diet, exercise, substance abuse, stress, and exposure to toxins are thought to modify gene expression. In other words, environment and lifestyle choices can make a difference in the expression of the genetic blueprint that identical twins start with. More research is needed into this fascinating area of study. One thing is certain, that the human body is a complex organism with many factors influencing growth, development, health and aging.

This should be good news! You do not have to accept that because a disease, mental health issue, personality characteristic, or any other trait that runs in your family that you are stuck with it. You can't change the genes you were born with, but you might be able to change your gene expression.

The concept of **epigenetics** introduces the possibility that how we live our lives can alter the expression of our genes, and perhaps the genes of our children. The term **brain plasticity** refers to the brain as an organ that can change and grow and develop. The brain changes as we learn new things, as we grow, and in response to injury. Just as skeletal muscle grows stronger with physical training and with proper nutrition, the brain "muscle" also grows with necessary nutrients provided by exercise, nutrition, and new experiences. **Neurogenesis** is the growth of new brain cells. DNA controls the process of neurogenesis, and a protein called brain-derived neurotrophic factor (BDNF) acts as a catalyst for creating new neurons. BDNF has been shown to be decreased in studies on Alzheimers patients. David Perlmutter, M.D., author of *Neurogenesis: How to Change Your Brain* writes, "Fortunately, many of the factors that influence our DNA to produce BDNF factors are under our direct control. The gene that turns on BDNF is activated by a variety of factors including physical exercise, caloric restriction, curcumin and the omega-3 fat, DHA. This is a powerful message. These factors are all within our grasp and represent choices we can make to turn on the gene for neurogenesis." This is indeed a powerful and exciting message that our lifestyle changes can actually transform our life.

Telomeres, the DNA protein "caps" on the end of our chromosomes, are the subject of numerous studies. Often compared to the plastic end on a shoelace which keeps the lace from fraying, telomeres inhibit chromosomal fraying. Obesity, chronic stress, smoking, depression and aging have been associated with shortened telomeres. A recent study in the Proceedings of the National Academy of Sciences found an association with chronic social stress in children and telomere length erosion (Mitchell, March 2014.) Molecular biologist and 2009 Nobel Laureate Elizabeth Blackburn identified the enzyme telomerase, which is responsible for lengthening and repairing telomeres. Dr. Dean Ornish, UCSF clinical professor of medicine, directed a 5 year study that followed 35 men with early stage prostate cancer to explore how specific comprehensive lifestyle changes could impact telomere length and telomerase activity. The findings were published in the journal *The Lancet Oncology*. "Our genes, and our telomeres, are not necessarily our fate," said Ornish. "These findings indicate that telomeres may lengthen to the degree that people change how they live. Research indicates that longer telomeres are associated with fewer illnesses and longer life." The participants in the study made lifestyle changes not only to their diet and activity levels, but they also made time for relaxation and community.

The study found these four lifestyle changes changed the participants lives: (1) eating a whole food plant based diet, high in fruits and vegetables and high in unrefined grains and low in fat and refined carbohydrates, (2) moderate exercise such as walking 30 minutes on most days of the week, (3) managing stress with a practice of mindfulness, meditation, and a discipline such as yoga, and finally (4) having a strong social support-a network of friends and family or a

> **Go4Life**, launched in October of 2012 by the U.S. Surgeon General and the National Institutes for Health, is a program targeting mature Americans. http://go4life.nis.gov/ is the website which encourages older Americans to make exercise a priority. Tips for success, workouts, videos, and success stories are available. Share with your favorite elder-and be sure and do the activity with them!

FACTORS CONTRIBUTING TO LIFESPAN

GENES

ENVIRONMENT

BEHAVIORAL TRAITS

supportive community-these things can lengthen your telomeres and perhaps lengthen your life as well. This is exciting research noting that not only can we stop telomere shortening, but we can actually increase telomere length with the significant but doable lifestyle changes noted in the study.

These provocative findings are not really a fountain of youth . . . but maybe? The fact remains that it is our choice how we live. We are impacted by these choices throughout our lifespan-from before we are born until the day we die. Choices have consequences, but our lives are in flux. It is never too late to make better choices in order to live life well.

Have you released any Irisin lately?

Called the "exercise hormone," irisin was first reported on in January 2012 issue of the Journal *Nature* by Bruce Spiegelman and colleagues at the Dana-Farber Cancer Institute in Boston. Irisin is released by skeletal muscle as a response to endurance exercise, and is thought to be how muscle helps regulate the activity of adipose tissue. "Irisin increases adipocyte mitochondrial biogenesis, and the expression of uncoupling proteins, leading to greater mitochondrial heat production and energy expenditure." (Crivello, 2014) Spielgelman and colleagues discovered in 2013 that when Irisin is released during aerobic activity, new growth of brain neurons (neurogenesis) occurs and genes involved in learning and memory are activated. In 2014, a research team in the UK at Aston University lead by Dr. James Brown found that Irisin is related to the molecular benefits of exercise on telomere length, indicating that irisin may be involved in slowing the aging process (Christopher Bergland, The Athlete's Way).

Prevention of Hypokinetic Conditions: Planning Your Activity Program

Most adults know that they should be active, and they may be aware that a more active life would make them feel better. The truth is that sometimes it is very difficult to know how to get started and how to incorporate activity into busy lives. *The most important thing is to get started.* Remember that lifestyle activity (defined on the next page) is easier to incorporate into a hectic schedule; walk during a coffee break, grab ten minutes to move around while on break, or jump rope for a study break. Planned exercise can be more of a challenge. The following are a few suggestions to help jump-start the new you.

1. Establish why you want to exercise.
2. Write down reasonable long-term goals.

3. Write down short-term goals that support the long-term goals.
4. Record the behaviors that need to change in order to support the goals. (A person wanting to quit smoking may want to quit working at a bar and work in a nonsmoking environment.)
5. Write in a log: feelings, food, activity and goal progress are all appropriate.
6. Develop a weekly plan for the activity that supports your goals.
7. Tell your friends and family about your goals and ask for their support; or ask them to join you.
8. Reward yourself when any goals are met (rewards should be non-food items, and should not be a day "off" from behaviors that promote your goals).
9. When goals are not met, check your log. What can you change to more effectively support your goals?
10. Periodically re-evaluate goals.

It has been firmly established that physical activity should be a part of our daily lives. Exercise enhances weight management and overall wellness by burning calories, speeding up metabolism, building muscle tissue, and balancing appetite with energy expenditure. More importantly, an active life decreases health risk and typically makes you feel good, and feel good about yourself. Look for opportunities to be active and have fun at the same time.

Creativity and Aging

Typically mental processes being to slow down with age; however, some characteristics such as creativity can flourish with age. There are many examples of great creative accomplishments by elderly artists. Michaelangelo completed his final frescoes for the vatican's Pauline Chapel at 75. Georgia O'Keeffe painted into her 90's despite failing eyesight. Benjamin Franklin invented bifocal glasses at 78 to correct his poor vision. Folks that have lived longer and have had more experience tend to be more comfortable in their own skin. Because mature adults seldom experience the adolescent need to "fit in," they are more likely to have the freedom to express themselves. This may enhance creative endeavors. So look forward to good health and artful aging in your golden years.

> **"Real Age"**
> Dr. Michael Roizen has developed a "real" age test. Log on to www.realage.com to take the free test. According to Roizen, exercising regularly can make your "real" age as much as nine years younger.

As Westerners living in an industrialized nation, we are fortunate to have choices. Nutrition choices at dinnertime and in between meals . . . Cheetos, Cheerios, or cheese? Choosing to be more active daily, such as gardening, taking the stairs instead of the elevator, walking rather than riding, can make a surprising dent in our energy reserves. The 1996 Surgeon General's Report (Satcher, 1996) encouraged all individuals to try and expend 150 calories extra each day, above and beyond a normal routine. **Lifestyle activity** is searching for opportunities to expend some extra energy, rather than searching for opportunities to conserve energy with convenient devices such as cell phones and electric pencil sharpeners. Students who walk across big campuses between classes rather than take a bus expend more energy. One day might not have a significant impact; however, at the end of the semester the cumulative effects of walking can add up to enhanced health.

> It is never too late to start exercising, but it is *always* too soon to stop!"
> -Walter Bortz

4 Lifestyle Choices That Keep You Young

1. Plant-based diet (high in fruits, vegetables and unrefined grains, and low in fat and refined carbohydrates)
2. Moderate exercise (walking 30 minutes a day, six days a week)
3. Stress reduction (gentle yoga-based stretching, mindfulness, breathing, meditation)
4. Social support network (friends, family, sense of community)

> "Those who think they have no time for bodily exercise will sooner or later have to find time for illness."
> Edward Stanley (1826–1893)

Examples of Lifestyle Activity: Looking for Opportunities to Expend More Calories

- Taking the stairs instead of the elevator.
- Parking farther from your destination to increase walking distance.
- Walking rather than riding.
- Vacuuming with vigor, taking big lunging steps.
- Doing sit-ups during the commercials of your favorite program.
- Playing Frisbee or planting a garden instead of watching TV.

Planned exercise is important for fitness benefits. It is important to find an activity that is enjoyable because that will ensure a more sincere commitment—one that may become permanent over a lifetime. Plan it! Don't leave your exercise to chance, because chances are, you won't have time. If you hate to run, don't train for a marathon. Set realistic goals. Research has shown that specific goal setting greatly enhances the chances of achieving said goals. Use the goal activity at the end of Chapter 1. Be reasonable and look for behaviors to change in support of your goals. Enlist the help and support of friends and family. Walking is the most popular activity in the United States, most likely because it is easy and requires no special training or equipment except a good pair of shoes. Choose an activity you enjoy and as the shoe company says, "Just do it!"

Lifestyle activity, planned exercise, stress management, and good nutrition choices can make a difference in whether or not an individual suffers from a hypokinetic condition. **Hypokinetic** literally means too little activity. Kraus and Rabb first coined the term hypokinetic in 1961. Hypokinetic diseases include the leading causes of death, such as coronary heart disease and cancer, as well as debilitating conditions such as low back pain, osteoporosis, obesity, diabetes, and mental health disorders. Simply changing an individual's lifestyle to one that includes more physical activity can reduce the incidence of many hypokinetic conditions. *Regular consistent activity can decrease the potential of contracting a hypokinetic disease.*

For example, expending an extra 500–1,000 calories per week can decrease health risk (see Figure 3.1). Expending an extra 1,000–2,000 calories per week can decrease overall health risk more and also moderately increase cardiovascular fitness. An expenditure of 2,000–3,500 calories per week can decrease overall health risk, as well as significantly increase cardiovascular fitness over time. Typically expending beyond 3,500 calories can increase risk of musculoskeletal injuries and burnout.

Caloric expenditure from both lifestyle activity and planned exercise has a significant impact on health. Participating in a little extra activity helps by decreasing overall health risk and by enhancing self-confidence. Expending more extra energy does the same as expending a little extra activity with further decreased risk, added fitness benefits, and potential weight loss, particularly if exercise is within the target heart rate zone (see Figure 3.2). However, doing too much activity can cause burnout, injury, or possibly an obsession with exercise. As in all areas of life, balance and common sense are important.

It has been previously noted that the CDC reported that lifestyle is the single greatest factor affecting longevity of life. This is especially critical to note for future generations as America's children are more sedentary and at higher risk for developing hypokinetic diseases than their parents or grandparents. Childhood obesity is a national epidemic; this is partially due to technology (think energy-saving remote controls and cell phones) and the ease with

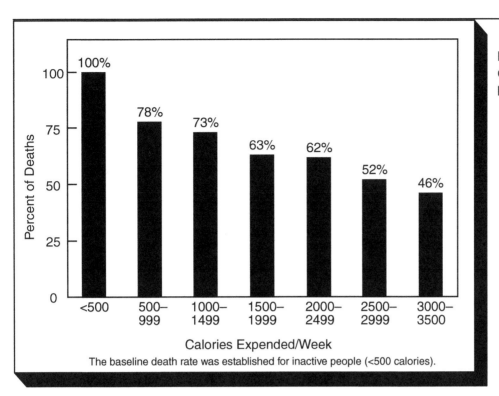

FIGURE 3.1
Deaths Decrease as Caloric Expenditure Increases

The baseline death rate was established for inactive people (<500 calories).

Source: Data from C. Bouchard et al. *Exercise Fitness and Health*. Champaign, IL: Human Kinetics Publishers, 1990.

which tasks are performed. To combat this trend, parents can help plan activities to ensure their children accumulate at least sixty minutes of activity a day. The best way to accomplish this is for parents to be good role models and to lead active lifestyles. Parents can plan active family outings and participate with children, as well as limit sedentary activities.

Documentation from many organizations and research facilities support the benefits of a healthy lifestyle. The following are just a few of the significant groups that have contributed to our current knowledge of lifestyle choices related to health. The "Rules" or Guidelines have changed over the years, but the documentation from all of the following organizations reveals that physical activity is a major key to a healthy lifestyle:

- 400 bc.—*Regimen*—Hippocrates makes exercise recommendations.
- 1961—*Hypokinetic*—term coined by Krause and Rabb relating to "too little activity."
- 1979—*World Health Organization* (WHO) classified obesity as a disease.
- 2010—*Healthy People 2010* (National Health Promotion and disease Prevention Objectives) developed statements by expert groups representing over 300 national organizations that include realistic health goals to be achieved by the year 2010, and now for the year 2020.
- 1992—*The American Heart Association* (AHA) identified inactivity as a major cardiac risk factor.
- 1995—*American College of Sports Medicine* (ACSM) established physical activity guidelines, updated periodically. Promotes and integrates scientific research, education, and practical applications of sports medicine and exercise science to maintain and enhance physical performance, fitness, health, and quality of life.
- 1996—*Surgeon General's Report*—landmark report that, after a thorough review of literature, traced the link between physical activity and good health.

Students who walk to class rather than ride are increasing their lifestyle activity as well as their metabolic rate.

- Ongoing—*Centers for Disease Control* (CDC) provides scientific and technical leadership and assistance to help states, national organizations, and professional groups reduce major risk factors associated with chronic diseases in the U.S.
- 2007—*Exercise Is Medicine* (EIM) launched by medical doctors in conjunction with ACSM, to integrate exercise as preventive medicine as a regular part of medical treatment.
- 2008—*U.S. Department of Health and Human Services* (HHS) released the current physical activity guidelines.
- 2011—*2011 Dietary Guidelines for Americans* (USDA Center for Nutrition Policy and Promotion) long-awaited, evidence-based guidelines encouraging Americans to reduce calorie consumption and increase physical activity.
- 2011—*Million Hearts Initiative* launched with HHS and public and private partners, a commitment to prevent 1 million cardiovascular events (including stroke) in 5 years.
- 2011—*2020 Health Impact Goal*—AHA/ASA (American Stroke Association) launched to improve health of all American by 20% while reducing deaths due to cardiovascular disease and stroke by 20%.
- 2012—*American Cancer Society Guidelines on Nutrition and Physical Activity for Cancer Prevention*—updated every 5 years; most cancer risk is due to factors that are not inherited, but lifestyle-related. Focuses on environment and community for support in choosing healthy behaviors.
- 2012—*Go4Life* is developed specifically to teach, aid, and encourage older Americans how to be active.
- 2012—Heart Disease and Stroke Statistics Update (AH)—updated yearly; comprehensive analysis of health data.
- 2012—*Inactivity Physiology* is added to the medical lexicon.

The fact that Americans need these guidelines at all is evidence that we are not following them. The health of most Americans in the past 30 years has declined as we have eaten more, as we have become more sedentary, and as we have had an explosion in technology. Regardless of what a government agency deems or what research says, the bottom line is that we all have to make the choice, daily, to move more. Our life depends on it.

Types of Hypokinetic Conditions

Cardiovascular Disease (CVD)

The cardiovascular system is responsible for delivering oxygen and other nutrients to the body. The major components of the cardiovascular system are the heart, blood, and the vessels that carry the blood. Cardiovascular disease (CVD) is a catch-all term that includes several disease processes including various diseases of the heart, stroke, high blood pressure, congestive heart failure, and atherosclerosis. The heart muscle may become damaged or lose its ability to contract effectively. The vessels that supply the heart with oxygen may become blocked or damaged and subsequently compromise the heart muscle. Finally, the peripheral vascular system (all of the vessels outside the heart) may become damaged and decrease the ability to provide oxygen to other parts of the body.

The great news is that between 1999–2009, deaths due to cardiovascular disease declined 32.7% (AHA, 2013). Americans are also on the whole, living longer, as life expectancy increases. CVD still claims one in three deaths in the United States. Many of the risk factors for CVD are lifestyle-related and therefore preventable, yet Americans, including women, are much more likely to die from CVD than anything else. The cost of CVD is very high, both in dollars and in productivity lost. Health Impact Goals 2020 and the Million Hearts Initiative are efforts to coordinate public, private, and governmental

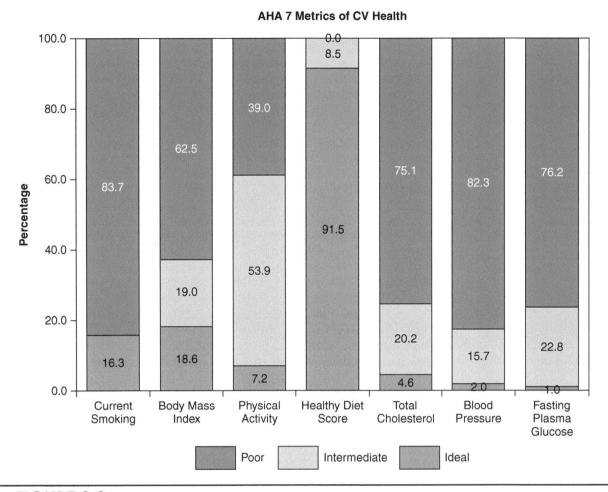

FIGURE 3.2

Prevalence (unadjusted) estimates for poor, intermediate, and ideal cardiovascular health for each of the 7 metrics of cardiovascular health in the American Heart Association 2020 goals, US children aged 12 to 19 years, National Health and Nutrition Examination Survey (Nhanes) 2007–2008 (available data as of June 1, 2011).

resources to focus on the positives, on prevention. As part of Health Impact Goals 2020, the AHA developed **7 Metrics of Cardiovascular Health** (see Figure 3.2). MyLifeCheck.org is a part of a campaign to increase awareness of positive attributes of health. Check it out.

Who Is At Risk for CVD?

The Surgeon General's Report (Satcher, 1996) placed **physical inactivity** as a significant risk factor for cardiovascular diseases and other health disorders. Most sedentary Americans are at risk as stated previously. There are an estimated 82,600,000 Americans that have some form of CVD. Many factors can predispose a person to be at risk for CVD. Sedentary living, habitual stress, smoking, poor diet, high blood pressure, diabetes, obesity, high cholesterol, and family history can all increase risk. Advancing age increases risk. Males typically have a higher risk than women until women are post-menopausal, then risk evens out. Misconceptions still exist that CVD is not a real problem for women. Because more women have heart attacks when they are older, the initial heart attack is more likely to be fatal. It is important for women to realize that CVD is an equal opportunity killer. Just like men, more women die from heart disease than anything else.

Certain populations have an inherently higher health risk such as African Americans and Hispanics. Genetic predisposition is a strong factor; familial tendencies toward elevated triglycerides, fat distribution (abdominal fat accumulation denotes a higher health risk than hip/thigh accumulation of fat), and high **low-density lipoprotein cholesterol (LDL-C)** levels increase risk. LDL-C is a blood lipid that indicates a higher cardiac risk. Saturated fat intake tends to increase LDL cholesterol. Dr. William Franklin of Georgetown University Medical School in Washington claims that anyone who has a close relative who has had a heart attack should begin monitoring his heart with regular stress tests when he is 45. If your father died in his 40's of a heart attack, then you should be concerned a decade earlier in your 30's. Variables such as age, gender, race, and genetic makeup may place you at a higher or lower risk but cannot be changed. These can be termed unalterable risk factors.

CVD Prevention

Since you cannot change your age, your sex, and who your parents are, focus on what you *can* change. These risk factors include but are not limited to: diet, drug use, smoking history, cholesterol levels, obesity, high blood pressure, and last but definitely not least, physical inactivity (see pages 16 and 19). This is a critical point, since activity level is a risk factor that can be easily modified and is often overlooked. Increasing an individual's activity level can prevent many of the diseases discussed in this chapter. Cardiovascular disease is the leading cause of death in the United States. "About every 25 seconds an American will experience a coronary event, and about every minute someone will die from one" (AHA, 2012) (see Figure 3.4). With this being the case, consider your own risk. How can you adjust your current lifestyle habits to decrease your risk? Read the Benefits of Exercise (on page 53) and Figure 3.4 to determine how exercise helps CVD.

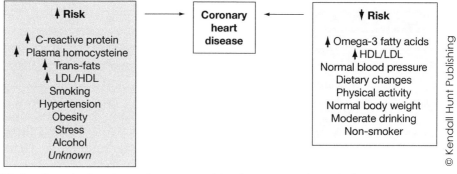

Make the choice to reduce your risk of coronary heart disease.

Claudette's Story

"I consider myself to be relatively healthy and I exercise for about ninety minutes every morning. I started having pain in my chest and face during my exercising, and finally went to the cardiologist. I never thought that the pain in my face could be related to my heart, so I was shocked when the tests showed that I had had a heart attack. I thought I was too young, but my father died of a heart attack when he was only 38, so I had family history as a risk factor. After my second heart attack, I knew that I needed to help get the message out. **Women need to know** that heart disease is their biggest health threat."

Source: National Heart, Lung and Blood Institute, National Health Institute.

The "ABCS" of heart disease and stroke prevention

Aspirin therapy
Blood pressure control
Cholesterol control
Smoking cessation

CVD and stroke are largely preventable for a significant part of the lifespan. High blood pressure, high cholesterol, and smoking continue to put people at risk of heart attack and stroke. To address these risk factors, the Centers for Disease Control and Prevention is focusing many of its efforts on the "ABCS" of heart disease and stroke prevention: **appropriate Aspirin therapy, Blood pressure control, Cholesterol control, and support for Smoking cessation** for those trying to quit and, even more generally, comprehensive tobacco prevention and control efforts. (CDC, 2012)

The American Heart Association projects that by 2030, 40.5% of the U.S. population will have some form of CVD, costing the healthcare system an estimated $1 trillion every year. (AHA, 2012)

Benefits of Exercise

Consistent physical activity affects cardiovascular disease by one or more of the following mechanisms:

- Improved cardiovascular fitness and health
- Greater lean (fat-free) body mass
- Improved strength and muscular endurance
- Stronger heart muscle
- Lower heart rate
- Increased oxygen to the brain
- Reduced blood fat including low-density-lipoprotein cholesterol (LDL-C)
- Increased protective high-density-lipoprotein cholesterol (HDL-C)
- Delayed development of atherosclerosis
- Increased work capacity
- Improved peripheral circulation
- Improved coronary circulation
- Reduced risk of heart attack
- Reduced risk of stroke
- Reduced risk of hypertension
- Greater chance of surviving a heart attack
- Greater oxygen carrying capacity of the blood

Exercise improves your body and mind more than you might expect.

Arteriosclerosis

Arteriosclerosis is a term used to describe the thickening and hardening of the arteries. Healthy arteries are elastic and will dilate and constrict with changes in blood flow, which allows proper maintenance of blood pressure. Hardened, non-elastic arteries do not expand with blood flow and can increase intrarterial pressure causing high blood pressure. Both high blood

FIGURE 3.3

The Benefits of Exercise
© ATurner, 2012, Shutterstock, Inc.

Increasing lifestyle activity by spending less time on the couch and doing something active daily will have a positive impact on your health.

pressure and arteriosclerosis increase the risk of an **aneurysm**. With an aneurysm, the artery loses its integrity and balloons out under the pressure created by the pumping heart, in much the same way as an old garden hose might if placed under pressure. If an aneurysm occurs in the vessels of the brain, a stroke might occur. Aneurysms in the large vessels can place a person at risk of sudden death. Maintaining normal elasticity of the arteries is very important for good health. Exercise helps to manage symptoms and the factors that contribute to cardiac risk.

Atherosclerosis Atherosclerosis is a type of arteriosclerosis. Atherosclerosis is the long-term buildup of fatty deposits and other substances such as cholesterol, cellular waste products, calcium, and fibrin (clotting material in the blood) on the interior walls of arteries (see Figure 3.6). The leading theory states that plaque develops when the endothelium (a thin layer of cells that line the interior vessel wall) is damaged due to major fluctuations in blood pressure, increased levels of blood triglycerides, cholesterol, and cigarette smoking. Conditions such as these accelerate the development of atherosclerosis. Due to this plaque development, the flow of blood within the artery decreases because the diameter of the vessel is decreased. This may create a partial or total blockage (called an occlusion) that may cause high blood pressure, a heart attack, or stroke. This process can occur in any vessel of the body. If it occurs outside of the brain or heart, it is termed peripheral vascular disease. Within the heart the gradual narrowing of the coronary arteries to the myocardium, or heart muscle, is called coronary artery disease. Atherosclerosis is a disease that can start early in childhood. With a family history of high cholesterol, it is important to check a child's cholesterol levels early in life. Besides heart disease and stroke, atherosclerosis can lead to kidney disease. The rate of progression of atherosclerosis depends on family history and lifestyle choices. Exercise helps manage symptoms as well as increase coronary collateral circulation. Collateral arteries are the vessels that form preceding the blockage as an artery slowly becomes occluded. Collateral vessels such as these can help lessen the severity of a heart attack when the artery becomes totally blocked. High cholesterol levels can increase risk of atherosclerosis, and low-density lipoprotein cholesterol is thought to contribute to the arterial occlusion. **Triglycerides** are another type of blood fat which at high levels is associated with high risk (see Figure 3.7). Regular physical activity has been shown to lower risk by lowering blood lipid (fat) levels.

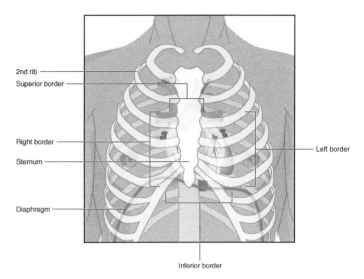

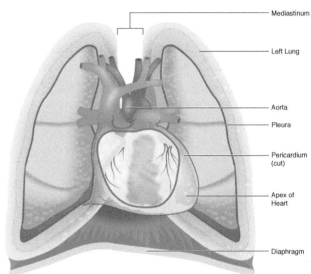

From *Human Anatomy and Physiology: A Functional Approach*, 2/e by Joseph Crivello. Copyright © 2013 by Kendall Hunt Publishing Company. Reprinted by permission.

Peripheral vascular disease is simply a term attributed to disease of the peripheral vessels. The lack of proper circulation may cause fluids to pool in the extremities. Associated leg pain, cramping, numbness, tingling, coldness, and loss of hair to affected limbs are common signs. The restrictions in blood flow are typically caused by years of arteriosclerosis and atherosclerosis in the vessels of the extremities. The risk factors are the same as those for cardiovascular disease. One difference is that the disease process may progress extensively before the affected person begins to notice any problems. The heart and brain are much more sensitive to compromised blood flow than are the extremities.

Hypertension, or high blood pressure, is often called the "silent killer" because typically there are no symptoms. Because hypertension is asymptomatic, it is important to get your blood pressure checked on a regular basis. In 2009, the estimated prevalence of hypertension (a blood pressure reading of 140/90 mm or higher) was one in 3 adults. High blood pressure is associated with a shortened life span. Interestingly, under the age of 45 more males typically have a higher blood pressure, while after age 55, more females tend to have a higher blood pressure (AHA, 2008). High blood pressure causes the heart to work harder. Chronic, untreated hypertension can lead to aneurysms in blood vessels, heart failure from an enlarged heart, kidney failure, atherosclerosis, and blindness. African Americans have highest incidence of high blood pressure at 44%.

The top number is the **systolic** reading, which represents the arterial pressure when the heart is contracting and forcing the blood through the arteries. The bottom number is the **diastolic** reading, which represents the force of the blood on the arteries while the heart is relaxing between beats. In 2003 new blood pressure guidelines were issued, with a new "prehypertensive" category identified (see Figure 3.9). A blood pressure reading of 115/75 is the new thresh-

> If you don't smoke, don't start. **If you do smoke, get help to quit now!** Many effective programs, nicotine patches, and other medications are available to help you quit. As soon as you stop smoking, your risk of heart disease starts to drop. In time your risk will be about the same as if you'd never smoked.

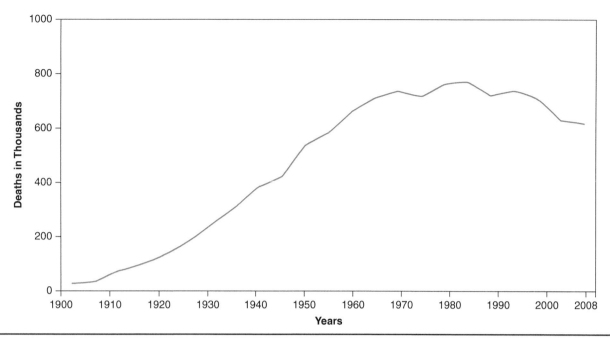

FIGURE 3.4

Deaths attributable to diseases of the heart (United States: 1990–2008)
Source: National Center for Health Statistics

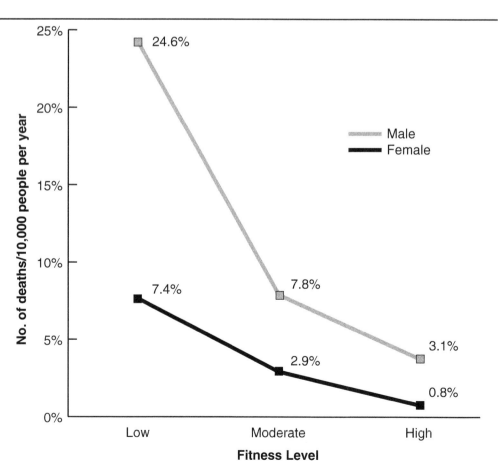

FIGURE 3.5

Relationship between Different Levels of Fitness and Death Due to Cardiovascular Disease among Men and Women

Source: Blair et al., Physical fitness and all-cause mortality: A prospective study of healthy men and women. *Journal of the American Medical Association* 262(17): 2395–2401, 1989. (Adapted from S. N. Blair, N. W. Kohl, III, R. S. Paffenbarger, Jr., D. G. Clark, K. H. Cooper, and L. W. Gibbons. Physical fitness and all-cause mortality: A prospective study of healthy men and women.)

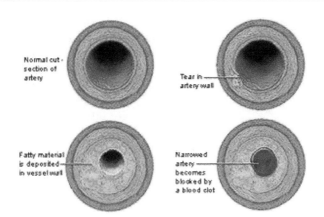

FIGURE 3.6

Atherosclerosis

Coronary artery disease occurs when a substance called plaque builds up in the arteries that supply blood to the heart (called coronary arteries). Plaque is made up of cholesterol deposits, which can accumulate in your arteries. When this happens, your arteries can narrow over time. This process is called atherosclerosis.

Source: Courtesy of the Center for Disease Control

FIGURE 3.7

Cholesterol and Triglyceride Levels and Risk of Heart Disease

Total Cholesterol Level	Category
Less than 200 mg/dL	Desirable level that puts you at lower risk for heart disease. A cholesterol level of 200 mg/dL or higher raises your risk.
200–239 mg/dL	Borderline high
240 mg/dL and above	High blood cholesterol. A person with this level has more than twice the risk of heart disease as someone whose cholesterol is below 200 mg/dL.
Cholesterol levels are measured in milligrams (mg) of cholesterol per deciliter (dL) of blood.	

LDL Cholesterol Level	Category
Less than 100 mg/dL	Optimal
100–129 mg/dL	Near or above optimal
130–159 mg/dL	Borderline high
160–189 mg/dL	High
190 mg/dL and above	Very High
mg/dL = milligrams per deciliter of blood	

HDL Cholesterol Level	Category
Less than 40 mg/dL	A major risk factor for heart disease.
40–59 mg/dL	The higher your HDL level, the better.
60 mg/dL and above	An HDL of 60 mg/dL and above is considered protective against heart disease.
mg/dL = milligrams per deciliter of blood	

Triglyceride Level	Category
Less than 150 mg/dL	Normal
150–199 mg/dL	Borderline high
200–499 mg/dL	High
500 mg/dL and above	Very high
mg/dL = milligrams per deciliter of blood	

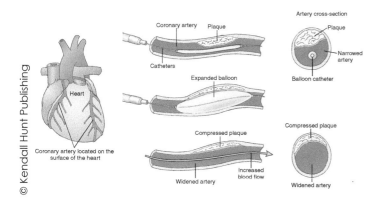

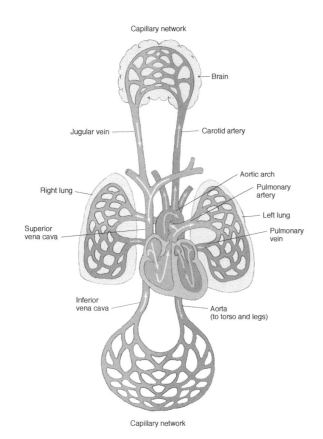

old above which cardiovascular complication can occur. The prehypertensive category includes a systolic pressure from 120–139 and a diastolic pressure from 80–89 as a warning zone. If your are considered prehypertensive, it is time to take action by modifying your lifestyle. Any reading consistently over 139/89 mm Hg is high blood pressure and indicates a high risk. With persons over 50 years old, a systolic reading of 140 or above is a more important CVD risk factor than the diastolic reading (JNC VII, 2003).

Hypertension cannot be cured, but it can be successfully treated and controlled. Most people with hypertension have additional risk factors for cardiovascular disease. Some of the risk factors for high blood pressure include Hispanic or African American heritage, older age, family history, a diet high in fat and sodium, alcoholism, stress, obesity, and inactivity. Exercise has been shown to help symptoms of high blood pressure in mild to moderate hypertension.

Heart Attack

A heart attack or **myocardial infarction** occurs when an artery that provides the heart muscle with oxygen becomes blocked or flow is decreased. The area of the heart muscle served by that artery does not receive adequate oxygen and becomes injured and may eventually die. The heart attack may be so small as to be imperceptible by the victim, or so massive that the victim will die. It is often reported that heart attack victims delay seeking medical help with the onset of symptoms. Every minute counts! In one study, men waited an average of three hours before seeking help. Women waited four hours. It is important to seek medical help at the first sign of a heart attack.

Teens, Sleep, and Blood Pressure In the News

A new study finds that teens who get too little sleep or erratic sleep may elevate their blood pressure. "Our study underscores the high rate of poor quality and inadequate sleep in adolescence coupled with the risk of developing high blood pressure and other health problems which may lead to cardiovascular disease," says Susan Redline, M.D., professor of medicine and pediatrics and director of University Hospital's Sleep Center at Case Western Reserve University in Cleveland, Ohio. Researchers say technology in bedrooms (phone, games, computers, music) may be part of the problem (AHA, 2008).

Women who smoke and take oral contraceptives are ten times more likely to have a heart attack (Payne and Hahn, 2000). "Smoking and oral contraceptives (OC) appear to act synergistically in increasing the risk of arterial thrombotic disease, particularly in heavy smokers and with old OC formulations," Ojvind Lidegaard reported in 1998. In addition to the classic symptoms of heart attack listed in the box on page 124, women were more likely than men to report throat discomfort, pressing on the chest, and vomiting.

Exercise is the cornerstone therapy for the primary prevention, treatment, and control of hypertension, according to the Position Stand *Exercise and Hypertension* released from the American College of Sports Medicine (ACSM). Adults with hypertension should seek to gain at least thirty minutes of moderate-intensity physical activity on most, if not all, days of the week, but they should be evaluated, treated, and monitored closely.

Each person may experience heart disease in a different way and unfortunately, a fatal sudden cardiac arrest may be the only symptom. Heart attack symptoms for women may be different than the classic symptoms that are commonly known such as chest, jaw, or left arm pain with shortness of breath and weakness. Women may experience more subtle symptoms such as fatigue, depression, back pain, or pain throughout the chest. Don't wait to get help, as time is critical when experiencing a heart attack.

Some findings suggest that **coronary collateral circulation** is increased with regular physical activity (Corbin and Welk, 2009). This increased vascularization may decrease the risk of having a heart attack, as well as increase the chances of survival if a heart attack does occur. This happens because the new vessels, which form as a result of exercise, can take over if a major coronary artery is blocked. Since 1951, the death rate from heart attacks has declined by 51 percent, yet more Americans die from coronary artery disease than from any other disease. Both treatment and prevention for heart attacks has increased due to revolutionary new surgical treatments, new drugs, and new information about the etiology of heart disease. Many of the drugs reserved for treating cardiac patients in the past are now used as aggressive prevention in high-risk patients. The AHA has developed Heart Attack Symptoms and Warning Signs.

Acting F.A.S.T. Is Key for Stroke

Acting F.A.S.T. can help stroke patients get the treatments they desperately need. The most effective stroke treatments are only available if the stroke is recognized and diagnosed within 3 hours of the first symptoms. If you think someone may be having a stroke, act F.A.S.T.[1] and do the following simple test:

F—Face: Ask the person to smile. Does one side of the face droop?
A—Arms: Ask the person to raise both arms. Does one arm drift downward?
S—Speech: Ask the person to repeat a simple phrase. Is their speech slurred or strange?
T—Time: If you observe any of these signs, call 9-1-1 immediately.

Note the time when any symptoms first appear. Some treatments for stroke only work if given in the first 3 hours after symptoms appear. Do not drive to the hospital or let someone else drive you. Call an ambulance so that medical personnel can begin life-saving treatment on the way to the emergency room.

Heart Attack: Warning Signs Common in Women

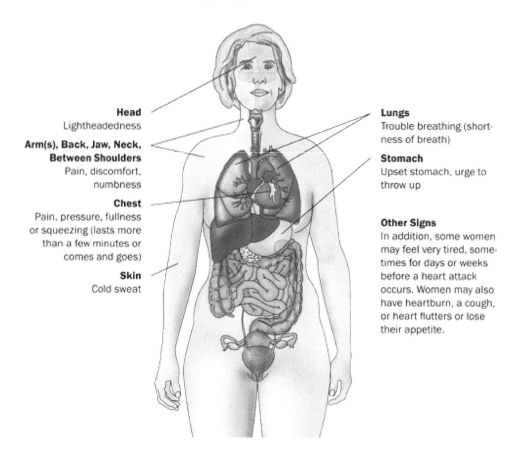

Head
Lightheadedness

Arm(s), Back, Jaw, Neck, Between Shoulders
Pain, discomfort, numbness

Chest
Pain, pressure, fullness or squeezing (lasts more than a few minutes or comes and goes)

Skin
Cold sweat

Lungs
Trouble breathing (shortness of breath)

Stomach
Upset stomach, urge to throw up

Other Signs
In addition, some women may feel very tired, sometimes for days or weeks before a heart attack occurs. Women may also have heartburn, a cough, or heart flutters or lose their appetite.

Source: Image courtesy of the Office on Women's Health, U.S. Department of Health and Human Services. Women may experience cardiac symptoms that are different from men.

FIGURE 3.8
Blood Pressure is known as the 'silent killer' because many people do not realize that they have high BP. Do you know your BP?

$\frac{117}{76}$ mm Hg

Read as "117 over 76 millimeters of mercury"

Systolic
The top number, which is also the higher of the two numbers, measures the pressure in the arteries when the heart beats (when the heart muscle contracts).

Diastolic
The bottom number, which is also the lower of the two numbers, measures the pressure in the arteries between heartbeats (when the heart muscle is resting between beats and refilling with blood).

Source: American Heart Association

Do you know the warning signs of a stroke? There is a public awareness campaign to increase knowledge of stroke warning signs and symptoms (see box on page 59). Stroke, or more recently called "**brain attack**," is the third leading cause of death affecting 795,000 Americans per year (AHA, 2013). This occurs when the vessels that supply the brain with nutrients become damaged or occluded and the brain tissue dies because of insufficient oxygen. The cerebral artery, the main supply of nutrients to the brain, can be narrowed due to atherosclerosis. The conditions that precipitate stroke may take years to develop. Stroke has the same risk factors as heart disease. Hypertension is the most notable risk factor. Like heart disease, conditions

Blood Pressure Category	Systolic mm Hg (upper #)		Diastolic mm Hg (lower #)
Normal	less than **120**	and	less than **80**
Prehypertension	120–139	or	80–89
High Blood Pressue (Hypertension) **Stage 1**	140–159	or	90–99
High Blood Pressue (Hypertension) **Stage 2**	**160** or higher	or	**100** or higher
Hypertensive Crisis (Emergency care needed)	Higher than **180**	or	Higher than **110**

* Your doctor should evaluate unusually low blood pressure readings.

FIGURE 3.9

This chart reflects blood pressure categories defined by the American Heart Association.
Source: American Heart Association

favorable to stroke also respond favorably to exercise. Ischemic (thrombosis and embolism) strokes are the most common form of stroke (87 percent) and occur as a result of a blockage to the cerebral artery (AHA, 2009). The process is similar to that which occurs in a heart attack. Intracerebral hemorrhage, or aneurysm, in which the vessel may rupture and cause bleeding inside the head and result in pressure on the brain, are 10 percent of strokes. Three percent of strokes are caused by hemorrhage. The least common form of stroke results from compression that can occur as a result of a hemorrhage or brain tumor. African Americans have the highest risk at 44% (AHA, 2012). African Americans also have a high incidence of stroke risk factors such as high blood pressure. On the average, someone in the United States has a stroke every forty seconds, and every three to four minutes someone dies of a stroke (AHA, 2012). One-third of all stroke victims die, one-third of stroke victims suffer permanent disability, and one-third of stroke victims gradually return to their normal daily routines (Bishop and Aldana, 1999). Stroke is also a leading cause of serious disability. Various studies have shown significant trends toward lower stroke risk with moderate and high levels of leisure time physical activity.

Make sure you know the risk factors for having a stroke. Do you know the signs and symptoms of a brain attack? (see table at end of chapter) Although it is much more common to suffer a stroke as you age, a stroke is possible at any age. If you suspect someone is having a stroke, alert medical personnel immediately. Time to treatment is critical.

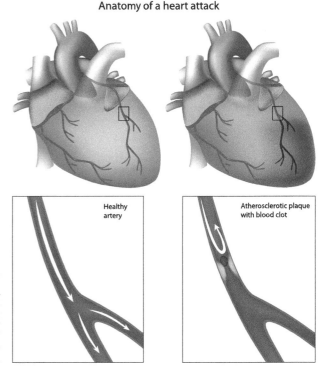

Anatomy of a heart attack

Healthy artery

Atherosclerotic plaque with blood clot

© Alila Medical Media, 2014. Used under license from Shutterstock, Inc.

Signs of a stoke could be sudden numbness or weakness in one side of the body, sudden confusion, sudden trouble seeing, sudden trouble walking or loss of balance or a sudden severe headache with no known cause.

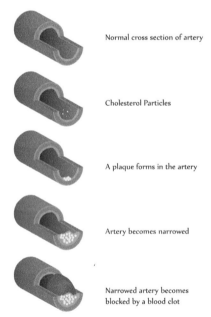

Risk Factors for Cardiovascular Disease

Controllable Risk Factors

- **Cigarette Smoking**—Smokers have two to four times the risk of developing cardiovascular disease than do nonsmokers (AHA, 2009). Cigarette smoking is the most "potent" of the preventable risk factors. Former U.S. Surgeon General C. Everett Koop claims that cigarette smoking is the number one preventable cause of death and disease in the United States and the most important health issue of our time. Smoking accounts for 50 percent of the female deaths due to heart attack before the age of 55 (Rosato, 1994).
- **Hypertension**—The AHA (2012) reports that approximately 76.4 million American adults and children have high blood pressure. Reports from the Harvard Alumni Study (1986) show that subjects who did not engage in vigorous sports or activity were 35 percent more likely to develop hypertension than those who were regularly active. Hypertension is the most important modifiable risk factor for stroke.
- **Cholesterol**—Dietary cholesterol contributes to blood serum cholesterol (cholesterol circulating in the blood), which can contribute to heart disease. Every 1 percent reduction in serum cholesterol can result in a 2–3 percent reduction in the risk of heart disease (AHA, 2009). To lower cholesterol, reduce intake of dietary saturated fat, increase consumption of soluble fiber, maintain a healthy weight, do not smoke, and exercise regularly.
- **Inactivity**—Physical inactivity can be very debilitating to the human body. The changes brought about by the aging process can be simulated in a few weeks of bed rest for a young person. Aerobic exercise on a regular basis can favorably influence the other modifiable risk factors for heart disease. Consistent, moderate amounts of physical activity can promote health and longevity. The Surgeon General's report (Satcher, 1996) states that as few as 150 extra calories expended daily exercising can dramatically decrease CVD risk.
- **Obesity**—Highly correlated to heart disease, mild to moderate obesity is associated with an increase in risk of CVD. Fat distribution can also predict higher risk. A waist-to-hip ratio that is greater than 1.0 for men and greater than 0.8 for women constitutes a higher risk because abdominal fat is more easily mobilized and dispersed into the bloodstream, thereby elevating serum cholesterol levels. A BMI over 30 is considered obese.
- **Diabetes**—At least 65 percent of diabetics die of some form of CVD (CDC, 1999). Exercise is critical to help increase the sensitivity of the body's cells to insulin. 18.3 million Americans have diabetes (AHA, 2012.)

Famed television host **David Letterman** was not overweight when he had quintuple bypass surgery at age 52 in 2000. Although Mr. Letterman didn't look like the typical person who has a heart attack, he had several risk factors going against him. He had a family history—his father, Harry, died of a heart attack in his 50's. Mr. Letterman had high cholesterol. Most likely his job would be considered high stress. David Letterman credits Dr. Wayne Isom, who operated on his heart, with saving his life. In an interview with another talk show host, Larry King (who coincidentally also was operated on by Dr. Isom for quadruple bypass surgery) asked Dr. Isom what was important in avoiding heart disease. Besides exercise, controlling stress, managing weight, and eating well, Dr. Isom said that attitude is very, very important. Post-heart surgery, the patient must decide for himself that he is going to get well. An important part of cardiac rehabilitation is a *positive attitude*.

Uncontrollable Risk Factors

- ***Age***—Risk of CVD rises as a person ages.
- ***Gender***—Men have a higher risk than women until women reach postmenopausal age. Remember that CVD is an equal opportunity killer!
- ***Heredity***—A family history of heart disease will increase risk.

Contributing Risk Factors

- ***Stress***—Although difficult to measure in concrete form, stress is considered a factor in the development and acceleration of CVD. Without stress-management techniques, constant stress can manifest itself in a physical nature in the human body. Stress contributes to many of today's illnesses.
- ***Triglycerides***—Most of the fat in the human body is stored in the form of triglycerides. Elevated triglyceride levels are thought to increase CVD risk by being involved in the plaque formation of atherosclerosis.

Obesity

Since 1979 the World Heath Organization (WHO) has classified obesity as a disease. "Obesity is a complex condition, one with serious social and psychological dimensions, that affects virtually all age and socioeconomic groups and threatens to overwhelm both developed and developing countries. As of 2000, the number of obese adults has increased to over 300 million" (WHO, 2008). "Globesity" may be the new term coined for the world's heavy populations. While malnutrition still contributes to an estimated 60 percent of deaths in children ages 5 and under globally, in the United States the excess body weight and physical inactivity that leads to obesity cause more than 112,000 deaths each year, making it the second leading cause of death in our county.

The figure above shows the prevalence of obesity among adults aged ≥20 years, by race/ethnicity and sex in the United States during 2009–2010, according to the National Health and Nutrition Examination Survey. Among adults aged ≥20 years in 2009–2010, 35.5% of men and 35.8% of women were obese. Among men, 38.8% of non-Hispanic blacks, 37.0% of Hispanics, and 36.2% of non-Hispanic whites were obese. Among women, 58.5% of non-Hispanic blacks, 41.4% of Hispanics, and 32.3% of non-Hispanic whites were obese.

Obesity causes, contributes to, and complicates many of the diseases that afflict Americans. Obesity is associated with a shortened life, serious organ impairment,

---

Health Risks of Obesity
Each of the diseases listed below is followed by the percentage of cases that are caused by obesity.

Colon cancer	10%
Breast cancer	11%
Hypertension	33%
Heart disease	70%
Diabetes	90%
(Type II, non-insulin-dependent)	

As these statistics show, being obese greatly increases the risk of many serious and even life-threatening diseases.

poor self-concept, and a higher risk of cardiovascular disease and diabetes, as well as colon and breast cancer.

Fat distribution is related to health risk (Canoy, 2007). "Apples" describe male-fat patterned distribution with fat accumulating mostly around the torso. "Pears" describe female-fat patterned distribution with fat accumulating mostly on the hips and upper thighs (see Figure 3.11). Apples have a higher health risk especially if they have visceral fat located around internal organs.

Childhood Obesity

> **Childhood Obesity:**
> How bad is it?
> About one in three children and teens in the U.S. is overweight or obese.
> Overweight kids have a 70–80 percent chance of staying overweight their entire lives.
> Obese and overweight adults now outnumber those at a healthy weight; nearly seven in 10 U.S. adults are overweight or obese.
> Reprinted with permission, www.heart.org, © 2010, American Heart Association, Inc.

We are a society of excesses. Unfortunately, everyone seems to be getting bigger—all ages, sexes, races independent of socioeconomic status, gender, or locale. The increase in overweight children causes the most concern. For some parents, childhood obesity has become a bigger concern than smoking or drug abuse. In the last 30 years, school age children (ages 6–11) have increased obesity rates from 4% to 20%.

The causes of **childhood obesity** are complex. As with adult obesity, the bottom line is that if there is a caloric intake surplus, weight will be gained. Infants can be overfed, toddlers can be pacified with candy, and teenagers love soft drinks and junk food. Overweight parents are more likely to have overweight children because the children learn eating and activity patterns from parents. To try to combat childhood obesity, adults should look at their own lifestyle habits: 44 oz sodas several times a day, fast food throughout each week, and lots of time spent playing video games or watching TV will increase the risk of obesity for children and adults alike. These behaviors are not the way to encourage kids to become healthy. Simple things such as eating balanced meals together at home, recreating as a family, and not having junk food always accessible at home are good places to start. Childhood obesity can have a negative impact on overall health risk including psychosocial consequences. Perhaps as individuals, families, schools, communities, and as a nation we should emphasize being more active. Overweight by some arbitrary standard may or may not mean cardiovascular or diabetic risk. Sedentary living, however, almost always predicates poor health in the future, if not sooner. "School aged youth should participate daily in 60 minutes or more of moderate to vigorous physical activity that is developmentally appropriate, enjoyable, and involves a variety of activities" (*Journal of Pediatrics*, 2005). How ironic that we should have to tell our kids to go out and play! Adults would do well to follow this same advice.

In regards to aging, if a young adult is overweight then the odds are he or she will be overweight or obese in middle age and older. Established patterns in life are difficult to change. Tim Spector (St Thomas Hospital in London, UK) reported in *The Lancet* that obesity may accelerate the aging process. Spector found dramatic differences in obese and lean women in the length of the telomeres on the end of their white blood cell chromosomes. The difference was over 8 years of aging. Spector also found that smokers were biologically older

---

The Surgeon General Encourages Americans to **Know Their Health History**—In the fall of 2008 the acting Surgeon General encouraged all Americans to take advantage of family gatherings to speak with family members to discuss, identify, and make a record of health problems that seem to run in the family. Doing this can offer insight into your health risk. Check out the Web-based tool "My Family Health Portrait" at www.hhs.gov/familyhistory/

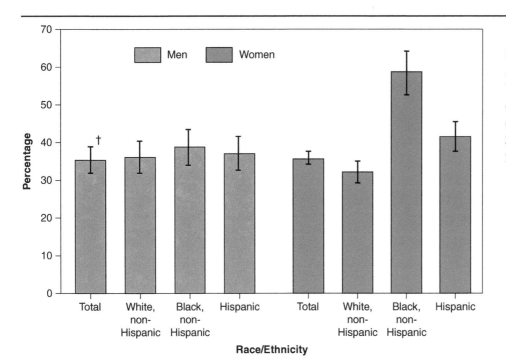

FIGURE 3.10

Prevalence of Obesity Among Adults Aged ≥20 Years, by Race/Ethnicity and Sex—National Health and Nutrition Examination Survey, United States, 2009–2010

Source: National Health and Nutrition Examination Survey, 2009–2010.

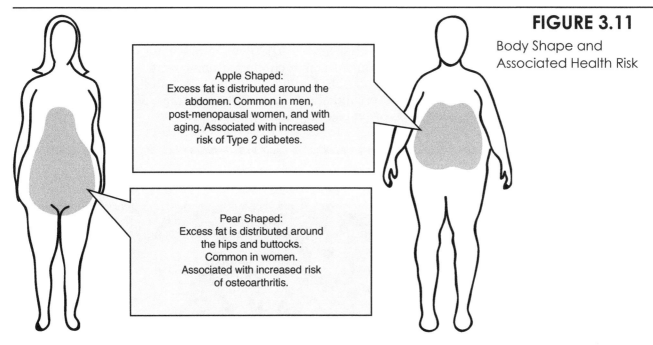

FIGURE 3.11

Body Shape and Associated Health Risk

Source: From National Institute of Diabetes and Kidney Disease.

than nonsmokers by 4.6 years, according to telomere length. Losing weight or quitting smoking can slow the loss of telomere length. Poor lifestyle choices can have a synergistic effect and contribute to aging, however it is prudent to focus on the fact that positive lifestyle choices also can have a positive synergistic effect on both quality of life and the aging process.

Causes of Obesity

Is it your genes or your fast-food lunches every day? Most likely it is both. Since you cannot change who your parents are, change your lifestyle habits. *Physical inactivity is certainly a major, if not the primary, cause of obesity in the United States today* (Wilmore, 1994). Most often caloric intake exceeds caloric expenditure. Glandular disorders affect 2 percent of the obese population. Genetically we are predisposed to a certain somatotype, fat distribution, size, and weight.

In every person, body weight is the result of many factors; genetic, metabolic, behavioral, environmental, cultural as well as socioeconomic influences (Surgeon General, 2005). An individual's lifestyle choices can help to modify these

Let's Move! America's Move to Raise a Healthier Generation of Kids

The health of American children should be a bipartisan initiative in every administration

Today's kids live differently than kids raised a generation ago. Most children, 30 years ago, walked to and from school, had homemade meals at home, and maybe one snack a day (3–6 snacks are common today). Fast food meals were a rarity. Kids played at recess and had gym class. Today, budget cuts require gym class, librarians, and after-school activities to be cut. Sweetened drinks are the norm for many kids and can be a source of extra calories. Parents are busy and the average American child spends 7.5 hours with electronic devices for entertainment. Less than 1/3 of high school students get the recommended amount of daily physical activity. In the last 30 years, our society has changed dramatically.

With the Let's Move! Campaign, the goal is to end the epidemic of childhood obesity in one generation. Mrs. Obama hopes kids can put "play" back into their lives by having the children experience "that exercise and eating good stuff" can be fun. The first lady quoted a young 7-year-old Penacook, New Hampshire resident, Caitlyn Habel, remarking that, "I like being able to play games before school because it's really fun and it helps me wake up my heart." 3/9/12 (whitehouse.gov/the press-office)

It has taken 3 decades to create an environment where childhood obesity rates have tripled. Programs such as Let's Move! strive to work with schools, families, communities, and the corporate sector to change the circumstances for this generation and also for generations to come. So choose to get involved in your community. Ask a kid to help you plant a community garden!

"The physical and emotional health of an entire generation and the economic health and security of our nation is at stake." —First Lady Michelle Obama at the Let's Move! Launch on February 9, 2010.

Stand2Learn

One simple thing can help a child build and maintain healthy bones and muscles, reduce risk of developing obesity and chronic diseases, reduce feelings of depression and anxiety, and promote feelings of well-being. That same thing can help improve academic achievement, and also improve academic behavior such as time on task, concentration, and attentiveness in the classroom. What is that one thing? Exercise. Savvy school districts should make sure kids are active every day at school! Perhaps those school districts should speak with some researchers at the Texas A&M Health Science Center involved in Stand2Learn.

In College Station, TX, a pilot study was conducted by researchers at College Hills Elementary School. Mark Benden, Ph.D., CPE, assistant professor at the TAMHSC-School of Rural Public Health, examined the effect giving students in several first grade classrooms the option to stand at a tall desk rather than sit during the day while in class.

"Students in classrooms with the standing-height desks are choosing to stand more than two-thirds of the time and are burning an average of 17 percent more calories overall than their classmates in traditional seated classrooms," Dr. Benden said. "More importantly, overweight and obese students are burning 32 percent more calories while working at standing desks than their peers who work in traditional seated classrooms." Besides expending more calories the students are also experiencing some fringe benefits which the teachers must like. "In addition to increasing physical activity, teachers in these classrooms note that standing desks seem to increase alertness and attentiveness of students while decreasing disruptive behavior."

Perhaps in the not too distant future classrooms will be full of stand up desks and the rest of us will all be working at treadmill desks.

Source: Courtesy of Stand2Learn

Standing for eight hours a day adds up to 160 calories, the equivalent of a half-hour walk. Over weeks and years, the energetic difference between mostly sitting and standing is staggering.
Daniel Lieberman, *The Story of the Human Body*

tendencies. Nineteen out of twenty overweight teenagers will be overweight adults (Texas A&M University Human Nutrition Conference, 1998).

Physiological Responses to Obesity

For an obese person, more blood vessels are needed to circulate blood. The heart has to pump harder, therefore increasing blood pressure. Extra weight can be tough on the musculoskeletal joints, causing problems with arthritis, gout, bone and joint diseases, varicose veins, gallbladder disease, as well as complications during pregnancy. Obese individuals often are heat intolerant and experience shortness of breath during heavy exercise. Obesity increases most cancer risks (Bishop and Aldana, 1999).

Cancer

Cancer is characterized by the uncontrollable growth and spread of abnormal cells. Cancer cells do not follow the normal code of DNA that is encrypted in noncancerous cells.

What Is Cancer?

Possibly in the future people will be able to go to the doctor for a simple blood test to determine whether they will have cancer or not. Unfortunately there seems to be no rhyme or reason for some cancer cases. Lifestyle choices, as well as heredity and also luck, play a big role in a person's risk of developing cancer. Even personality can influence if a person is prone to cancer. With health promotion and prevention, fewer people may develop cancer, and more cancer patients may survive.

Can Cancer Be Prevented?

It is theorized that 80 percent of cancers can be prevented with positive lifestyle choices. Avoiding tobacco and over-exposure to sunlight are two major examples. Eating a varied diet, consuming antioxidants, having a positive attitude, and participating in regular physical activity are simple choices that can have a large impact on cancer prevention. Thirty-five percent of the total cancer death toll is associated with diet (Rosato, 1994), and fit individuals may

TABLE 3.2 ♦ Physical Activity and Cancer

Cancer Type	Effect of Physical Activity
Colon	Exercise speeds movement of food and cancer-causing substances through the digestive system, and reduces prostaglandins (substances linked to cancer in the colon).
Breast	Exercise decreases the amount of exposure of breast tissue to circulating estrogen. Lower body fat is also associated with lower estrogen levels. Early life activity is deemed important for both reasons. Fatigue from therapy is reduced by exercise.
Rectal	Similar to colon cancer, exercise leads to more regular bowel movements and reduces "transit time."
Prostate	Fatigue from therapy is reduced by exercise.

Can You Make a Difference?

Get involved in your local school district as an activist for good health. When school budgets are tight, P.E. teachers are often the first to be let go. Often we think one person can't make a difference. Molly Barker didn't let that stop her from forming a grassroots organization that targets the emotional and the mental fitness as well as developing the physical fitness of young girls. The program, called *Girls on the Run,* is a twelve-week program that culminates in the participants running or walking a 5-km road race. The road race is secondary to what the girls experience in the twelve weeks leading up to the race. Positive preteen emotional development is the focus. The girls might warm up by running/walking around a track, and then have focused girl talk. They discuss positive people in their lives. Issues such as pressures to look a certain way, anorexia, bullying, nutrition, the role of women in society, and what makes each girl special are contemplated as a group. *Girls on the Run* is now in over one-hundred cities across Canada and the United States. One person can make a difference. Be a good role model for your kids, your siblings, your relatives, for any children that you come in contact with. Like Molly Barker, choose to be involved not only with your own health, but also with the health of your community.

---

have a decreased risk of reproductive organ cancers (Bishop and Aldana, 1999). Cancer is the second leading cause of death in the United States, accounting for about 23 percent of all deaths yearly (Hoeger et al., 2009).

Does Exercise Help?

Recognition of the potential of exercise to prevent cancer came in 1985 when the American Cancer Society began recommending exercise to protect against cancer. Regular activity has been shown to reduce risk of colon cancer (see Table 3.2). Active people have lower death rates from cancer than inactive people—50 to 250 percent lower. Colon, breast, rectal, and prostate cancers each have an established link with inactivity.

Exercising early in life also seems to have an impact in reducing risk of breast cancer in post-menopausal women. A study at the USC Norris Cancer Center reported that one to three hours of exercise a week over a woman's reproductive lifetime (between the teens and age 40) may result in a 20 to 30 percent risk reduction for breast cancer. Exercise that averaged four or more hours per week resulted in a 60 percent reduction! A woman starting to exercise in her 20's or 30's can also experience reduced risk. Active females, such as a dancer or track athlete, may put off the age of onset of menstruation, and if they continue to be active, they may experience earlier menopause than their inactive counterparts. This results in a lower lifetime exposure to estrogen, which also reduces cancer risk. Ironically lower estrogen levels may contribute to osteoporosis.

It is also thought that exercise can boost immunity that can help kill abnormal cancer cells (Bishop and Aldana, 1999). Dr. Steven Blair at the Institute for Aerobics Research in Dallas, Texas, has done long-term epidemiological studies that show rate of death due to cancer is significantly lower in patients with elevated levels of fitness. It must also be noted that people who are active tend to also participate in other healthy behaviors, such as eating a varied diet low in fat and high in fiber. These other behaviors may also influence cancer risk and help those with cancer lead more fulfilling and productive lives. The American Cancer Society reports that people with healthy lifestyles (non-smokers, regular physical activity, and sufficient sleep) have the lowest cancer mortality rates.

> "Exercise is a known remedy for the weakness and low spirits that cancer patients experience during their recovery. It boosts energy and endurance, and also builds confidence and optimism. But, within the past five years, several medical investigations have revealed a surprising new fact: Exercise may also help prevent cancer" (Rosato, 1994).

Diabetes

Diabetes is a disorder that involves high blood sugar levels and inadequate insulin production by the pancreas or inadequate utilization of insulin by the cells (Wilmore, 1994). Type II diabetes will be discussed in this chapter.

Who Gets Diabetes?

Eighty percent of the adults who develop Type II diabetes are obese (Surgeon General, 2005). The mortality rate is greater in diabetics with CVD—68% of people with diabetes die from some form of CVD. Each year, 1.6 million new cases of diabetes are diagnosed (AHA, 2012). Diabetes is the seventh leading cause of death in people over 40 (Corbin and Welk, 2009). Due to the surge in childhood obesity in the decade of the 90's, children are more at risk for diabetes. Diabetes is one of the most important risk factors for stroke in women.

Can Diabetes Be Prevented?

Research shows that changing lifestyle habits to decrease risk for heart disease also decreases risk for diabetes. "According to research, a seven percent loss of body weight and 150 minutes of moderate-intensity physical activity a week can reduce the chance of developing diabetes by 58 percent in those who are at high risk. These lifestyle changes cut the risk of developing type II diabetes regardless of age, ethnicity, gender, or weight." Type II diabetes may account for 90–95 percent of all diagnosed cases of diabetes (AHA, 2012).

Does Exercise Help?

Exercise plays an important role in managing this disease, as exercise helps control body fat and improves insulin sensitivity and glucose tolerance. Exercise does not prevent Type II diabetes; however, exercise does help manage the disorder.

Aging and Diabetes

Having diabetes is an additional challenge to the aging adult. There is even more reason to stay vigilant about eating well, staying active and monitoring blood glucose. Having a diabetes management checklist including medication dosage, medical appointments, blood glucose measurements, meal planning and exercise is recommended.

Metabolic Syndrome

Moderate and vigorous activity is associated with a lower risk of developing **metabolic syndrome**. Metabolic syndrome is a "cluster" of cardiovascular risk factors including overweight or obesity (waist circumference above 102 cm for men or above 88 cm for women), high blood pressure (above 130/85 mm Hg or current drug treatment for hypertension), elevated triglycerides (150 mg/dL or higher), low levels of high-density lipoprotein (below 40 mg/dL in men and below 50 mg/dL in women), and high fasting glucose levels (100 mg/dL or higher) (AHA, 2009). Having three or more of these risk factors puts you at higher risk of developing CVD or diabetes. In studies done at the Cooper Institute in Dallas, the risk of metabolic syndrome for men with moderate fitness was 26 percent lower, and for men with high fitness the risk was 53 percent lower compared to their lower fitness counterparts. For women the risk was 20 percent and 63 percent lower, respectively. It is clear that to prevent metabolic syndrome, improving cardiovascular fitness through regular physical activity is critical. Also called Syndrome X, the prevalence of metabolic syndrome goes up with age. This set of symptoms is very similar to prediabetes.

Low Back Pain

Low back pain is characterized by chronic discomfort in the lumbar region of the back. Chronic back pain may be the result of an injury; however, back pain is most often due to a lack of fitness. The National Safety Council data indicates that the back is the most frequently injured of all the body parts, with the injury rate double that of other body parts. Intervertebral disks can suffer degeneration from overuse, which is more common in men than in women. Backache is the second leading medical complaint when visiting a physician (Corbin and Welk, 2009).

Who Suffers from Low Back Pain?

More than eight out of ten Americans will suffer some back-related pain at some point in their lifetime (Corbin and Welk, 2009). Low back pain is epidemic throughout the world and is the major cause of disability in people aged 20 through 45 in the United States. Ninety percent of back injuries occur in the lumbar region (Donatelle and Davis, 2007). Thirty to 70 percent of all Americans have recurring back problems; two million Americans cannot hold a job as a result. Back pain is the most frequent cause of inactivity in individuals under the age of 45 (Corbin and Welk, 2009). Improper lifting, faulty work habits, heredity, diseases such as scoliosis, and excess weight are other causes of low back pain. Undue psychological stress can cause back pain via tight muscles and constricted blood vessels (Hoeger et al., 2009).

Can Low Back Pain Be Prevented?

Lack of activity is the most common reason for low back pain, so movement is critical to good back health. Staying active, using common sense regarding lifting heavy objects, and managing weight all are important in low back pain prevention. Decrease occupational risks. Use caution, as it is often employees new to a job who injure their back. Another factor in low back pain is poor posture while sitting, standing, or walking.

Does Exercise Help?

Exercise helps with enhancing posture, balance, strength, and flexibility. Strengthening abdominal muscles, which are the complimentary muscle group to the lower back, helps support the spine. Stretching the hip flexors and the hamstrings are important to help tilt the anterior portion of the pelvis back. Low back pain and tight hamstrings are highly correlated. Excess weight around the torso and abdominal region pulls the pelvis forward, causing potential strain in the lumbar region. In general, strengthening the "core" (all the muscles from the shoulders and the hips) helps prevent back pain. **Sarcopenia** is the loss of skeletal muscle mass and functional strength often associated with aging. Around 50 years of age, muscle deteriorates about 5% every decade thereafter. The upside is that "mature" muscle responds positively to a regular strength training and flexibility program to help manage weight over a lifetime.

Osteoporosis

Osteoporosis is a disease characterized by low bone density and structural deterioration of bone tissue, which can lead to increased bone fragility and

---

How much calcium does a college student need?
1,300 mg daily

- Good sources of calcium: low-fat dairy products, dark green leafy vegetables, broccoli, tofu, sardines, and salmon.
- Calcium-fortified foods: cereals, breads, orange juice, and some antacids.

Diabetic Walkers Gain Fitness

Diabetics who **walked moderately for thirty-eight minutes** (4,400 steps or 2.2 miles) did not lose weight; however, they showed significant effects: risk of heart disease decreased; cholesterol improved; triglycerides improved; and they saved $288.00 in health costs per year.

Diabetics who **walked ninety minutes** (10,000 steps or 5 miles) saw bigger benefits: the number of walkers needing insulin therapy decreased by 25 percent; those receiving insulin therapy reduced the dosage by an average of eleven units per day; cholesterol, triglycerides, blood pressure, and heart disease risk decreased; and they saved over $1,200.00 per year.

Diabetics in the control group that **walked 0 minutes** saw health care costs rise $500.00; insulin use, cholesterol, blood pressure, triglycerides, and heart disease risk all increased. This study was conducted for two years. (Sullivan et al., 2009)

© Tyler Olson, 2012, Shutterstock, Inc.

---

increased risk of fractures to the skeletal structure. Osteoporosis is sometimes called the "silent disease" because there are often no symptoms as bone density decreases. The *Dallas Morning News* (August 26, 2001) describes osteoporosis as an "epidemic of young women with old bones." Many young women delay the onset of menstruation due to high activity levels, which in turn lowers body fat and estrogen levels. Your physician may recommend a bone mineral density test. The test is noninvasive, painless, and safe (Otis and Goldingay, 2000).

Bone is living, growing tissue. With adequate nutrition and activity, bone formation continues to occur throughout a lifetime. Old bone is removed through **resorption**, and new bone is formed through a process called **formation**. Bones need to be fed and cared for just as the rest of our body. Childhood and teenage years are when new bone is developed more quickly than the old bone is resorbed. Bones become stronger and denser until peak bone mass is attained at approximately age 30. Thereafter, bone loss exceeds bone formation. **Osteopenia** is low bone mass that precedes osteoporosis. Note that weak, porous and brittle bones are a condition, not a disease. *Adequate calcium intake, minimal exposure to sunlight for vitamin D, and regular physical activity are critical for young adults because the higher peak bone mass is at age 30, the less likely it is that osteoporosis will develop in later years.*

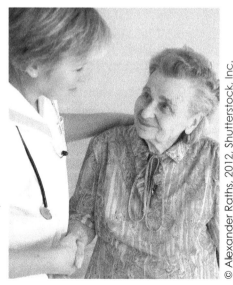

An estimated 10 million Americans—8 million women and 2 million men have osteoporosis.

For 34 million Americans, osteoporosis is a major public health threat (Surgeon General, 2004). Considered to afflict mostly women, this disease can affect males as well. Of the women with osteoporosis, 80 percent are post-menopausal. One out of two women and one out of eight men over 50 will get osteoporosis in their lifetime.

Risk increases with age. Have you observed older women who seem to slump? Many women with low bone density have **kyphosis** (also called dowager's hump), or a rounding of the upper back. The head tilts forward because often the cervical vertebrae in the upper spine actually suffer compression fractures. This keeps older women from being able to stand up straight or to get a full breath. Small, thin-boned women are at higher risk, and there may also be a genetic factor. If there are people in a family with weak, thin bones then relatives with the same

© Alexander Raths, 2012, Shutterstock, Inc.

Bone Loss during Spaceflight

In the 1980's, NASA scientists observed a dramatic spike in calcium excreted by astronauts after the first seven days of spaceflight. Researchers have since confirmed that humans lose bone mineral density during spaceflight at a rate ten-fold faster than does a post-menopausal woman; the lack of gravitational forces, even with daily exercise during space missions, has a very dramatic effect on bone mass. There are also changes in the cross-sectional geometry of long bones, for instance, the femoral neck near the hip joint, that further reduce bone strength and increase risk of a hip fracture should that astronaut fall soon after returning to earth. Some of this bone loss may be due to reduced blood flow to bone with the shifts in body fluids while in microgravity, according to a study by Dr. Michael Delp at Texas A&M University. This reduced blood flow in turn may affect in various ways the activity of bone cells responsible for bone formation and bone resorption, altering the balance in favor of resorption (Colleran et al., 2000). Related studies conducted by Dr. Susan Bloomfield at Texas A&M University demonstrated that this bone loss is not uniform across the skeleton but focused in trabecular ("spongy") bone sites (for instance, in the ends of the long bones) (Bloomfield et al., 2002). This is the same type of bone that is lost first with the development of osteoporosis here on earth. Another potential contributor to bone loss in astronauts might be reduced caloric intake, quite common during busy missions. Restricting caloric intake by 40 percent causes reductions in trabecular bone formation rate similar in magnitude to that observed with the unloading of microgravity (Baek et al., 2008). This finding has important implications for the many Americans who attempt long-term restriction of caloric intake to achieve weight loss.

Colleran, P.N., M.K. Wilkerson, S.A. Bloomfield, L.J. Suva, R.T. Turner, and M.D. Delp. Alterations in skeletal perfusion with simulated microgravity: a possible mechanism for bone remodeling. *J. Appl. Physiol.* 89: 1046–1054, 2000.

Bloomfield, S.A., M.R. Allen, H.A. Hogan, and M.D. Delp. Site-and compartment-specific changes in bone with hindlimb unloading in mature adult rats. *Bone* 31: 149–157, 2002.

Baek, K., A.A. Barlow, M.R. Allen, and S.A. Bloomfield. Food restriction and simulated microgravity: effects on bone and serum leptin. *J. Appl. Physiol.* 104: 1086–1093. 2008.

Shape Up America!

Try using digital technology—there are numerous free apps and some pricey high tech options for tracking your fitness progress. If you like gimmicks and gadgets, the cheapest option is probably a pedometer. A pedometer is a small device that clips on your belt and counts your steps. The average American walks approximately 900–3,000 steps per day in daily normal activities. The former U.S. Surgeon General, C. Everett Koop, developed Shape Up America! in 1994 to highlight health risks of obesity. The extra 150 calories or thirty minutes per day recommended by the Surgeon General's 1996 Report may not be enough for you to reach your fitness or weight loss goals. Studies indicate that a sufficient goal would be to walk 10,000 steps per day. Shape Up America! challenges you to walk 10,000 steps. Without a conscious effort, 10,000 steps would be a difficult task. Be sure and log your steps each day to work up to 10,000. Give it a try; if you like the latest thing, a pedometer is much cheaper than an ab roller or a treadmill!

Current Recommendations to Decrease Osteoporosis Risk
- Engage in daily weight-bearing aerobic activity
- Weight training (the ACSM recommends ten–twelve reps, two sets two times weekly)
- Vitamin D (wellbalanced diet and adequate exposure to sunlight)
- Estrogen replacement therapy (for some women, especially post-menopausal women)

body type may have an inherently higher risk. Post-menopausal Caucasian and Asian women are at the highest risk. It is unknown why these particular groups are more susceptible to osteoporosis. African Americans have bone that is 10 percent more dense than Caucasians (Greenberg et al., 1998). Others at risk include those with poor diets, especially if calcium and vitamin D are low over a long period of time. It is estimated that 75 percent of adults do not consume enough calcium on a daily basis (Bishop and Aldana, 1999). An inactive lifestyle contributes greatly. A history of excessive use of alcohol or cigarette smoking can also increase risk.

Another growing group of high-risk individuals is the eating disordered. Many active young women suffer stress fractures, which can be a sign of osteoporosis. If a person is extremely active with a low percentage of body fat, then hormone levels may be askew. Prolonged **amenorrhea** (absence of menstruation) can signal low body fat or an eating disorder. If symptoms such as amenorrhea, disordered eating, or abuse of exercise are suspected, then it would be prudent for the physician to order or for the athlete to consider asking for a bone density test.

Can Osteoporosis Be Prevented?

The good news is that osteoporosis can be both prevented and treated. Regular physical activity reduces the risk of developing osteoporosis. A lifetime of low calcium intake is associated with low bone mass (www.osteo.org). Adequate calcium intake is critical for optimal bone mass. Growing children, adolescents, and pregnant and breast-feeding women need more calcium. It is estimated by the National Institutes of Health that less than 10 percent of girls age 10–17 years are getting the calcium they need each day. A varied diet with green leafy vegetables and plenty of dairy will help ensure good calcium intake. Many calcium-fortified foods are now available. A varied diet will also ensure adequate intake of vitamin D, which aids in prevention. It is also advisable to limit caffeine and phosphate-containing soda, which may interfere with calcium absorption. Prolonged high-protein diets may also contribute to calcium loss in bone. A high-sodium diet is thought to increase calcium excretion through the kidneys. For post-menopausal women, some physicians consider hormone replacement therapy to help strengthen bones. Weightbearing exercise such as walking, running, tennis, and basketball is an excellent way to strengthen bones to help prevent osteoporosis.

Does Exercise Help?

The stress caused by working against gravity during activity strengthens and causes bones to be more dense, just as any other living tissue. Weight-training is highly recommended to keep the bones strong and to build bone mass. Consider the muscle atrophy experienced when a person is confined to bed rest, or a limb that is in a cast for a period of time. Bones deteriorate just as muscles deteriorate without the stimulation of movement. An interesting current topic of study is the effect of zero gravity in space on bone mass. It appears that even a short duration in space can impact bone density (see Bone Loss during Spaceflight, on page 137).

Physical activity is presented as the only known intervention that can potentially increase bone mass and strength in the early years of life and reduce the risk of falling in older populations according to a new Position Stand from the American College of Sports Medicine (ACSM). The official ACSM position stand encourages the adoption of specific exercise prescriptions designed for various ages to best capitalize on the chances to accrue and preserve bone throughout the various stages of life (Surgeon General, 2004).

If your dad or brother had a heart attack before age 55, or if your mom or sister had one before age 65, you're more likely to develop heart disease. This does not mean you will have a heart attack. It means you should take extra good care of your heart to keep it healthy.

References

Alters, S. and Schiff, W. *Essential Concepts for Healthy Living* (5th ed). Sudbury, MA., Jones and Bartlett. 2009.

American College of Sports Medicine (ACSM). *Exercise and Hypertension.*

American Heart Association (AHA). *Poor Teen Sleep Habits May Raise Blood Pressure, Lead to CVD.* News release December 10, 2008.

American Heart Association (AHA). *Heart Disease and Stroke Statistics—2012 Update.* 2012. www.aha.com

Canoy, M. P. et al. Abdominal Fat Distribution Predicts Heart Disease, *Circulation*, 2007. Castelli, W. P., Chair, Women, smoking and oral contraceptives; Highlights of a consensus conference, Montreal, November 1997.

Center for Health and Health Care in Schools, School of Public Health and Health Services, George Washington University Medical Center. *Childhood Overweight: What the Research Tells Us.* September 2007 Update. *www.healthinschools.org*

Bergland, Christopher, 4 *Lifestyle Choices That Will Keep You Young*, The Athlete's Way http://www.psychologytoday.com/em/133701

Biology of Aging, National Institutes on Aging, Publication No. 11-7561, Nov. 2011.

Crivello, Joseph, *Human Anatomy and Physiology, A Functional Approach*, Kendall Hunt Publishing Co. Dubuque, Iowa, 2014.

Speigelman, Bruce, et al., *A PGC1-α-dependent myokine that drives brown-fat-like development of white fat and thermogenesis Nature* 481, 463–468 Published 11 January 2012.

American College of Sports Medicine *Strength Training for Bone, Muscle and Hormones* Current Comment, 2014

Neurogenesis: How to Change Your Brain http://www.huffingtonpost.com/dr-david-perlmutter-md/neurogenesis-what-it-mean_b_777163.html#es_share_ended retrieved June 2014.

Mitchell, et. al., *Social disadvantage, genetic sensitivity, and children's telomere length*, Proceedings of the National Academy of Sciences, vol.111 no. 16.

Fernandez, Elizabeth, *Lifestyle Changes May Lengthen Telomeres, A Measure of Cell Aging* http://www.ucsf.edu/news/2013/09/108886/lifestyle-changes-may-lengthen-telomeres-measure-cell-aging retrieved June 2014.

Corbin, C. and Welk G. *Concepts of Physical Fitness* (15th ed). New york: McGraw-Hill. 2009.

Donatelle, R. J. and Davis, L. G. *Access to Health* (10th ed). Boston: Benjamin Cummings. 2007.

Flegal, K. M., Carrol, M. D., Kuczmarski, R. J., and Johnson, C. L. Overweight and Obesity in the United States: Prevalence and Trends, 1960–1994. *International Journal of Obesity and Related Metabolic Disorders*, 22: 39–47. 1998.

Frye, D. W. Contracting Officer. NHANES Iv, Central Lipid Laboratory for National Health and Nutrition Survey. October 1999.

Gaesser, G. Obesity, Health, and Metabolic Fitness, *www.thinkmuscle.com/articles* Gibbs, W. W. Obesity: An Overblown Epidemic? *Scientific American*, May 23, 2005. Greenberg, J. et al. *Physical Fitness and Wellness* (2nd ed). Boston: Allyn and Bacon. 1998.

Hafen, B. Q., Karren, K. J., and Frandsen, K. J. *First Aid for Colleges and Universities* (7th ed). Boston: Allyn and Bacon. 1999.

The Heart Truth for Women: Women and Heart Disease, *www.hearttruth.gov*

Hoeger, W. W. K., Turner, L. W., and Hafen, B. Q. *Wellness Guidelines for a Healthy Lifestyle* (4th ed). Belmont, CA: Thomson Wadsworth. 2009.

Koop, C. E. Shape Up America! http://www.shapeupamericastore.org, 1994.

National Center for Health Statistics, U.S. Department of Health and Human Services, Centers for Disease Control and Prevention. Hyattsville, MD.

National Institutes on Aging, NIH, www.nia.nih.gov 2008.

National Institutes of Health, http://www.nhlbi.nih.gov/actintime/rhar/md.htm

National Institutes of Health. *Sixth Report on the Joint National Committee on Prevention, Detection, Evaluation and Treatment of High Blood Pressure.* 1997.

National Institutes of Health. Osteoporosis and Related Bone Disease, http://www.osteo.org

Lumey LH, Stein AD. Offspring birth weights after maternal intrauterine undernutrition: a comparison within sibships. *Am J Epidemiol* 1997; **146:** 810–819.

Ochoa, L. W., editor. Women's Health and Wellness, an Illustrated Guide (26th ed).

Skokie, IL: Lippincott Williams & Wilkins. 2002.

Otis, C. L. and Goldingay, R. *The Athletic Woman's Survival Guide.* Champaign, IL: Human Kinetics Publishers. 2000.

Paffenbarger, R. et al. Physical Activity and Physical Fitness as Determinants of Health and Longevity. In C. Bouchard et al. *Exercise Fitness and Health.* Champaign, IL: Human Kinetics Publishers. 1990.

Payne, W. A. and Hahn, D. B. *Understanding Your Health* (6th ed). St. Louis, MO: Mosby. 2000.

Powers, S. K. and Dodd, S. L. *Total Fitness and Wellness.* San Francisco: Pearson Benjamin

Cummings. 2009.

Rosato, F. *Fitness to Wellness: The Physical Connection* (3rd ed). Minneapolis: West. 1994. Satcher, D. *Surgeon General's Report on Physical Activity and Health.* Atlanta, GA: CDC. 1996.

Sesso, H. D. and Paffenbarger, R. S. The Harvard Alumni Health Study, Harvard School of Public Health. Boston: 1956.

Seventh Report of the Joint National Committee on Prevention, Detection, Evaluation, and Treatment of High Blood Pressure (JNC vII). *Hypertension,* December 2003.

Springen, K. and Seibert, S. Artful Aging, *Newsweek,* January 17, p. 57. 2005.

Sullivan, P. W. et. al. Obesity, Inactivity, and the Prevalence of Diabetes and Diabetes-related Cardiovascular Comorbidities in the U.S., 2000–2002, *Diabetes Care,* 28: 1599–1603, 2009.

Surgeon General's Call to Action to Prevent and Decrease Overweight and Obesity, www.surgeongeneral.gov

Surgeon General's Report on Bone Health and Osteoporosis: What It Means to You. Washington, DC: U.S. DHHS. October, 2004.

Texas A&M University Human Nutrition Conference. College Station, TX. 1998. Weinttraub, A. yoga: It's Not Just An Exercise. *Psychology Today.* November, 2000. Wilmore, J. H. Exercise, Obesity, and Weight Control, *Physical Activity and Research*

Digest. Washington D.C.: President's Council on Physical Fitness & Sports. 1994.

World Health Organization (WHO). *Controlling the Obesity Epidemic.* Geneva: Author December 2008.

www.exerciseismedicine.com

www.osteo.org

Contacts

American College of Sports Medicine (ACSM)
http://www.acsm.org

American Heart Association
http://www.americanheart.org

American Medical Association
http://www.ama-assn.org

Franklin Institute of Science: interactive multimedia tour of the heart
http://www.fi.edu/biosci/heart.html

Dr. Koop's Community: health improvement info
http://www.drkoop.com

Stayhealthy.com: comprehensive Internet resources continuously updated
http://www.stayhealthy.com/

Go Ask Alice: sponsored by Columbia University Health Service; question & answer format
http://www.alice.columbia.edu/index.html

Centers for Disease Control and Prevention: Information and national health statistics plus more
http://www.cdc/gov

National Health Information Center: 100 organizations listed here to provide answers to health-related questions
http://nhic-nt.health.org/

Weight-control Information Network
www.win.niddk.nih.gov

> 1 WIN Way
> Bethesda, MD 20892-3665
> (toll-free number) 877-946-4627

Recommended Reading

The Roadmap to 100: The Breakthrough Science of Living a long and Healthy Life by Walter M. Bortz (St. Martin's Press, 2010)

The Athletic Woman's Survival Guide by Carol Otis & Roger Goldingay (Human Kinetics Publishers, 2000)

Strong Women Stay Young by Miriam E. Nelson (Bantam, 1997)

Activities

Notebook Activities

Self-Assessment of Cardiovascular Fitness
Healthy Back Test
Is Your Blood in Tune?
Is Osteoporosis in Your Future?

Name _____ Section _____ Date _____

NOTEBOOK ACTIVITY

Healthy Back Test

These tests are among the ones used by physicians and therapists to make differential diagnoses of back problems. You and your partner can use them to determine if you have muscle tightness that may make you "at risk" for back problems. Discontinue any of these tests if they produce pain or numbness, or tingling sensations in the back, hips, or legs. Experiencing any of these sensations may be an indication that you have a low back problem that requires diagnosis by your physician. Partners should use great caution in applying force. Be gentle and listen to your partner's feedback.

Test 1—Back to Wall

Stand with your back against a wall, with head, heels, shoulders, and calves of legs touching the wall as shown in the diagram. Try to flatten your neck and the hollow of your back by pressing your buttocks down against the wall. Your partner should just be able to place a hand in the space between the wall and the small of your back.

- If this space is greater than the thickness of his/her hand, you probably have lordosis with shortened lumbar and hip flexor muscles.

❏ **Pass**   ❏ **Fail**

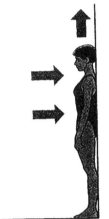

Test 2—Straight Leg Lift

Lie on your back with hands behind your neck. The partner on your left should stabilize your right leg by placing his/her right hand on the knee. With the left hand, your partner should grasp the left ankle and raise your left leg as near to a right angle as possible. In this position (as shown in the diagram), your lower back should be in contact with the floor. Your right leg should remain straight and on the floor throughout the test.

- If your left leg bends at the knee, short hamstring muscles are indicated. If your back arches and/or your right leg does not remain flat on the floor, short lumbar muscles or hip flexor muscles (or both) are indicated. Repeat the test on the opposite side. (Both sides must pass in order to pass the test.)

❏ **Pass**   ❏ **Fail**

Test 3—Thomas Test

Lie on your back on a table or bench with your right leg extended beyond the edge of the table (approximately one-third of the thigh off the table). Bring your left knee to your chest and pull the thigh down tightly with your hands. Your lower back should remain flat against the table as shown in the diagram. Your right thigh should remain on the table.

- If your right thigh lifts off the table while the left knee is hugged to the chest, a tight hip flexor (iliopsoas) on that side is indicated. Repeat on the opposite side. (Both sides must pass in order to pass the test.)

❏ **Pass**   ❏ **Fail**

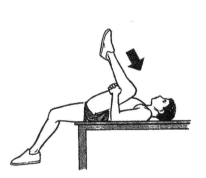

Test 4—Ely's Test

Lie prone; flex right knee. Partner gently pushes right heel toward the buttocks. Stop when resistance is felt or when partner expresses discomfort.

- If pelvis leaves the floor or hip flexes or knee fails to bend freely (135 degrees) or heel fails to touch buttocks, there is tightness in the quadriceps muscles. Repeat with left leg. (Both sides must pass to pass the test.)

❏ **Pass**   ❏ **Fail**

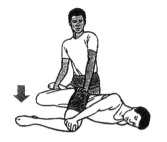

Test 5—Ober's Test

Lie on left side with left leg flexed ninety degrees at the hip and ninety degrees at the knee. Partner places right hip in neutral position (no flexion) and right knee in ninety-degree flexion; partner then allows the weight of the leg to lower it toward the floor.

- If there is no tightness in the iliotibial band (fascia and muscles on lateral side of leg), the knee touches the floor without pain and the test is passed. Repeat on the other side. (Both sides must pass in order to pass the test.)

  ❏ Pass ❏ Fail

Test 6—Press-Up (Straight Arm)

Perform the press-up.

- If you can press to a straight-arm position, keeping your pubis in contact with the floor, and if your partner determines that the arch in your back is a continuous curve (not just a sharp angle at the lumbosacral joint), then there is adequate flexibility in spinal extension.

  ❏ Pass ❏ Fail

Test 7—Knee Roll

Lie supine with both knees and hips flexed ninety degrees, arms extended to the sides at shoulder level. Keep the knees and hips in that position and lower them to the floor on the right and then on the left.

- If you can accomplish this and still keep your shoulders in contact with the floor, then you have adequate rotation in the spine, especially at the lumbar and thoracic junction. (You must pass both sides in order to pass the test.)

  ❏ Pass ❏ Fail

Healthy Back Ratings

Classification	Number of Tests Passed
Excellent	7
Very good	6
Good	5
Fair	4
Poor	1–3

Name _____ Section _____ Date _____

NOTEBOOK ACTIVITY

Is Osteoporosis in Your Future?

Risk Factors You CANNOT Control:

1. Are you female?
2. Do you have a family history of osteoporosis?
3. Are your ancestors from the British Isles, northern Europe, China, or Japan?
4. Are you very fair-skinned?
5. Are you small-boned?
6. Are you over age 35?
7. Have you had your ovaries removed, or did you have an early menopause?
8. Are you allergic to milk and milk products?
9. Have you never been pregnant?
10. Do you have cancer or kidney disease?
11. Do you have to take chemotherapy, steroids, anticonvulsants, or anticoagulants?

Risk Factors You CAN Control:

12. Do you smoke?
13. Do you drink alcohol?
14. Do you avoid milk and cheese in your diet?
15. Do you get very little exercise?
16. Do you drink a lot of soft drinks?
17. Is your diet high in protein?
18. Do you consume a lot of caffeine (five or more cups of coffee per day or equivalent)?
19. Are you amenorrheic (without a monthly period)?
20. Do you get less than 1,000 mg of calcium a day?
21. Is your body weight very low?
22. Do you go on extreme or crash diets?
23. Do you have a high sodium (salt) intake?

If you answered "yes" to three (3) of the above questions, you are at risk for osteoporosis and may want to ask your doctor to give you a bone density screening test. The more questions you answered "yes" to, the higher your risk of developing osteoporosis in the future.

Many clinical studies suggest that osteoporosis is a preventable disease. As you can see from the quiz, you can do several things right now to help prevent osteoporosis in your future.

Source: Adapted from Marion Laboratories, Inc.

Chapter 4
Complementary and Alternative Medicine

OBJECTIVES

Students will be able to:
- Identify components of holistic self-care.
- Compare and contrast conventional, complementary and alternative health care systems.
- Identify the therapies contained in each major CAM domain.
- Identify pros and cons associated with each major CAM practice.
- Integrate the benefits of mindfulness into everyday life.
- Participate in a self-guided relaxation/meditation practice.

Adaptation of *Health and Fitness: A Guide to a Healthly Lifestyle, 5/e* by Laura Bounds, Gayden Darnell, Kirstin Brekken Shea and Dottiede Agnor. Copyright © 2012 by Kendall Hunt Publishing Company.

> "The doctor of the future will give no medicine, but will interest his patients in the care of the human frame, in diet, and in the cause and prevention of disease."
>
> –Thomas Edison

Introduction

by Amy V. Nowak, Ph.D. TAMU

As an adult, making informed healthcare decisions is key to achieving higher levels of health and wellness. To become an informed healthcare consumer, it is important to be aware of the many choices available to you in healthcare today. Learning about each form of healthcare, as well as its risks and benefits, will allow you to make the best choices that meet the needs of your health situation.

Americans value choice in any venue and healthcare is no different. When buying a car, you want a quality vehicle to meet your needs, a company that can give you a good product, and a salesperson that you trust, and who is dedicated to helping you meet your goals. The same goes for healthcare. You want to find a healthcare option that fits with your healthcare needs, a treatment which will be effective, and a healthcare provider whom you trust and who is dedicated to helping you. With healthcare, you have the option of choosing one or a combination of healthcare approaches and providers to best meet your needs.

There are two main camps of healthcare today. You are probably very familiar with what is called conventional medicine. When you go to a conventional clinic or hospital, doctors and nurses work to diagnose an illness and then treat the symptoms with medication, surgery, or radiation. The roots of conventional medicine date back to the mid-1800's with the discovery of the germ and its relationship to illness. While this is the main form of healthcare used in the United States and similar developed nations around the world, more and more people are turning to other types of healthcare that do not fit in the mainstream of conventional medicine.

The other main category of medicine is a group of traditional systems and practices, which are currently called complementary and alternative medicine (CAM). The terms complementary and alternative are actually designations of how traditional medical practices, some of which developed thousands of years ago, are used in relation to conventional care. Systems and practices are considered **complementary** when used in conjunction with conventional care and **alternative** when used instead of conventional care. The emphasis in CAM is holistic, in the sense that its purpose is to treat the whole person and support the body's natural ability to heal itself. The increasing use of CAM is expected to continue as people seek out options in healthcare to best meet their needs.

Medicine has come a long way since the days of the snake-oil salesmen of the early nineteenth century. In the 1800's, homeopaths, midwives, naturopaths, and an assortment of lay healers used herbs and nostrums to combat illness. Thanks to the wonders of modern conventional and emergency medicine, many of the ill and injured can survive what fifty years ago would have meant certain death. This is surely being played out in the modern landscape of the war-torn Middle East. Due to improved body armor, field medical procedures, and medevac capabilities, wounded soldiers are surviving what they would not have survived in the Vietnam War or World War II. Conventional medicine can work mini-miracles in acute trauma care, the treatment of bacterial infections and life-threatening diseases. Life saving antibiotics and other drugs have revolutionized the medical field. What conventional medicine has failed to do is prevent the lifestyle-related hypokinetic diseases that plague Western society. Conventional (also called Western, allopathic, or biomedical) medicine developed from the evidence-based scientific method. Traditionally, alternative medicine (also called natural, unconventional, or unorthodox in the past) has been based on anecdotal evidence, word of mouth, testimonials, or even the placebo effect. Scientists in the bio-medical research community are recognizing that more and more Americans are choosing complementary and alternative medicine (hereafter referred to as CAM), and therefore funding to test the safety and efficacy of CAM approaches is increasing. *Part of the attraction of the CAM modalities may be their identification with prevention rather than cure, and consequently CAM has come to be identified with wellness and self-care.*

Alternative practitioners emphasize a wholesome diet rich in organic fruits, vegetables, nuts, seeds, fiber, pure water, and organically raised meat products.

Today, alternative medicine is also called holistic, complementary, or integrative. Refer back to the wellness dimensions from Chapter 1; a holistic practitioner considers the physical, emotional, mental, social, occupational, environmental, and spiritual factors associated with the individual as a "whole person." "Practitioners of alternative medicine approach healing from a holistic perspective where the primary goal is the creation and maintenance of optimum health in body, mind, and spirit. In addition to the comprehensive care they provide to achieve that goal, they also serve as teachers, instructing their patients in effective methods of selfcare. Such methods not only assist patients in their journey back to wellness, but also help them prevent disease from occurring in the first place "(Goldberg, 2002).

Alternative practitioners emphasize **holistic self-care.** A wholesome **diet** minimizing intake of processed food with foods rich in organic fruits and vegetables, nuts, seeds, fiber, pure water, and organically raised meat products is recommended. **Exercise** is critical to maintaining physical health. Adequate **sleep** is necessary to allow the regenerative processes in the body to work. Keeping the **environment** at home and work healthy may mean adding indoor plants, air filters, humidifiers, and avoiding toxic chemicals and secondhand smoke. Peace of mind and contentment are part of **good mental health.** Spiritual health is also considered an important part of self-care. **Spiritual health** can be gained through prayer, meditation, or even giving of yourself through volunteerism. *In alternative or holistic care, the patient takes an active role and is responsible for looking at all aspects or his/her health.*

How many Americans are using CAM? According to a 2007 National Health Interview Study (NHIS) four out of ten adults use some form of CAM therapy on a regular basis. Twelve percent of children (ages 0–18 years) use CAM (see Figure 4.1). Native Alaskan and American Indians were the most likely to use some form of CAM, followed by white adults. Caucasian college-educated women in a higher income bracket use CAM more than other segments of

FIGURE 4.1

CAM Use by U.S. Adults and Children

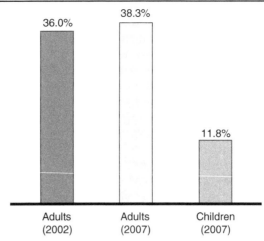

Source: National Institute of Health.

the population. However, CAM is practiced by all types of people across racial, cultural, and socioeconomic lines (see Figure 4.2). The most commonly used CAM therapies are non–vitamin, non–mineral, natural products (for instance, fish oil, echinacea, DHA, glucosamine, and ginseng), deep breathing, meditation, yoga, massage, chiropractic care, and diet-based therapies (see Figure 4.3). Chiropractic is used the most by patients with back pain.

Note: The 2007 NHIS did not include folk medicine practices (i.e., covering a wart with a penny and then burying it) or religious healing, as in prayer for oneself or for others. The 2002 NHIS did include prayer in its survey.

Figure 4.4 shows several accepted and widely used treatments that are rooted in CAM. Indeed, even exercise prescribed as a healing modality was once considered "alternative." "In an age of M.R.I. scans and spinal fusion surgery, a treatment as low tech as exercise can seem to some patients rudimentary or even dangerously illogical" (Ryzik, 2005). Cardiovascular exercise helps with increased circulation and flexibility, and core muscle strength focuses on supporting the spine which can help prevent future pain. Perhaps in the future many more CAM modalities will become mainstream (see Table 4.1).

FIGURE 4.2

CAM Use by Race/Ethnicity among Adults 2007

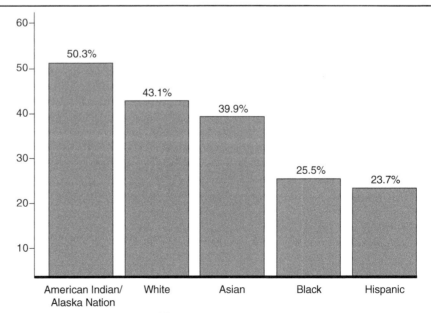

Source: National Institute of Health.

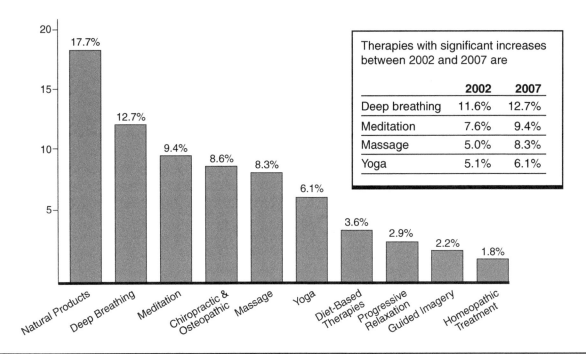

FIGURE 4.3
Ten Most Common CAM Therapies among Adults 2007
Source: Center for Disease Control

TABLE 4.1 ♦ Currently Used CAM Modalities
CAM Modalities Now in Mainstream Medicine
Codeine for pain
Digitalis for heart failure
Quinine for malaria
Aspirin for fever
Behavioral therapy for headache
Hypnosis for smoking cessation
Exercise for diabetes
Support groups for breast cancer
Low-fat, low-cholesterol diets

Alternative Healthcare Systems

Alternative healthcare systems are holistic "whole person" systems. Whole person systems refer to treating more than just a patient's symptoms. The CAM practitioner often interviews the patient in an attempt to determine the patient's history, eating habits, lifestyle choices, and so on. Some patients report that they appreciate the fact that their practitioner often regards self-care, positive lifestyle habits, behaviors, quality of life, and the combined role of the mind, body, and spirituality in health, disease, and healing as being very important (WHCCAMP, 2002). Typically the CAM practitioner works out of a small facility and spends a fair amount of time with their patient, which may be more attractive to the patient than the short fifteen-minute appointment most people get with their busy conventional doctor. Included in the Alternative Healthcare System domain of CAM is ayurvedic medicine, homeopathic medicine, Native American medicine, and traditional Chinese medicine (including acupuncture and Chinese herbal medicine). Interestingly, the CAM therapies with the most acceptance by the medical community are some of the

> "The competent physician, before he attempts to give medicine to the patient, makes himself acquainted not only with the disease, but also with the habits and constitution of the sick man."
> —Cicero

> "The art of healing comes from nature and not from the physician. Therefore, the physician must start from nature with an open mind."
> —Paracelsus

Essential Oils

by Mike Hanik, M.S. TAMU

Essential oils are the natural, aromatic, volatile liquids found in shrubs, flowers, roots, trees, bushes, and seeds. Essential oils are also capable of readily changing from liquid to vapor at normal temperatures and pressures. Each essential oil may consist of hundreds of different and unique chemical compounds that defend plants against insects, environmental conditions, and disease. They also help plants grow, live, and adapt to its environment. Essential oils are extracted from aromatic plant sources by steam distillation. This allows the essential oils to be highly concentrated. It often takes an entire plant to produce a single drop of essential oil.

Essential oils often have a pleasant aroma and their chemical makeup may provide many health benefits. This is why essential oils have been used throughout history by many cultures for medicinal and therapeutic purposes. Recently there has been a renewed interest in studying the many benefits of essential oils.

Peppermint oil has been shown to have a calming and numbing effect. It has been used to treat headaches, skin irritations, anxiety associated with depression, nausea, and diarrhea. In test tube experiments, peppermint kills some types of bacteria, viruses, and fungus suggesting it may have antibacterial, antiviral, and antifungal properties.

Another common essential oil is lavender oil. Research has confirmed that lavender produces calming, soothing, and sedative effects when its scent is inhaled. Lavender oil may also help alleviate insomnia, anxiety, and fatigue.

Prior to using essential oils it is important to learn about the chemistry and safety of the oils. In order to experience the benefits of essential oils, high quality of pure essential oils must be used. A person will not experience the benefits of essential oils if they use diluted, adulterated, or synthetic oils.

It is important to research the companies that supply the essential oils. Learn how the company selects, grows and harvests its plants. Where the plant is grown, the quality of the soil, and time of day that the plant is harvested can all impact the quality and makeup of the essential oil. The distillation process is also important in making high quality essential oils. In order to get the best quality essential oils, proper temperature and pressure, length of time, equipment, and batch size must be closely monitored throughout the distillation process. Like any other product, it is important to fully research and understand essential oils in order to make an informed decision.

FIGURE 4.4

CAM Domains and Their Related Pratices

Source: White House Commission on Complementary and Alternative Medicine Policy (2008)

most infrequently used by patients—hypnotherapy, acupuncture, and biofeedback.

Ayurveda is thought to be the oldest medical system known. In Hindu mythology Ayurveda is considered the medicine of the gods. In Sanskrit, ayurveda is "knowledge of life," with life being defined as mind, body, and spiritual awareness. "Ayurveda is based on the belief that the natural state of the body is one of balance. We become ill when this balance is disrupted, with specific conditions or symptoms indicating a particular disease or imbalance. Ayurveda emphasizes strengthening and purifying the whole person, whereas in conventional medicine, the focus is on a set of symptoms or an isolated region of the body" (Alternative Medicine Foundation, 2005).

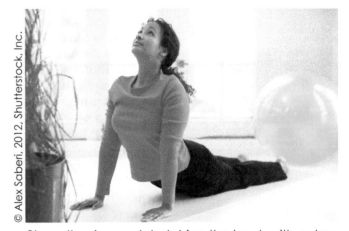

Strengthening and stretching the back with cobra.

Ayurvedic teaching states that every living thing in the universe is made of these five elements: earth, water, fire, air, and space. The elements combine to determine a dosha, or metabolic type. A person's personality and character determine which of three doshas they are: Vata, Pitta, or Kapha. The practitioner can determine which dosha a patient is, then prescribe botanics, exercise, yoga, and scash or massage therapy according to the person's particular dosha type. Ayurvedic practitioners mainly diagnose by observation and by touch. Caution with botanicals is advised as a 2008 study determined that a type of ayurveda that uses Rasa shastra (herbal medicines mixed with minerals and metals) may be cause for concern. Twenty-one percent of the medicines tested (obtained from the Internet) had unsafe levels of lead, arsenic, and mercury (Saper, Phillips, Sehgal, et al., 2008).

Homeopathy is based on a three pronged theory that *like cures like,* treatment is very individualized, and less is more. Homeopathic practitioners give very diluted forms of the substance that causes the symptoms of the disease in healthy people to the ill in the hopes that it will help support the body's natural healing power. The World Health Organization (WHO) has cited homeopathy as one of the systems of traditional medicine that should be integrated worldwide with conventional medicine in order to provide adequate global care in the twenty-first century (Goldberg, 2002). Homeopaths use low cost herbals, chemicals, and minerals.

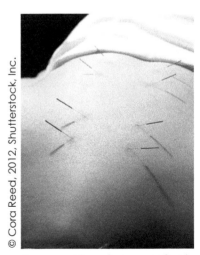

Acupuncture is an ancient medical art using the insertion of very fine needles into the body.

Naturopathy is based on the motto "Vis Medicatrix Naturae," which is Latin for *helping nature heal*. Naturopaths emphasize restoring health rather than curing disease. Naturopaths utilize many different healing "tools" found in nature, such as magnets, water, heat, crystals, the sun, herbal medicine, manipulation, light therapy, electrical currents, and more. Naturopaths argue that Americans should return to a more natural and to a simpler way of life. Some naturopaths contend that we should go so far as to cease fluoridation of water and eliminate the addition of preservatives to food. There are three naturopathic training schools in the United States and one in Canada. Although these schools have a four year program emphasizing humanistic medicine, the naturopath is not an M.D.

Traditional oriental medicine (TOM) is a comprehensive system that dates back to the Stone Age. Also called traditional Chinese medicine (TCM), it includes acupuncture, acupressure, herbal medicine (discussed under Biological-Based Therapies), oriental massage, and qi gong (discussed under Energy Therapies). **Acupuncture** is an ancient medical art using the insertion of very fine needles on the body in order to affect physiological functioning in the body. The needles are placed on the body at points that correspond to twelve meridians throughout the body. Manipulation of the needles, electrical stimulation, heat, and burning herbs (moxibustion) can be used in acupuncture. Before making a diagnosis, the practitioner talks with, and asks questions about, the patient. Typically the acupuncturist will check the pulse and the tongue of the patient to help diagnose the problem. A reputable acupuncturist will use disposable needles or sterilize reusable needles in an autoclave. With the millions of people treated with acupuncture, there have been relatively few complications reported to the U.S. Food and Drug Administration (FDA). Acupuncturists have a Master of Traditional Oriental Medicine and are required to be state-licensed.

All TOM recognizes an energy force that flows through the body called qi (pronounced chi). Qi consists of the spiritual, physical, mental, and emotional aspects of life. Yin and yang are the vital forces of life that run throughout the twelve meridians within the body. Stimulation of points on the meridians is thought to activate the qi, which restores the body's equilibrium and allows the free flow of qi. The body is considered a flowing, self-healing system. Pain and discomfort can be the result of stagnation of energy which needs to be brought back into balance. Patients may experience calm and peacefulness as well as rejuvenation when their qi has been restored.

U.S. medical doctors became more interested in acupuncture in 1971 when James Reston, a well respected New York Times columnist, had to undergo emergency surgery while in China. Doctors there eased his post-surgery pain

---

Can acupuncture give the athlete an edge in competition? It is possible that acupuncture treatment can be a positive adjunct to training, just like massage or physical therapy. Needles placed at sites of inflammation may reduce time out of training due to injury or swelling of tissue. There is little research, but Whitfield Reaves (2008) has used pre-performance needling and found personal benefit. Ear (auricular) acupuncture has been used during an athlete's competition, with small "tacks" kept in the ear. Acupuncture points don't work for everyone, but perhaps some sports acupuncture can make your next run a little more enjoyable.

with acupuncture. There have been numerous studies done in the United States on the effectiveness of acupuncture. In December 2004 results of the largest randomized, controlled phase III clinical trial of acupuncture ever conducted were published in the *Annals of Internal Medicine*. The study was conducted on 570 patients with osteoarthritis of the knee. The results showed that "acupuncture reduces pain and functional impairment of osteoarthritis of the knee" (NIH, 2004). Dr. Brian M. Berman, M.D. of the University of Maryland School of Medicine directed the study and concluded that acupuncture is an effective complement to conventional arthritis treatment. According to a CDC 2002 survey, 2.1 million Americans have used acupuncture.

Acupressure is similar to acupuncture, but without the needles. The practitioner applies pressure to critical points along the meridian lines to balance yin and yang. There are different pressure points corresponding to specific parts of the body. The pressure releases muscular tension and promotes circulation of blood and qi to promote healing. Gradual steady penetrating pressure for up to three minutes is common (Gach, 1990). Simple acupressure techniques can be practiced on oneself. For example, between the forefinger and thumb is an acupressure point for headaches. **Shiatsu** is a type of acupressure massage using fingers, elbows, fists, and so on to apply pressure to restore the flow of energy in the body.

> "When qi gathers, the physical body is formed, when qi disperses, the body passes on."
> –Ancient Chinese Proverb

Manipulative and Body-Based Therapies

Manipulative and body-based therapies in CAM use movement or manipulation of part of the body (see Figure 4.5).

Chiropractic is a medical treatment defined as the science of spinal manipulation. Chiropractic is the most commonly used form of CAM in the United States with 18 million Americans visiting the chiropractor each year. A June 2005 Consumer Reports Survey of 34,000 readers reported that of those interviewed with back pain, more went to the chiropractor than used prescription drugs. Practiced in earnest in the United States since 1895, chiropractic can trace its roots back to Galen and Hippocrates who laid their hands on patients for manipulation. Chiropractic is considered to be the oldest indigenous CAM practices in the United States (NIH Lecture Series,

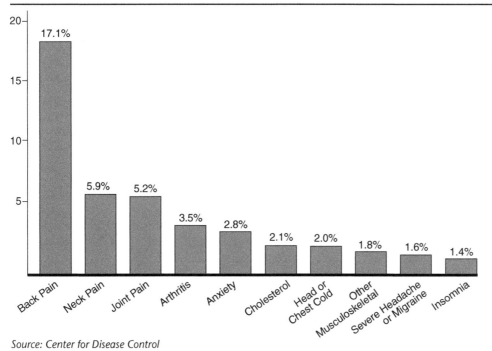

FIGURE 4.5

Diseases/Conditions for Which CAM Is utilized in the U.S.

Source: Center for Disease Control

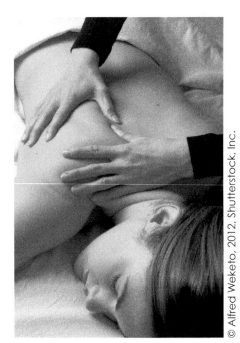

Massage involves manipulation of muscle and connective tissue to enhance function and promote relaxation and well-being.

2002). Chiropractic has been seen in the past to be in competition with conventional medical treatment. Early in the last century, state medical boards used their power to restrict chiropractic practice. The chiropractors successfully brought an antitrust suit against the American Medical Association for allegedly trying to eliminate chiropractic practice in the United States. Today the medical doctors and the doctors of chiropractic enjoy a better working relationship. Chiropractors train for up to six years with in-depth courses in anatomy, physiology, nutrition, and pathology. They also have clinical training.

Chiropractors manipulate the spine, often with high-velocity, low-amplitude spinal adjustments, to align the spine in order to let energy flow through the nervous system. It is unclear exactly how chiropractic works, however scientific evidence supports the use of chiropractic to treat acute or chronic back pain (NIH Lecture Series, 2002). Safety is always a concern, and the apparent risk for lumbar vertebrae adjustment is one in 1 million. Back and head complaints are the most common reason patients visit the physician or chiropractor, which results in $100 billion annually in lost productivity.

Many Americans suffer from neck pain. A recent study looked at the effectiveness of spinal manipulation, home exercise and medication to reduce acute neck pain. The spinal manipulation group received adjustments and mobilization exercises for the spine. The home exercise group received detailed instructions for gentle exercises for the neck and shoulders to be done 6 to 8 times per day. The third group received non-steroidal anti-inflammatory drugs (NSAIDS) as well as acetaminophen. Narcotic medications were an option for those who could not tolerate the NSAIDS. Both the exercises and the spinal manipulations groups were equal in improvement with participant rated pain. At 12 weeks, 82% of the exercise and manipulation groups experienced a 50% reduction in pain. Of the medication group, 69% noted improvements of at least 50% reduction in pain. The findings were similar at 26 and 52 weeks. "Additionally, the spinal manipulation group reported greater global improvement, participant satisfaction and function than the medication group" (NCCAM, 2012). The results from this study suggest that movement-based therapies can be effective in chronic pain management, especially before invasive techniques like surgery are considered. The *Clinical Journal of Pain* also reported results from a 2009 study that chronic neck pain patients experienced benefits from therapeutic massage (NCCAM, 2009).

Massage involves manipulation of muscle and connective tissue to enhance function of those tissues and to promote relaxation and well-being. Massage is growing in popularity as the use and acceptance of massage therapy increases. Many Fortune 500 companies are including massage as a benefit for their employees. Even small companies that offer on-site fifteen-minute massage are seeing the benefit in lower employee absenteeism due to headache, fatigue, and back pain (AMTA, 2005). Deep tissue, Swedish, myofascial release, petressage (kneading), sports massage, and trigger point therapy are just a few of the popular types of massage today.

Reflexology is based on the fact that the feet and hands represent a microcosm of the body and that specific parts of the foot and hand correspond or "reflex" to other parts of the body. Working with the feet has been used in many ancient medical practices; however William Fitzgerald developed modern reflexology in the early 1900's in England.

Craniosacral therapy has its origins in the 1800's with Andrew Still M.D. The current form of craniosacral therapy was developed by osteopathic physician John E. Upledger at Michigan State University as a therapy that uses gentle touch to evaluate the physiological functioning of the craniosacral system. The craniosacral system is comprised of the membranes and the cerebrospinal fluid that surrounds and protects the spinal cord. Imbalance in the cerebral and spinal systems may cause sensory or motor dysfunction (IAHE, 2005). As relaxing as a massage, this therapy is typically used by people experiencing chronic pain who have not found relief with other therapies.

Biological-Based Therapies

Biological-based therapies use substances found in nature such as food, vitamins, minerals, herbal products, animal-derived products, probiotics, amino acids, whole diets, and functional foods. Some biological-based therapies are evidenced-based. For example, the FDA now fortifies some foods with folic acid to deter potential neural tube defects in developing fetuses. There are other biological-based therapies that are as of yet unproven. An example is the use of shark cartilage as a treatment for cancer. The consumer should be informed and use common sense and do a little research before spending money and making important decisions regarding healthcare. Drugs are monitored by the FDA, but biological-based systems are measured for truth in advertising by the Federal Trade Commission (FTC). The following biological-based therapies are just a few of the options for consumers today.

Macrobiotics is more than a diet, it is a discipline based on a philosophy of balance in accordance with the universe. It involves managing or changing diet to enhance health or for spiritual benefit. Macrobiotics is characterized by excluding meat and concentrating heavily on whole grains. Besides modifying diet, basic macrobiotic practices emphasize an active life, a positive mental outlook, and regularly eating small portions. There are numerous testimonials from cancer patients that have recovered from a stage IV cancer diagnosis using the macrobiotic diet. The National Cancer Institute has funded a clinical study to determine the effects of a macrobiotic diet on cancer therapy (www.clinicaltrials.gov/ct/gui/c/alb/show/ NCT00010829). As you recall, evidence-based science needs clinical trials to provide scientific evidence in order for a therapy or treatment to have wide acceptance.

Herbals and dietary supplements are a hot trend in the industry, making manufacturers four billion dollars richer each year. Herbal therapy has been around for several thousand years. It is likely the oldest and most widely used therapy with roots in traditional Oriental medicine and the ayurvedic tradition. Herbs are substances derived from trees, flowers, plants, seaweed, and lichen. Herbs are prepared in several different forms: tinctures which contain grain alcohol for preservation, freeze-dried extracts, and standardized extracts. Herbs are contained in some manufactured drugs; drugs can also contain a synthetic copy of the herb. *Many plant extracts can be very beneficial, however it is prudent to remember that herbs are drugs and should be consumed only as prescribed.* Even if the consumer is using the herbal remedy correctly, there may be an adverse interaction with food, over-the-counter drugs (OTC), vitamins and minerals, or prescriptions drugs. Recent studies done by NCCAM (NIH, 2002) found that St. John's wort reduces the action of a common AIDS drug called Indinavir (see Figure 4.6). St. John's wort, commonly used for mild to moderate depression, clears 50 percent of all pharmaceutical drugs from the human body (Markowitz et al., 2003).

An herbalist is a practitioner who bases most of his therapy on the medicinal qualities of plant and herbs. Herbs are prescribed so much in some parts of Europe that they might not even be considered alternative. There are volumes of testimonials, lots of anecdotal evidence, and many cultural traditions supporting herbal therapy. Gingko biloboa is purported to help with memory. St. John's wort helps with depression (some studies support this, some refute it). Saw

> "Let thy food be thy medicine and thy medicine be thy food."
> —Hippocrates (460–377 B.C.)

> "No illness which can be treated by the diet should be treated by any other means."
> –Moses Maimonides (1135–1204)

Many plant extracts can be beneficial, however it is prudent to remember that herbs are drugs and should be consumed only as prescribed.

FIGURE 4.6

Biologically Based Systems: St. John's Wort Lowers Blood Levels of HIV Protease Inhibitor Indinavir

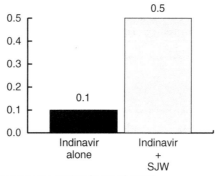

Note: HIV Inhibition threshold = 0.1. Indinavir Level (ug/ml)

palmetto helps manage an enlarged prostate. Butterbur, bee pollen, and stinging nettle may help with allergy symptoms. Evening primrose oil helps to manage PMS. The list is endless! When working with an experienced herbalist it is important to try and regulate the quality of the product you are getting. Using caution, especially when self-prescribing, is important because: safety is assumed, not proven; products are not standardized; products can be contaminated; you may have an allergic reaction, some herbs or certain amounts of the herb can be toxic, and the herbs can interact with drugs. Purity, standardization, and quality of the herbs can be an issue in consistency and the amount of the herb in the product. Another reason to use caution is that sometimes we get the sense that if a little works, perhaps a little more will work better. Toxic levels of drugs and herbs can be dangerous. If your friend takes 200 mg of a drug or herb, then you might do the same with disastrous consequences. You may be a nonresponder for that substance and get no result, or you may tolerate the substance and need a larger amount. Another issue is the amount of product actually contained in the packaging. See Figure 4.7 on ginseng.

An example of an unsafe drug is ephedra, derived from the Chinese herb Ma Huang. Traditionally Ma Huang has been used in China to treat asthma and other ailments associated with respiration. Ephedra was confirmed to be a factor in the death of Orioles pitching prospect Steve Belcher in February of 2003. Steve was taking ephedra to give him energy and to assist him with weight loss. The facts that Steve used ephedra, it was hot, and he was exercising

Did You Know?

National Consumers League Food and Drug Interaction Brochure

NCL is your consumer healthcare advocate.

Medicines are powerful. The drugs your physician prescribes to you can help your health. A drug's effectiveness can be rendered ineffective or enhanced by food, drink, herbs (botanicals), and other drugs in your diet. Log on to this Web site, write, or call the NCL to obtain this important food and drug interaction brochure.

www.nclnet.org

National Consumers League (nonprofit membership organization)

1701 K Street, NW, Suite 1200

Washington DC 20006 (202) 835-3323

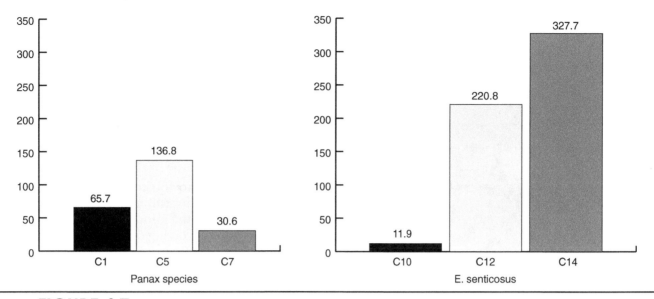

FIGURE 4.7
Variability in Commercial Ginseng Products
Adapted from Harkey, *American Journal of Clinical Nutrition,* 2001.

combined to cause his death. The FDA has since banned the use of ephedra. If you are wondering about the safety of a particular herb, a very informative website that contains warnings and safety information is the USFDA Center for Food Safety and Applied Nutrition's Dietary Supplements: Warnings and Safety Information (www.cfsan.fda.gov/~dms/ds-warn.html).

Functional foods are foods that contain compounds like phytochemicals that are beneficial beyond basic nutrition, especially when eaten on a regular basis as part of a varied diet. This is a relatively new classification of foods and the specific benefits are still being determined. It is possible that functional foods act synergistically with other foods and antioxidants. Isoflavones in soy products, omega-3 fatty acids in cold water fish, essential fatty acids and fiber in ground flaxseed, and probiotic yeasts found in some yogurts are examples of other functional foods. If you eat a wholesome diet with plenty of fruits and vegetables and whole grains, then most likely you are getting functional foods in your diet (Corbin et al., 2009).

Vitamins and minerals are two of the six essential components for life in the human diet. As with herbs, a little is good but more is not always better. Use caution when megadosing on vitamins and minerals to avoid toxicity and health risks. As always, it is good to consult your physician or an educated CAM practitioner. We will not discuss the particulars of vitamins and minerals here; however, remember that often the best way to get vitamins and minerals

> "The revolution we call mind-body medicine was based on this simple discovery: Wherever thought goes, a chemical goes with it. This insight has turned into a powerful tool that allows us to understand, for example, why recent widows are twice as likely to develop breast cancer, and why the chronically depressed are four times more likely to get sick. In both cases, distressed mental states get converted into the bio-chemicals that create disease."
> —Deepak Chopra, M.D. (1993)

There are numerous meditative techniques. Almost all techniques involve focus on breathing.

is through a daily balanced diet. For those that have chronically deficient diets, a multivitamin and mineral supplement is most likely a good recommendation. For evidence-based information on herbal and supple ments, log onto *HerbMed* for herbal information or to *Medline Plus* for other supplement information; the address is found at the back of the chapter in the list of resources.

Mind-Body Medicine

Mind-body medicine taps into the connection between the physical body and the very powerful mind. Mind-body therapies are designed to enhance the mind's capacity to influence the body. Creative therapy involving dance, art, and music, as well as prayer and mental healing, are considered to be in the mind-body category.

Meditation is a way to deal with the effects of chronic stress. The person meditating is able to let go by focusing primarily on taking time (twenty minutes is great) to relax, shutting out external stimuli. There are numerous meditative techniques, and there really is no "right" way to meditate. Almost all techniques involve focus on breathing. The most important factor is to commit the time to just do nothing. This can be a challenge in our busy lives today. Try simply to find a quiet place, eliminate distractions, close your eyes, repeat a word or phrase that is meaningful for you, and say that word over and over with each exhalation. Allow yourself to just let go. "For 30 years, research has told us that meditation works beautifully as an antidote to stress," says Daniel Goleman, author of *Destructive Emotions* (2003). More and more conventional health professionals are recommending meditation as a way to deal with chronic stress as well as chronic pain. Experiencing the calming effect of meditation, if only for ten minutes each day, creates a period of physical relief that can enhance immune function. Over time, the benefits of meditation can have a cumulative effect, improving the well-being of the meditator (Heistand, 2005). With practice, it can calm the body and quiet the mind. The benefits are numerous. Try committing to the above Meditation Exercise technique daily for one week to see if it has a positive effect on your life.

Meditation practice can alter brain activity. Andrew Newberg is a University of Pennsylvania neurologist who has studied changes in brain activity during meditation. Using radioactive dye with functional brain imaging, Newberg demonstrated that the brains of Tibetan Buddhist monks blocked out information from the part of the brain that orients the body in space and time. The monks focused their energy inward, while blocking out any external stimuli (Pure Insight, 2005).

Previous studies have determined that meditation has a positive influence on symptoms of stress and anxiety. Another more recent study used brain imaging to look at specifically which regions of the brain were affected by mindfulness meditation. In the study conducted by researchers from University of Massachusetts Medical School and the Bender Institute of Neuroimaging in Germany, the meditation group had increases in gray matter concentration in the left hippocampus. The hippocampus is an area of the brain involved in learning, memory, and emotional control. This is one of the first studies to actually look at what is going on in the brain during meditation.

Applied psycho-neuro-immunology is based on research into psycho-neuroendocrino-immunolgy, the science of how our experiences are encoded neurologically and about how this affects our immune and hormonal systems. How our bodies' encounter, adapt, and react to stress directly affects our immune system. This approach treats the whole body. Our minds can keep us in a dark hole of hopelessness and helplessness with a condition like depression,

Meditation Exercise

The most important thing is to allow yourself to do nothing.

Easy steps to meditation:

1. Eliminate distractions.
Turn off the cell phone, decide not to answer the door, let family or friends know you are unavailable for the next ten or fifteen minutes. Go to the bathroom, get a drink of water, and generally take care of any physical distractions that might arise. You might find it helpful to set a timer.

2. Just sit.
Get comfortable—consider sitting on a folded blanket to allow for less stress on the knees. Take a deep breath and allow your spine to extend and your ribs to lift. Maintain a tall posture as you soften your shoulders. Dim lights, a candle, incense, and appropriate music are nice but not necessary.

3. Let go.
Practice silence. Close your eyes and quietly observe the thoughts that come to your mind. Acknowledge them and then let them go. Let go of the outside stimuli so you can focus inward.

4. Listen.
Listen to the sounds of life in and around you. Become receptive to the sounds that are obvious, but also to the sounds that you normally don't hear because your attention is elsewhere. Hear without judgment; just observe. Notice your awareness of the present as it deepens.

5. Use your senses.
Cultivate an awareness of the present moment through sensations. Be attentive to where your body is connecting to the earth. What do you feel? Softness, hardness, coolness, warmth, pressure, and ease. How does your body change with each inhalation and exhalation? Settle into the present moment using your breath and your senses. Begin to focus on your inner self.

6. Simply breathe.
Attend to your breath. Try not to change your breath, just observe it how it is. Use all of your senses to increase awareness of how your body responds to your breath. Relax into your breath. Follow the rhythm of your breath with each inhalation and exhalation. When your mind wanders, just refocus and come back to the breath.

7. Mantra.
Saying a simple word, a phrase, a prayer, or anything meaningful to you over and over again as a mantra can coax you into a contemplative state. Repeat the mantra softly and slowly in an undulating rhythm with your breath, like riding a wave.

8. Practice kindness.
In your quiet state, consider someone who might be in need of some understanding and goodwill. Focus on this person. "In your mind's eye, send this person love, happiness, and well-being. Soften your skin, open the floodgates of your heart, and let gentle goodwill pour forth."

With consistent practice, meditation can make a difference in your life.

Adapted from Meditation 101 by Claudia Cummins, **Yoga Journal.** www.yogajournal.com/practice, 2009

---

which can cause a breakdown of our immune system. The reverse can be true as well. "The profound power of the mind that causes this rundown and eventual loss of resistance can also naturally be used positively to tune-up and boost the immune system to maximum level: to repel viruses, bacterium and other micro-organisms and to speed up healing" (AAAPNI).

Prayer and Spirituality—*Newsweek* recently did a cover story on the growth of spirituality in America today (Adler, 2005). The article greatly contrasts with the cover story of a 1966 *Time* article entitled "Is God Dead?" Spirituality is experiencing resurgence among Americans: 55 percent of Americans consider themselves spiritual and religious, and 24 percent consider themselves spiritual but not religious. Young people especially seem to be searching for greater meaning in a rigorous form of faith and prayer. Wanting more than

Beginning Your Own Spiritual Journey

Whether a person's quest for spiritual health takes the form of a love for nature, a weekly visit to a place of religious worship, or some other guise, it is clear that spirituality benefits overall health. While it is possible to achieve spiritual health in many ways, the following ideas have helped a number of people on their spiritual path:

Relaxation and Meditation

"There is no greater source of strength and power for me in my life now than going still, being quiet, and recognizing what real power is," says Oprah Winfrey on the segment of her daily television show called "Remembering Your Spirit." Many people take the time to sit quietly and to meditate; for example, more than five million people worldwide practice transcendental meditation, one popular relaxation technique.

Time in Nature

For Henry David Thoreau, who fled civilization to live on Walden Pond, nature was the temple of God and the perennial source of life. A powerfully spiritual moment—and one we have all experienced—is the instant we are confronted with earth's perfection and are filled with awe. The scientist Carl Sagan wrote about his time-in-nature experience: "The wind whips through the canyons of the American Southwest, and there is no one to hear it but us." The crisp, clean smell of the woods after rainfall, the soothing rhythm of crickets on a summer night, the beauty of freshly fallen snow—these experiences inspire unspeakable awe and humility because of the small but rich part that we, as individuals, play in the larger scheme of the universe.

Intimacy with Others

Loving selflessly is part of spiritual experience. Living life with passion and allowing ourselves to "feel" may be the greatest element of the spiritual journey. Experiencing emotion through a poignant musical passage, feeling the grief of a lost love, and surrendering to love's beauty are all part of human spirituality. By giving, sharing, and loving, we become whole and experience all that we are capable of feeling.

Spiritual Readings

Ranging from inspirational self-help books available at the local bookstore to traditional religious works, the written word has provided insight and guidance throughout human history, during its times of joy and darkest moments. For some it's the Bible; for others, it may be the Quran; and for still others, it may be a contemporary book such as *Spiritual Healing: Scientific Validation of a Healing Revolution* by Daniel J. Benor, M.D. (Vision Publications, 2001). To find books that will foster your personal growth and healing, listen to what others recommend and then search for whatever will move you or speak to you.

Prayer

Prayer may be the oldest spiritual practice and the most popular one in America. Almost all world religions include a form of prayer. Says George Lucas, who plays on religious themes such as good and evil in his blockbuster *Star Wars* series, "Religion is basically a container for faith. And faith is a very important part of what allows us to remain stable, remain balanced." The mental and emotional release, along with a sense of connection to a transcendent dimension, may be at the heart of prayer's effectiveness.

From *Psychology Today* (September/October 1999), 48; The Transcendental Meditation Program (see http://www.tm.org).

what traditional religious services offer, they want to experience God in their daily lives. Many are drawing on the influence of Eastern religions to enhance their traditional doctrines. Meditations, centering prayer, silent contemplation, as well as disciplines like yoga, are often used in addition to traditional church services (Adler, 2005).

More than half of the medical schools in the country now offer an elective course on "Spirituality and Medicine." Questions abound about the role of prayer in healing the sick. Eighty-four percent of Americans polled believe praying for the sick can improve their chance of recovery (Kalb, 2004). Anecdotal evidence says prayer works. Science demands concrete evidence; however, prayer is hard to measure.

Yoga is a mind/body/spiritual discipline that is rooted in the ancient Hindu religion traced back 5,000 years. Yoga has become popular in the United States in the last several decades. According to the *New York Times,* 16 million Americans currently practice yoga. There are many different styles of yoga—gentle, meditative, powerful, and relaxing yoga to name of few. Movement in yoga can be rigorous and intense, or gentle and calming. Physicians are recognizing the benefits that a yoga practice can have on strengthening the physical body and quieting the mind. In addition to the many physical benefits, the practice of yoga with emphasis on breathing can be a great stress management tool. Movement, breathing, chanting, and sound are a prelude to meditation and conscious relaxation that regular yoga practice can provide. Conscious relaxation gives our minds a break from the daily chatter and unending stimuli to which we are continuously exposed.

Guided imagery is a concept that has been used successfully with people suffering from post-traumatic stress disorder. Psychologist Kathleen Reyntjiens, Ph.D., reports that imagery was an integral part of her recovery efforts with those recently traumatized by Hurricane Katrina. "As our previous research with imagery has indicated, these simple self-regulation techniques—especially imagery and conscious breathing—are helping to minimize distress, anxiety, hypervigilance, anger, sadness, and insomnia, and allow all of us to be more effective, efficient, kind and caring neighbors in our survival and clean-up efforts" (Naparstek, 2005).

Mindfulness is a concept that includes strategies and activities that help us be more in the present moment. This helps us connect more intently with ourselves, others, and with nature. Peacefulness can be the result of being mindful throughout the day. Mindfulness is often thought of in conjunction with spirituality.

Feldenkrais method is "a form of somatic education that uses gentle movement and directed attention to improve movement and enhance human functioning. Through this method, you can increase your ease and range of motion, improve your flexibility and coordination, and rediscover your innate capacity for graceful, efficient movement" (The North American Feldenkrais Guild). Feldenkrais is excellent for dancers, athletes, and others, as well as those limited by neuromuscular pain or neurological dysfunction. Moshe Feldenkrais, an Israeli engineer, developed this technique.

Somatic movement re-educates the neuromuscular system toward greater health and well-being. Through hands-on movement work by the somatics practitioner, "people can learn to manage stress, relieve back pain, breathe more freely, heal from trauma to the neuromuscular system, and speed recuperation after illness or surgery" (Brockport, 2005). The Feldenkrais method is a type of somatic movement education. Meditation, visualization, craniosacral therapy, and myofascial release techniques are often practiced by the somatic movement practitioner.

Animal-assisted therapy is the use of companion animals to help people with special needs. Evidence is mounting that spending time with a loved pet not only has emotional and psychological benefits, but physiological benefits as well. The act of petting and caring for a loved animal can reduce blood

> "Yoga is the stilling of the restlessness of the mind."
> —Yoga Sutras

Evidence is mounting that spending time with a loved pet has emotional, psychological, and physiological benefits.

pressure and heart rate and improve survival rates from heart disease (Arkow, www.animal therapy). Close to half of the psychologists responding to a survey indicated prescribing a pet to combat loneliness or depression. According to Phil Arkow, instructor of the Animal Assisted Therapy course at Camden County College in Blackwood, New Jersey, elderly people who have pets visit physicians 16 percent less than those who do not; dog owners in particular make 21 percent fewer visits. "A pet is an island of sanity in what appears to be an insane world. Friendship retains its traditional values and securities in one's relationship with one's pet. Whether a dog, cat, bird, fish, turtle, or what have you, one can rely upon the fact that one's pet will always remain a faithful, intimate, non-competitive friend—regardless of the good or ill fortune life brings us" (Dr. Boris Levinson, child psychologist).

Energy Therapies

Energy therapies engage the use of energy fields that surround the body and penetrate the body. The science behind energy fields has yet to be proven.

Qi gong combines movement, meditation, and regulation of breathing to enhance the flow of vital energy (qi), improve blood circulation, and enhance immune function (Donatelle, 2004). Qi gong literally means the skill of attracting vital energy. Those that practice qi gong call it a "self-healing art" that uses visualization and imagery with movement and meditation.

Reiki (pronounced ray-key) is a type of energy work that utilizes touch and visualization. Reiki is based on ancient Tibetan teachings and is said to date back thousands of years. Today reiki is practiced using the Eastern concept of the five chakras in the body, as well as using the organs and glands from Western anatomy.

Therapeutic touch is purported to induce the relaxation response, alleviate pain, and to speed the healing process. In therapeutic touch, the patient is not actually touched. In one study people were wounded on their arms. The control group had conventional therapy, while the other group experienced therapeutic touch. The entire second group experienced quicker healing (Wirth, 1990).

Bioenergy practitioners use psychotherapy, grounding exercise, and deep breathing to assist in releasing muscular tension, pain, and illness. Pain and illness are thought to be caused by suppressed emotions and behaviors (AMFI, 2005).

Ultimately the responsibility lies with the patient to secure quality health care. As time goes on, more CAM modalities will be studied and the results will help guide consumers to which therapies are best for each individual person.

There are many more CAM therapies than are mentioned in this chapter. Conventional physicians and those they work closely with want the same things as most CAM practitioners—for patients to have good health and wellness. "The effectiveness of the healthcare delivery system in the future will depend upon its ability to make use of all approaches and modalities that provide a sound basis for promoting optimal health. People with better health habits have been shown to survive longer and to postpone and shorten disability" (WHCCAMP, 2002). Certainly many CAM practices will be useful in contributing to the nation's health goals. The modern patient is more informed and involved in his or her own health. Most likely the marriage of essential conventional practices with complementary and alternative therapeutics will be the way of the future.

Using the Internet for Credible Medical Information

You may find the Internet a valuable resource for researching potential CAM therapies. Using a typical search engine could net you thousands, if not more, hits. This information is intended as a resource to help you sift through the "junk mail" of sorts and to determine what is actual credible information.

Distinguish between different sorts of Web resources:

Information sites—have a domain ending .edu (education), .gov (government), .org (nonprofit organization), .net (technical services)—typically very reliable.

Advice and referral sites—look for the credentials, reputation, or experience to see if the source is credible.

Activist sites—typically promoting a particular cause rather than providing information.

Mindfulness in Everyday Life

Being mindful means focusing attention on what you're experiencing from moment to moment. It's a daunting challenge in a hectic world, but science has begun to establish that it's a worthwhile habit to cultivate. You can start by getting a sense of how much time you spend not being mindful. See if you recognize any of these statements from a questionnaire developed at the University of Rochester:

- I find it difficult to stay focused on what's happening in the present.
- I snack without paying much attention to what I'm eating.
- It seems I'm "running on automatic" without much awareness of what I'm doing.
- I rush through activities without being really attentive to them.
- I tend to walk quickly to get where I'm going without paying attention to what I experience along the way.
- I find myself listening to someone with one ear and doing something else at the same time.
- I tend not to notice physical tension or discomfort until they really grab my attention.

If these sound familiar, there's plenty of room for increasing mindfulness in your daily life. Take note of times when your thoughts are creating stress or distracting you from the present moment. The Mind/Body Medical Institute suggests that you slow down as you go about everyday activities, doing one thing at a time and bringing your full awareness to both the activity and your experience of it. Here are some tips for integrating mindfulness:

- Make something that occurs several times during the day, such as answering the phone or buckling your seat belt a reminder to return to the present—that is, think about what you're doing and observe yourself doing it.
- Pay attention to your breathing or your environment when you stop at red lights.
- Before you go to sleep, and when you awaken, take some "mindful" breaths. Instead of allowing your mind to wander over the day's concerns, direct your attention to your breathing. Feel its effects on your nostrils, lungs and abdomen. Try to think of nothing else.
- If the present moment involves stress— perhaps you're about to speak in public or undergo a medical test—observe your thoughts and emotions and how they affect your body.
- Find a task you usually do impatiently or unconsciously (standing in line or brushing your teeth, for example) and do it mindfully.

Being mindful doesn't mean you'll never "multitask," but you can make multitasking a conscious choice. It doesn't mean you'll never be in a hurry, but at least you will be aware that you are rushing. Although upsetting thoughts or emotions won't disappear, you will have more insight into them and become aware of your choices in responding to them.

From *Harvard Women's Watch*, Vol. 11, #6, February 2004.

Chat groups—can be unreliable but also a valuable extension of your community; a great way to connect with others sharing the same experience, such as a rare disease.

Individual testimonials—interesting but not always reliable or authoritative.

Commercial sites—the majority of Web sites are commercial and typically set up to sell products, therefore the information is usually biased.

TIP: You can avoid commercial Web sites by including "NOT .com" in your search string.

Other questions to consider:
Who is the group providing the Web site? How is the Web site funded?
How is the information selected and presented? What are the qualifications of the author?
Does the site follow ethical practices? (The Web site should be current; check their privacy policy if they ask information from you.)

Reference: The Alternative Medicine Foundation

References

Acupuncture Relieves Pain and Improves Function in Knee. *Osteoarthritis NIH News,* December 20, 2004 press release.

Adler, J. In Search of the Spiritual. *Newsweek,* August 29, 2005. Alternative Medicine Foundation www.amfoundation.org

American Massage Therapy Association. www.amtamassage.org Phone (847) 864-0123

Arkow, P. *Animal Assisted Therapy: A Premise and a Promise.* http://www.animaltherapy. net/Premise%20%26%20Promise.html

Association for the Advancement of Applied Psychoneuroimmunology. http://hometown.aol.com/AAAPNI

Barnes, P. M., Bloom, B., and Nahin, R. *CDC National Health Statistics Report #12.* Complementary and Alternative Medicine Use Among Adults and Children: United States, 2007. December 10, 2008.

Chopra, D. *Ageless Body, Ageless Mind.* Harmony Books. 1993.

ClinicalTrials.gov; US National Library of Medicine, US National Institute of Health. www. clinicaltrials.gov/ct/gui/c/alb/show/NCT00010829

International Alliance of Healthcare Educators (IAHE). Craniosacral Therapy/ Somatoemotional Release: Education for better patient care. 2005. http://www.iahe.come/ html/therapies.cst.jsp

Donatelle, R. J. *Access to Health* (8th ed). Pearson/Benjamin Cummings. 2004. Feldenkrais Educational Foundation of North America (FEFNA) 3611 SW Hood Ave. Suite 100 Portland, OR 97239, USA http://www.feldenkrais.com

Gach, M. R., *Acupressure's Potent Points.* Bantam Books. 1990.

Goldberg, B. *Alternative Medicine, The Definitive Guide* (2nd ed). Berkeley, CA: Celestial Arts. 2002.

Goleman, D. *Destructive Emotion: A Scientific Dialogue with the Dalai Lama.* 2003. Alternative Medicine Foundation Information. How to Assess Credibility on the Web. 2009. www.amfoundation.org/assess.htm

Information on Clinical Trials being conducted; www.clinicaltrials.gov/ct/gui/c/alb/show/NCT00010829

Kalb, C. Faith and Healing, *Newsweek,* November 10, 2004.

Markowitz, J. et al. Effect of St John's Wort on Drug Metabolism by Induction of Cytochrome P450 3A4 Enzyme. *JAMA,* (290),1500–1503. 2003. Naparstek, B. *Health Journeys.* 2005.

National Institutes of Health. NCCAM Online Continuing Education Lecture Series, Manipulative and Body-Based Therapies: Chiropractic and Spinal Manipulation. 2002. http://nccam.org/main.php

Psych-Neuro-Immunology http://hometown.aol.com/AAAPNI/

Pure Insight. 2005. http://pureinsight.org/PI/index/html

Saper, R. B., Phillips, R. S., Sehgal, A., et al. Lead, mercury, and arsenic in U.S. and Indian-manufactured Ayurvedic medicines sold via the Internet. *Journal of the American Medical Association*, 300(8):915–923. 2008.

Somatic Movement Studies http://www.brockport.edu/~dance/somatics/techniques.htm, 2005.

Reaves, W. Acupuncture and the Athlete, *ACSM Fit Society*, Fall 2008.

Ryzik, M. Z. Exercising That Back Pain Away. *New York Times*, September 15, 2005. Wirth, D. P. The Effects of Non-Contact Therapeutic Touch on the Healing Rate of Full Thickness Dermal Wounds. *Subtle Energies*, Vol. 1, No. 1, 1990.

USFDA Center for Food Safety and Applied Nutrition's Dietary Supplements: Warnings and Safety Information. www.cfsan.fda.gov/~dms/ds-warn.html

White House Commission on Complementary and Alternative Medicine Policy (WHCCAMP). Final Report. 2002. *(electronic version)*

Contacts

Acupuncture and Oriental Medicine Alliance

www.acupuncturealliance.org

Alternative Medicine: Health Care Information Resources

http://hsl.mcmaster.ca/tomflem/altmed.html This is an extremely thorough resource of alternative medicine for the informed consumer.

Alternative Medicine Foundation

www.amfoundation.org

American Massage Therapy Association

www.amtamassage.org

Phone (847) 864-0123

American Yoga Association

www.americanyogaassociation.org

The Art of Living Foundation www.artofliving.org Information on the science of the breath and it's healing qualities. Worldwide organization.

Center for Mindfulness in Medicine, Health Care, and Society; University of Massachusetts Medical School

www.umassmed.edu/cfm

Complementary and Alternative Medicine: From Promises to Proof. NIH

Craniosacral Therapy Association of North America

FDA/Center for Food Safety and Applied Nutrition

http://www.cfsan.fda.gov/~dms/ds-warn .html Warnings and safety information regarding dietary supplements.

Heistand, C. http://www.orgsites.com/ca/acco/_pgg4.php3 very thorough site using CAM with cancer patients.

HerbMed http://www.herbmed.org Evidence-based herbal resource http://nccam.nih.gov/news/images/campractice.htm (CAM category picture)

Medline Plus http://medlineplus.gov/ An excellent resource that is a service of the U.S. Library of Congress and the National Institute of Health.

MedWatch The FDA Safety Information and Adverse Event Reporting Program

http://www.fda.gov/medwatch

Mind/Body Medical Institute www.mindbody.harvard.edu

Movement Educators www.movementeducators.com This site features a free Mindful Movement Lesson in the Fledenkrais method.

National Institute of Ayurvedic Medicine (NIAM)

www.niam.com

National Association of Chiropractic Medicine (NACM)

www.chiromed.org

National Institutes for Health National Center for Complementary and Alternative Medicine (NCCAM)

http://nccam.org/main.php

Nutrition Science News

http://exchange.healthwell.com/nutritionsciencenews/ information and research on natural medicine

Psych-Neuro-Immunology

http://hometown.aol.com/AAAPNI/

Psycho-Neuro-Immunolgy

http://www.alpha-cs.co.za/PsychoNeuroImm.htm source of CD's to listen to harness the power of the mind to help health and well-being.

Qigong Association of America

http://www.qi.org/

Resources for Body, Mind and Spirit

www.healthjourneys.com Bellaruth Naparstek's very informative site. Ms. Napastek is a guided imagery pioneer and creator of desktopspa.com, (800) 800–8661.

Shiatsu: Japanese Massage

http://www.rianvisser.nl/shiatsu/e_index.htm This site includes a "do-in" link where you can learn specific exercises.

Somatic Movement Studies

http://www.brockport.edu/~dance/somatics/techniques.htm

Core movement patterning, Somatic Release and Contact Unwinding-educational/therapeutic techniques developed in East-West Somatics.

Tips for the Savvy Supplement User: Making Informend Decisions and Evaluating Information

http://www,cfsan.fda.gov/~dms/ds-savvy.html Dietary supplement info

University of Michigan Integrative Medicine

http://www.med.umich.edu/umim/

Source of the Healing Foods Pyramid.

USFDA Center for Food Safety and Applied Nutrition's Dietary Supplements: Warnings and Safety Information.

www.cfsan.fda.gov/~dms/ds-warn.html

White House Commission on Complementary and Alternative Medicine Policy, March 2002.

WHO Guidelines on Developing Consumer Information on Proper Use of Traditional, Complementary and Alternative Medicine.

Yoga Journal Periodical and online at www.yogajournal.com Excellent resource for all things related to yoga, relaxation, stress management and meditation.

Recommended Books For Further Reading

Ageless Body, Timeless Mind. By Deepak Chopra (Harmony Books, 1993)

Awakening the Spine. By Vanda Scaravelli (HarperCollins Publishers, 1991)

Peace Is Every Step: The Path of Mindfulness in Everyday Life. By Thich Nhat Hanh (Bantam Books, 1992)

Relax and Renew By Judith Lasater (Rodmell Press, 1995)

Wherever You Go There You Are: Mindfulness Meditation in Everyday Life. By Jon Kabat-Zinn (Hyperion, 1995)

The Spectrum: A Scientifically Proven Program to Feel Better, Live Longer, Lose Weight, and Gain Health. By Dean Ornish (Ballantine Books, 2007)

Chapter 5
Exercise Science and Sports Medicine Topics

OBJECTIVES

Students will be able to:
- Identify and explain key components necessary to safely maintain or improve cardiovascular fitness, muscular fitness, and flexibility.
- Explain the importance of cardiovascular conditioning for health maintenance and disease prevention.
- Integrate body measurements and fitness values to determine current health status.
- Recognize and explain the importance of having strong core musculature to overall health.
- Identify the key components of a cardiovascular and/or a muscular training program.

Adaptation of *Health and Fitness: A Guide to a Healthly Lifestyle, 5/e* by Laura Bounds, Gayden Darnell, Kirstin Brekken Shea and Dottiede Agnor. Copyright © 2012 by Kendall Hunt Publishing Company.

"Physical Fitness is not only one of the most important keys to a healthy body, it is the basis of dynamic and creative intellectual activity. The relationship between the soundness of the body and the activities of the mind is subtle and complex. Much is not yet understood. But we do know what the Greeks knew: That intelligence and skill can only function at the peak of their capacity when the body is healthy and strong; that hardy spirits and tough minds usually inhabit sound bodies."

—**President John F. Kennedy,**
"The Soft American," Sports Illustrated, December 26, 1960

President Obama commits forty-five to ninety minutes, six days a week, for cardio, weights, and basketball.

"It gives you more mental endurance and more energy to think clearly," said President Barack Obama. You might think the president was talking about the newest energy drink on the market, but he was referring to exercise. "He is quietly confident and competitive" was a description of President Barack Obama by U. S. Senator Bob Casey, D-Pa. Although it may sound like the way Mr. Obama ran his marathon election campaign, Senator Casey was actually describing Obama on the basketball court. The president commits forty-five to ninety minutes six days a week for cardio, weights, and basketball. He has good blood pressure and he eats fairly well, so President Obama told *Men's Health* magazine that "The main reason I do it is just to clear my head and relieve me of stress." The Obama children stay active with organized sports like soccer and gymnastics while wife Michelle Obama enjoys hitting the gym at least three days a week for ninety minutes doing cardio and lifting weights. The first lady calls exercise "therapeutic." The president also enjoys healthy snacks of nuts, seeds, and raisins. His vice happens to be an on-again off-again smoking habit. Perhaps the president will not only be a good example of living a fit lifestyle, but he might encourage many who may try to quit the smoking habit alongside him. The White House is, after all, a nonsmoking area.

Before President George W. Bush took office for his first term, President Clinton was known for jogging through the streets of Washington. In his first term President George W. Bush jogged for fitness, but in his second term he traded in his running shoes for a mountain bike. The biking allowed him to work his heart without stressing his knees. President John F. Kennedy wrote in 1960 that hardy minds and tough spirits usually inhabit sound bodies (see quote at the beginning of this chapter). Members of the Oval Office have a nonpartisan request, which is for Americans to increase their activity to hopefully decrease the increased prevalence of obesity and heart disease in America (see Figure 5.2 on page xx). If the president of the United States finds time to exercise, perhaps our excuse of "I don't have time" is not valid.

Why Is Physical Activity Important?

Just like a car can rust and fall into disrepair when it isn't driven for long periods of time, the human body can deteriorate and fall into disrepair when it isn't taxed beyond a resting state. Every system in the body, from the large muscles to the mitochondria in the cells, "runs" more optimally when the brain and the body are stimulated. Activities like walking, running, dancing, and playing tennis work the muscles and in turn the actions of the muscles stimulate bones to increase in density. The heart and the entire cardiorespiratory system increase activity to elevate heart rate, blood pressure and cardiac output to meet the demand for increased oxygen to the working muscles. Capillary circulation increases. The contractility of the cardiac (heart) muscle increases. Positive impact is seen in attitude, self-esteem, ability to focus, and risk of depression. The list goes on and on. We live better and most of us feel better when involved in some sort of moderate activity on most days of the week. Cars run better when driven, and people live better when active. For all ages, in all walks of life, regular activity can enhance quality of life. This chapter discusses options when considering your personal fitness program.

Clinical, scientific, and epidemiological studies indicate that physical activity has a positive effect on the delay in development of cardiovascular disease (ACSM, 2012). In a landmark report in 1996, the Surgeon General recommended that all Americans accumulate thirty minutes of activity on most, if not all, days of the week. Recent recommendations state that thirty minutes might not be enough (see Figure 5.2 on page xx). The 2010 dietary guidelines recommend most Americans should bump activity time to sixty minutes daily and up to ninety minutes if a recent significant weight loss is to be maintained. In 2002, the Institute of Medicine issued a statement that all Americans, regardless of age, weight, size and race, should achieve a total of sixty minutes of moderately intense physical activity daily. In December 2003, the National Sports and Physical Education (NASPE) changed the previous recommendation (1998) to be increased to "at least sixty minutes, and up to several hours of physical activity per day" for children 5 to 12 years of age.

In 2007 the American College of Sports Medicine (ACSM) and the American Heart Association (AHA) together released updated physical activity guidelines that emphasize thirty minutes of moderate activity five times weekly for most Americans. An alternative would be three twenty minute vigorously intense bouts of activity. Strength training is also recommended. Both ACSM and AHA endorse recommendations by the Institute of Medicine that in order to lose weight when these guidelines are already being met, "an increase in activity is a reasonable component of a strategy to lose weight" (ACSM/AHA, 2007). *For individuals that are overweight and want to avoid becoming obese, exercise sessions of forty-five to sixty minutes may be prudent. Formerly obese individuals who are trying to avoid regaining weight should consider up to ninety minutes of aerobic activity on most days of the week.*

How do we define moderate or vigorous activities? Examples of each follow. **Moderate activities** (You can still speak while doing them) are activities like line dancing, biking with no hills, gardening, tennis (doubles), manual wheelchair wheeling, walking briskly, and water aerobics. **Vigorous activities** are aerobic dance or step aerobics, biking hills or going faster than ten miles per hour, dancing vigorously, hiking uphill, jumping rope, martial arts, racewalking, jogging or running, sports with continuous running such as basketball, soccer and hockey, swimming laps, and singles tennis. With vigorous activities it would be difficult to carry on a conversation due to the intensity of the exercise. Try to do all of these activities for a minimum of ten minutes at a time. It is important to recognize that moderate activity is beneficial to everyone, while vigorous activity may not be appropriate for everyone.

Regardless of recommendations, children, adults, and the elderly all benefit from regular, consistent physical activity (see Figure 5.1). Choosing to seek opportunities to move such as walking, biking, swimming, gardening, and other activities can impact risk of disease as well as quality of life. In this chapter, we will discuss why you should exercise and give suggestions to help ensure your success.

FIGURE 5.1
Evidence abounds that everyone, all ages, can benefit from an enhanced quality of life with regular physical activity

Health Benefits of Physical Activity—A Review of the Strength of the Scientific Evidence

Adults and Older Adults

Strong Evidence
- Lower risk of:
  - Early death
  - Heart disease
  - Stroke
  - Type 2 diabetes
  - High blood pressure
  - Adverse blood lipid profile
  - Metabolic syndrome
  - Colon and breast cancers
- Prevention of weight gain
- Weight loss when combined with diet
- Improved cardiorespiratory and muscular fitness
- Prevention of falls
- Reduced depression
- Better cognitive function (older adults)

Moderate to Strong Evidence
- Better functional health (older adults)
- Reduced abdominal obesity

Moderate Evidence
- Weight maintenance after weight loss
- Lower risk of hip fracture
- Increased bone density
- Improved sleep quality
- Lower risk of lung and endometrial cancers

Children and Adolescents

Strong Evidence
- Improved cardiorespiratory endurance and muscular fitness
- Favorable body composition
- Improved bone health
- Improved cardiovascular and metabolic health biomarkers

Moderate Evidence
- Reduced symptoms of anxiety and depression.

Source: U.S. Department of Health & Human Services.

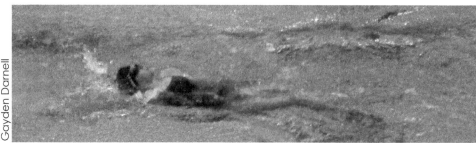

Swimming is an excellent lifetime fitness activity.

FIGURE 5.2

Determining Your Waist to Hip Ratio

Equipment
1. Tape measure
2. Partner to take measurements

Preparation
Wear clothes that will not add significantly to your measurements.

Instructions

Stand with your feet together and your arms at your sides. Raise your arms only high enough to allow for taking the measurements. Your partner should make sure that the tape is horizontal around the entire circumference and pulled snugly against your skin. The tape shouldn't be pulled so tight that it causes indentations in your skin. Record measurements to the nearest millimeter or one-sixteenth of an inch.

Waist. Measure at the smallest waist circumference. If you don't have a natural waist, measure at the level of your navel.
Hip. Measure at the largest hip circumference.

Calculating Your Ratio

You can use any unit of measurement (for example, inches or centimeters), as long as you're consistent. Waist-to-hip ratio equals waist measurement divided by hip measurement.

Determine Your Relative Risk

Find the risk category that corresponds to your ratio and age group on the appropriate figure below.

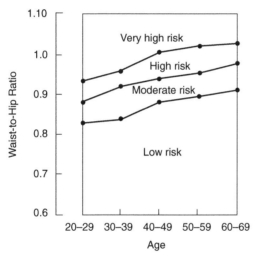
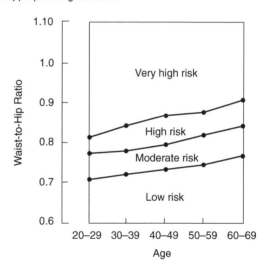

Source: from *Fit & Well* by Fahey, Insel and Roth, Mayfield Publishing.

Measuring Health Risk

There are two simple measures of overall health risk: **Waist-to-Hip Ratio** and **Body Mass Index (BMI).** See Figure 5.2 to determine your Waist-To-Hip ratio, and see the activity Body Mass Index Calculator in Chapter 7 to determine your BMI. Increasing lifestyle activity by spending less time on the couch and doing something daily will have a positive impact on your health. It is true that overweight and obesity are associated with complications and high risk factors such as hypertension, high blood cholesterol, and diabetes. *A growing body of evidence, however, indicates physical inactivity is more critical than excess weight in determining health risk.* Longitudinal studies such as the ongoing research by epidemiologist Steven Blair, previously of the Cooper Institute in Dallas, Texas, and information from the ongoing Harvard alumni study indicate that lifestyle is more significant than weight. **Fitter people have lower death rates regardless of weight** (see Figure 3.5). Indeed, the mortality rate for low fit males is more than 20% higher than for those that are high fit. While this effect is smaller for women, the decrease in mortality rate for high

Waist-to-Hip Ratio

Recent evidence from a study done at the University of Manchester in the United Kingdom indicates that **abdominal obesity** is a strong independent risk factor for heart disease. "A large waist with large hips is much less worrisome than a large waist with small hips." The conclusion of the study determined that the simple waist-to-hip ratio is a strong predictor of heart disease (AHA, 2007).

fit females is more than 6% compared to those who are low fit. Previously sedentary Harvard alumni (Sesso and Paffenbarger, 1956) who became active reduced their all-cause mortality rate by 23 percent. The alumni who lost weight (but were not active) did not improve their mortality rate. Improvements in metabolic fitness (glucose tolerance, blood pressure, and cholesterol) are often seen with just moderate amounts of physical activity. The good news is that overweight Americans don't need to go on a crash diet, buy a gym membership, or totally give up Twinkies. Increasing lifestyle activity and walking regularly, spending less time on the couch, and doing something active daily can have a positive impact on health.

Fitness or Fatness

Dr. Steven Blair is convinced we are too focused on obesity and overweight. Physical activity is much more crucial than a high BMI. People who are active, yet have a high BMI, have lower death rates than those who have a normal BMI but are sedentary. It is clear that if a sedentary person begins an exercise program, blood glucose and cholesterol could improve, yet that person might not loose any weight. *"Fitness is a more important indicator of health outcomes than fatness"* says Steven Ball, University of Missouri exercise physiologist.

Cardiovascular Fitness

Cardiovascular fitness refers to the ability of the heart, lungs, circulatory system, and energy supply system to perform at optimum levels for extended periods of time. **Cardiovascular endurance** is defined as the ability of the body to perform prolonged, large-muscle, dynamic exercise at moderate to high levels of intensity. The word **aerobic** means "in the presence of oxygen" and is used synonymously with *cardiovascular* as well as *cardiorespiratory* when describing a type of exercise.

Complete fitness is comprised of health-related fitness and skill-related fitness. **Health-related fitness** consists of cardiovascular fitness, muscular strength, muscular endurance, flexibility, and optimal body composition. The components of health-related fitness affect the body's ability to function efficiently and effectively. Optimal health-related fitness is not possible without regular physical activity. Most health clubs and fitness classes focus primarily on the health-related fitness components. **Skill-related fitness** includes agility, balance, coordination, reaction time, speed, and power. These attributes are critical for competitive athletes. Skill-related fitness is not essential in order to have cardiovascular fitness, nor will it necessarily make a person healthier. Balance is, however, important for seniors. Staying active

Basketball: Fitness, fun and competition at any age!
Louisiana
The Tigerettes Basketball Team, age 65+

The Tigerettes are five-time women's basketball gold medalists from Baton Rouge, Louisiana. Off the court, they are charming Southern belles and caring grandmothers. On the court, they transform into winner-take-all competitors who will dive, push, and elbow their way to another title. Learn more about the Tigerettes through the PBS documentary Age of Champions by visiting http://ageofchampions.org.

> Five components of health-related fitness:
> 1. cardiovascular fitness,
> 2. muscular strength,
> 3. muscular endurance,
> 4. flexibility, and
> 5. body composition.

> **Physical inactivity** is listed as the fourth leading preventable risk factor for global mortality rates-*World Health Organization*.

helps seniors maintain strength and balance, which can be critical in avoiding injuries.

Cardiovascular fitness is often referred to as the most important aspect of physical fitness because of its relevance to good health and optimal performance. **Muscular fitness** is important because of its effect on efficiency of human movement and basal metabolic rate. **Flexibility** is important for everyone, athletes and non-athletes alike, especially as a person ages. Knowing how to exercise correctly for effectiveness and reduced risk of injury is also important. Physically active individuals can expect to experience a positive impact on glucose regulation, blood pressure, blood cholesterol, bone density, body weight, and their outlook on life. Small amounts of activity for sedentary individuals can have a positive impact on overall health risk. For those who are overweight or obese, a loss of 5 to 10 percent of body weight can have a significant impact on body composition and overall health risk.

Exercising with friends can be enjoyable and help to keep you motivated and committed to your fitness program.

How important is participation in physical activity in achieving and maintaining good health? Since 1992, the American Heart Association has considered **inactivity** as important a risk factor for heart disease as high blood cholesterol, high blood pressure, and cigarette smoking. The U.S. Centers for Disease Control and Prevention (CDC) and the American College of Sports Medicine (ACSM) reported that 300,000 lives are lost each year due to inactivity (AHA, 2006). In 1996 the U.S. Surgeon General's Report made several definitive statements regarding physical activity and its impact on one's health (see Figure 5.3). Evidence is mounting that physical activity is an integral part of good health. Lack of exercise has spawned a whole new field of study termed **inactivity physiology**.

> $\dot{Q}=SV \times HR$ Cardiac output is the stroke volume times the heart rate during a one minute interval

There are many benefits associated with aerobic exercise. When one is aerobically fit, there is an overall reduction in the risk of coronary artery disease, i.e., stroke, blood vessel diseases, and heart diseases. Related to this reduction, there is a decrease in resting **heart rate** due to the improved efficiency of the heart. There is also an increase in **stroke volume.** The amount of blood pumped by the left ventricle of the heart with each beat is increased (stroke volume) which in turn increases total cardiac output. Cardiac output (Q) is the total volume of blood pumped during one minute.

A decrease in **systolic blood pressure,** which is the highest arterial blood pressure attained during the heart cycle, and a decrease in **diastolic blood pressure,** or lowest arterial pressure attained during the heart cycle, also occurs. There is also an increase in collateral circulation, which refers to the number of functioning capillaries both in the heart and throughout the body. Increased capillarization is an adaptation to regular aerobic activity. Delivery of oxygen to the working muscles and removal of metabolic wastes is more efficient with increased collateral circulation. In addition to the specific physiological changes just listed, other benefits of a mental,

FIGURE 5.3
1996 U.S. Surgeon General's Report: Physical Activity and Health

1. Males and females of all ages benefit from regular physical activity.
2. Significant health benefits can be obtained by moderately increasing daily activity on most, if not all, days of the week.
3. Additional health benefits can be gained through greater amounts of physical activity.
4. Physical activity reduces the risk of premature mortality in general, and of coronary heart disease, hypertension, colon cancer, and diabetes mellitus in particular.
5. More than 60 percent of American adults are not regularly physically active.
6. Nearly half of American youth 12 to 21 years of age are not vigorously active on a regular basis.
7. *People of all ages should try to accumulate thirty minutes of activity of moderate intensity (e.g., brisk walking) on most, if not all, days of the week.*

emotional, and physical nature will increase with regular aerobic exercise. Some of the potential benefits that have been documented when aerobic fitness levels increase are:

- a decrease in percent body fat,
- an increase in strength of connective tissues,
- a reduction in mental anxiety and depression,
- improved sleep patterns,
- a decrease in the speed of the aging process,
- an improvement in stress management,
- an increase in cognitive abilities.

> **Stress and the Heart**
> A strong heart will be more efficient than a weak heart when demands are imposed on it through stress. Exercise makes the heart stronger. Sympathetic nerve stimulation is responsible for the fight-or-flight response experienced as a result of emotional or physical stress. High fitness levels decrease the impact of this stress on the heart.

Heart rate becomes elevated during exercise because of the increase in demand for oxygen in the muscle tissues. Oxygen is attached to hemoglobin molecules and is transported in the blood. The heart pumps at a faster rate to meet the increased demand for oxygen. The heart is a muscle, and like other muscles, it becomes stronger due to the stress of exercise. Through regular exercise, the heart will increase slightly in size and significantly in strength, which results in an increased stroke volume. The primary difference is seen in the increased thickness and strength of the left ventricle wall. As a result of exercise, blood plasma volume increases, which allows stroke volume to increase. These two factors will cause resting heart rate to decrease, exercising heart rate will become more efficient, and there will be a quicker recovery to a resting heart rate after exercise ceases. Lack of exercise can contribute to many cardiovascular diseases and conditions, including **myocardial infarction,** or heart attack; **angina pectoris,** a condition caused by insufficient blood flow to the heart muscle that results in severe chest pain; and **atherosclerosis,** a build-up of fatty deposits causing blockage within the blood vessel.

Closely associated with the function and efficiency of the heart is the function and efficiency of the **circulatory system**. Blood flows from the heart to arteries and capillaries where oxygen is released and waste products are collected and removed from the tissues. The deoxygenated blood then makes the return trip to the heart through the venous system (As a result of aerobic exercise, blood flow to the skeletal muscles improves due to an increase in stroke volume, an increase in the number of capillaries, and an increase in the function of existing capillaries. This provides more efficient circulation both during exercise recovery and during daily activities.

Blood flow to the heart muscle is provided by two coronary arteries that branch off from the aorta and form a series of smaller vessels. With regular aerobic activity, the size of the coronary blood vessels increase and collateral circulation improves. These small blood vessels can supply oxygen to the cardiac muscle tissue when a sudden block occurs in a major vessel, such as during a heart attack. Often the degree of developed collateral circulation determines one's ability to survive a myocardial infarction, or heart attack. It appears that a regular exerciser might survive a heart attack due to collateral circulation within the heart, as the smaller collateral vessels take over when the primary artery becomes occluded, or blocked.

The lungs, air passages, and muscles involved in breathing that supply oxygen and remove carbon dioxide from the body are known as the respiratory system. During exercise, pulmonary ventilation, which is the movement of gases into and out of the lungs, increases in direct proportion to the body's metabolic needs. At lower exercise intensities, this is accomplished by increases in respiration depth. At higher intensities, the rate of respiration also increases. Although fatigue in strenuous exercise is frequently referred to as feeling "out of breath" or "winded," it appears that the normal capacity for pulmonary ventilation does not limit exercise performance (McArdle, Katch, and Katch, 1999). In a normal environment, one inhales sufficient amounts of oxygen. The breathing limitation is in the efficiency of the oxygen exchange at the cellular level. *The primary benefit of aerobic exercise to the respiratory system is an increase in strength and endurance of the respiratory muscles, not an increase in lung volume.* Maximal pulmonary ventilation volumes are dependent on body size (Wilmore and Costill, 1999). Muscles that elevate the thorax such as the diaphragm are referred to as muscles of inspiration. Muscles of expiration, including the abdominal muscles, depress the thorax. Regular aerobic training will result in an increase in both the strength and endurance of these muscles, and will also result in more efficient respiration.

Aerobic and Anaerobic Exercise

Aerobic exercise is activity that requires the body to supply oxygen to support performance over a period of time. Aerobic exercise is characterized by the use of the large muscle groups in a rhythmic mode with an increase in respiration and heart rate. *Aerobic* literally means "with oxygen." Walking, the most common form of exercise in the United States, is an aerobic activity. Other aerobic exercises include running, swimming, biking, cardiokickboxing, rowing, jump-roping, and any activity that fits the above criteria. As with most exercise, the rate of energy expenditure varies with an individual's skill level and intensity of exercise. *Aerobic activities of low intensity are ideal for the beginning or sedentary exerciser because they can be maintained for a longer period of time and have been shown to be effective in promoting weight loss and enhancing cardiovascular health.* Many activities are too intense to be maintained more than a few minutes; these activities are considered anaerobic.

Anaerobic literally means "in the absence of oxygen." **Anaerobic exercise** is exercise performed at intensity levels so great that the body's demand for oxygen exceeds its ability to supply it. Anaerobic activities are usually short in duration, high intensity, and result in the production of blood lactate. The energy for anaerobic activity is primarily from carbohydrates stored within the muscles, called **glycogen,** which is in limited supply. Fatigue rapidly sets in when glycogen stores are depleted. Examples of anaerobic activities include strength training, sprinting, and interval training. Sprinting requires so much energy that the intensity of the activity cannot be maintained for a long period of time. Anaerobic training can enhance the body's ability to cope with the effects of lactic acid and fatigue, thus promoting greater anaerobic fitness.

Interval training has been used for years, but now there is a new name for it–boot camp. The concept is simple: work for a shorter amount of time, but work harder. Working with a coach or personal trainer is recommended so that you also work smarter. Interval training involves high intensity cardiovascular exercise alternating between short rest or active rest periods. People who tend to get bored just running for 40 minutes often enjoy the variety and change in routine. The increased challenge of a 40-minute interval workout means increased energy expenditure. It is common to see the acronym HIIT, for High intensity interval training. Research indicates that 3 times per week is likely best for this type of training to limit injury and in order to allow for full recovery

Bicycling is an aerobic exercise that uses the large muscle groups in a rhythmic mode with an increase in respiration and heart rate.

between sessions (Kravitz). The basic variables manipulated when designing an interval training program include:

1. Duration (time/distance) of intervals
2. Duration of rest/recovery phase
3. Number of repetitions of intervals
4. Intensity (speed) of intervals
5. Frequency of interval workout sessions

Working hard with other people can be motivating due to the supportive and/or competitive nature in a class. A fringe benefit of interval training is that you will not only become more aerobically fit, but you might get stronger, faster, and more powerful. The anaerobic energy system is trained as well as the aerobic energy system in an interval type class. Due to increased caloric expenditure in a high intensity class, participants may experience weight loss success. Remember to warm up thoroughly, progress slowly, and stretch after activity.

Exercise Prescription, or How to Become FITT

To improve cardiovascular fitness, one must have a well-designed regimen of cardiovascular exercise. In order for improvement to occur, specific guidelines must be adhered to when designing a personal exercise program. As will be discussed later, the following guidelines apply not only to aerobic exercise, but to other components of physical fitness as well. The **FITT** acronym is easy to remember when identifying an appropriate cardiovascular exercise prescription: **frequency, intensity, time,** and **type.**

Frequency

Frequency refers to the number of exercise sessions per week. The American College of Sports Medicine recommends exercising three to five days per week at a moderate to vigorous level of intensity.

For individuals with a low level of aerobic fitness, beginning an exercise program by working out two times a week will result in an initial increase in aerobic fitness level. However, after some time, frequency and/or intensity will need to be increased for improvement to continue.

Intensity

Intensity refers to how hard one is working, and it can be measured by several techniques. These techniques include **measuring the heart rate** (see Figure 5.6) while exercising, **rating of perceived exertion (RPE),** and the **talk test**. To use heart rate as a measure of intensity, one's target heart rate range needs to be calculated before exercising. **Target heart rate range** is the intensity of training necessary to achieve cardiovascular improvement (see Figure 5.7). This target heart rate range indicates what an individual's heart rate should be during exercise. Calculation of target heart rate range using the Karvonen formula is done by multiplying maximum heart rate (220 minus one's age) by a designated intensity percentage. The American College of Sports Medicine guidelines for intensity recommends working between 55 and 90 percent of maximum heart rate, or between 50 and 85 percent of heart rate reserve (maximum heart rate minus resting heart rate). For individuals who are very unfit, the recommended range is 55 to 64 percent of maximum heart rate or 40 to 49 percent of heart rate reserve (Pollock, Gaesser, Butcher, Despres, Dishman, et al., 1998). Use of a heart rate monitor is also useful for specific heart rate and intensity feedback. Programs can be designed using heart rate to alleviate boredom and increase efficiency of the workout.

Reduce Belly Fat with Sprints
A 2012 study by Dr. Steve Boutcher, University of New South Wales, reported in the *Journal of Obesity* that one hour of sprinting (3 20 minute bouts on an exercise bike) for men produced the same visceral (abdominal) fat loss as continuous jogging for 7 hours per week. High intensity anaerobic work like sprinting will cause an increase in muscle mass whereas low intensity jogging typically does not result in significant muscle mass increases.

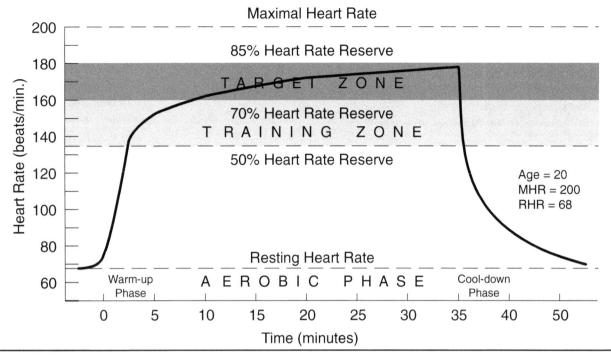

FIGURE 5.4

Cardiovascular Exercise Prescription Guidelines Tracking the Heart Rate Response Through a Typical Cardiovascular Exercise Session

FIGURE 5.5

Target Heart Rate Zones for Individuals of Ages 20 through 70. The zones cover 70–90 percent of maximum heart rate, which is indicated above the zones for selected ages

Source: From *Total Fitness and Wellness*, 5th Edition, by Powers and Dodd, Pearson Benjamin Cummings Publishers.

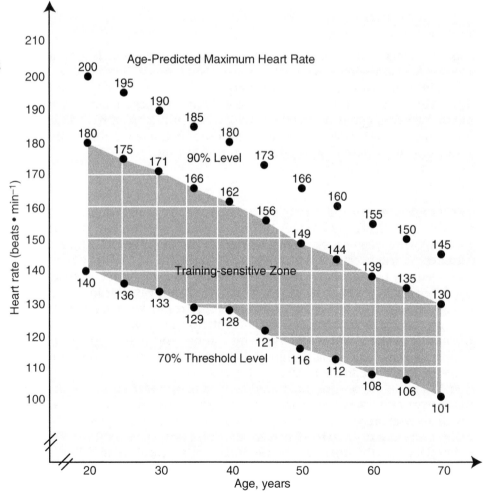

Another technique for measuring exercise intensity is through subjective self-evaluation or psychophysiological scale of how hard one is working. The body and the mind are called on to perceive one's effort. Gunner Borg designed the **Rating of Perceived Exertion Scale (RPE)** (see Table 5.1) in the early 1950s. It is a numbered scale from 6 to 20, with the lowest numbers being "very, very light" exercise, the highest numbers being "maximal" exercise, and the numbers in between representing a gradual increase in exercise intensity, from low to high. Using this scale, a rating of 10 corresponds roughly to 50 percent of maximal heart rate and a rating of 16 corresponds roughly to 90 percent of maximal heart rate. A person estimates his or her exercise intensity level by taking into consideration or "perceiving" how they feel, how much sleep they have had, whether or not they have eaten, whether or not they are ill, and so on. This scale is a useful tool for estimating exercise intensity when exact measures are not needed, and it is often used in a clinical setting as well as fitness classes and health clubs. Perceived exertion is also useful when a person is taking medication that can alter the heart rate. When using RPE, it is important to remember there is no right or wrong rating.

TABLE 5.1 ♦ The Borg RPE Scale

Score	Degree of Exertion
6	No exertion at all
7	Extremely light
8	
9	Very light
10	
11	Light
12	
13	Somewhat hard
14	
15	Hard (heavy)
16	
17	Very hard
18	
19	Extremely hard
20	Maximal exertion

Source: © Gunner Borg, 1970, 1985, 1998.

A third, and probably the easiest, technique to measure exercise intensity is the **talk test.** If you are exercising and must laboriously breathe rather than participate in a conversation, the exercise intensity is too high and training heart rate has probably been exceeded. Exercise at this intensity will be difficult to maintain for long periods of time. On the other hand, if you are able to sing, intensity level is probably insufficient for improvement in your fitness level.

Time

The third factor to be considered when designing a cardiovascular exercise workout is **time,** or duration. For benefits to be accrued in the cardiovascular system, exercise duration should be a minimum of twenty minutes of continuous exercise or several intermittent exercise sessions of a minimum of ten minutes each. Some beginning exercisers may not be capable of exercising continuously for twenty minutes at a prescribed intensity. While a minimum of twenty minutes is recommended, a duration of ten minutes can certainly be beneficial to people who are at a low fitness level and just beginning an exercise program. Duration and exercise intensity are interdependent, having an inverse relationship. As exercise effort increases, duration typically decreases. Distance runners exert a moderate effort for a long period of time, called long slow distance training. Sprinters exert a maximal effort for a brief period of time. Duration at a lower intensity is optimal for beginning exercisers. Exercise intensity levels should remain within recommended guidelines, while maximum duration is only limited by the participant's available fuel for energy and mental determination to keep going.

Type

Another factor that should be considered in determining a cardiovascular exercise prescription is **type,** or mode, of exercise. The choice of exercise modality is up to each individual, but one must keep in mind the specific requirements of cardiovascular exercise: use the large muscle groups via continuous and rhythmic movement, and exercise for a duration of twenty to thirty minutes or more in the target heart rate range a minimum of three to five times per week. Common types of aerobic exercise include running, walking, swimming, step aerobics, cross-country skiing, biking, or using a machine such as a rower, stairstepper, or treadmill. However, these are certainly not the only types of exercise available. Any exercise that meets the requirements of intensity and duration is acceptable. Some sports can provide aerobic exercise, depending on the nature of the sport, the position being played, and the skill level of the player. For example, an indoor soccer player could get an aerobic workout by playing a game, provided there is constant movement and training heart rate was maintained in the target heart rate range. Many sports provide an excellent way for people to expend a lot of energy and burn a significant number of calories, but the "play some and rest some" nature often prevents them from being good aerobic exercise. In order to achieve longterm cardiovascular fitness, it is good to pick a variety of activities, and to find activities that are pleasurable to the individual.

To achieve long-term cardiovascular fitness, select a variety of activities you find pleasurable.

Components of an Exercise Session

The sequence of a cardiovascular workout should be as follows: warm-up, easy optional stretch, workout, cool down, and stretch again with more intensity for increased flexibility.

Warm-Up

A good cardiovascular workout follows a specific sequence of events. First and foremost, to prepare the body and increase the comfort level for a cardiovascular workout, a warm-up is crucial. The purpose of the warm-up is to prepare the body and especially the heart for the more vigorous work to come. The warm-up should increase body temperature, increase heart rate, increase blood flow to the muscles that will be used during the workout, and include some rhythmic movement to loosen muscles that may be cold and/or tight. A warm-up should raise the pulse from a resting level to a rate somewhere near the low end of the recommended heart rate training zone. Beginning vigorous exercise without some kind of warm-up is not only difficult physically and mentally, but it can also contribute to musculoskeletal injuries.

Easy Stretch

It is important to warm up the muscles that will be used during the workout. Use caution when stretching prior to the workout, holding the stretches for a brief time and not stretching with much intensity. Stretching prior to the workout is optional. Many athletes choose to go through rhythmic limbering, with dynamic movements focusing on functional movement patterns that will be used in the race, sport, or fitness class rather than holding a static stretch.

Activity

When doing cardiovascular exercise, the aerobic component should be from fifteen minutes to sixty minutes, depending on the individual's fitness level and goals. A gradual increase in intensity and duration is recommended for beginners. It is important to pay attention to your body's signals to slow down or perhaps stop.

It is important to warm up muscles before stretching.

Cool Down and Stretch

After an aerobic exercise session, heart rate should be lowered gradually by slowly reducing the intensity of the exercise. Sudden stops are not recommended and can lead to muscle cramps, dizziness, and blood pooling in the legs. After a gradual cooldown such as walking, static stretching of the muscle groups is needed and highly recommended. *When muscle core and body core temperature are elevated, it is an optimal time to stretch for increased flexibility.* Warmer muscle core temperature increases the pliability of the muscle, allowing it to lengthen better.

Principles of Fitness Training—The Rules

There are specific principles that can be applied to any exercise program. Understanding these exercise principles will increase a person's chance of success with his/her exercise program.

Overload and Adaptation

The principle of overload and adaptation states that in order for a body system to become more efficient or stronger, it must be stressed beyond its normal working level. In other words, it must be overloaded. When this overload occurs, the system will respond by gradually adapting to this new load and increasing its work efficiency until another plateau is reached. When this occurs, additional overload must be applied for gain. The cardiovascular system can be overloaded in more than one way. For example, a person has been running for a few months and is running a distance of three miles in thirty minutes. The runner never goes farther than three miles and never runs faster than a ten-minute per mile pace. For this individual, some techniques of overloading would be: to increase distance, to run the same distance at a faster pace, or to add hills or sprint segments to the run. In terms of weight training, any time a person adds more weight to the bench press or increases the number of repetitions, that person is using the principle of overload. *In order for improvements to be realized, overload must occur.* The principle of overload and adaptation applies to muscular strength, cardiovascular and muscular endurance, and flexibility training.

Specificity

The principle of specificity refers to training specifically for an activity, or isolating a specific muscle group and/or movement pattern one would like to improve. For example, a 200-m sprinter would not train by running long, slow distances. Likewise, a racewalker would not train for competition by swimming. Workouts must be specific to one's goal with respect to the type of exercise, intensity, and duration. The warm-up should also be specific to a particular activity. Cross training, defined as using several different types of training, has recently increased in popularity. The benefits of cross training are to prevent injury from overuse and to decrease boredom. The principle of specificity does not negate participation in cross training activities; rather, it indicates that the primary training protocol should be in one's chosen activity.

Individual Differences

The principle of individual differences reminds us that individuals will respond differently to the same training protocol. Some individuals may be what is called a "low responder" to an exercise stimulus. It is not clear why individuals vary in response to exercise, but initial fitness level, age, gender, genetic composition, and previous history will also cause individual responses to specific activities to differ. Coaches, athletic trainers, and personal trainers should be especially aware of this principle when designing workouts in order to achieve maximum performance levels. It is also critical that individuals

A Complete Physical Activity Program

There are three principle components to a rounded program of physical activity: aerobic exercise, strength training exercise, and flexibility training. It is not essential that all three components be performed during the same workout session. Try to create a pattern that fits into your schedule and one to which you can adhere. *Commitment to a regular physical activity program is more important than intensity of the workouts*. Therefore, choose exercises you are likely to pursue and enjoy.

ACSM/AHA recent position stand *Guideline for adults under 65* states that **aerobic training** should be moderate intensity for thirty minutes five times per week or three times per week for twenty minutes with vigorous intensity. Remember that *if your schedule is tight, it is better to exercise for a shorter period of time than not at all*. Typical forms of aerobic exercise are walking and running (treadmills), stair climbing, bicycling (bicycle ergometers), rowing, cross-country skiing, and swimming. Many devices contain combinations of these motions.

For general purposes, **strength training** should be done two or three times per week. Strength training is performed with free weights or weight machines. For the purposes of general training, two or three upper body and lower body exercises should be done. Additionally, abdominal exercises are an important part of strength training.

Flexibility training is important and frequently neglected, resulting in increased tightness as we age and become less active. Stretching is most safely done with sustained gradual movements lasting a minimum of fifteen seconds per stretch. Strive to stretch every day.

realize that body type is genetically determined. Body fat distribution and metabolism are individual. Lifestyle and activity can affect one's physique; however, a large-framed person will never be a small-framed person and vice versa. Focusing more on enhanced health rather than trying to change one's body type is prudent.

The inevitable process of losing cardiovascular benefits with cessation of aerobic activity is known as the reversibility principle. The old adage "if you don't use it you lose it" applies here. Physiological changes will occur within the first two weeks of detraining and will continue for several months. Bed rest causes this detraining process to greatly accelerate. Consider the muscle atrophy that occurs with disuse when a cast is removed from a body part that has been immobilized for several weeks. The reversibility principle is clearly the justification for off-season programs for athletes and immediate initiation of physical rehabilitation programs for individuals with limited mobility or for those individuals recovering from injury.

There are many ways to measure a person's level of cardiovascular fitness. Over time, the body adapts to regular activity by not working as hard when given the same workload. An example would be running the mile. A person may have a goal of running the mile in eight minutes. At first an eight-minute mile may be a challenge, but with continuous practice, an eight-minute mile can be achieved, and may actually become easier. The body has adapted by allowing the pace to be maintained with less apparent effort. Working heart rate, the heart rate during activity, is lower. Oxygen delivery is increased to the working muscles. Respiration rate is less as the respiratory muscles become stronger. Muscles become stronger with use, creating ease of movement. Body composition typically becomes more favorable, which contributes to efficiency of movement. *Powers and Howley have shown that in general twelve to fifteen weeks of endurance exercise results in a 10 percent to 30 percent improvement in VO2 Max.*

VO_2 **Max** is the nomenclature for **maximum oxygen uptake**, the measure of the maximum amount of oxygen that an individual can utilize per minute of physical activity. VO_2 Max is expressed as milliliters of oxygen per kilogram of body weight per minute (ml/kg/min). As aerobic capacity increases, so does VO_2 Max. VO_2 Max is considered the best indicator of cardiovascular fitness. Unfortunately, measurement in a laboratory takes time, equipment, and technicians to administer the tests. Other ways to measure cardiovascular fitness include:

- 1.5-mile run (see 1.5-mile run activity (Notebook Activity on page xxx)
- 1-mile walk (below)
- various submaximal tests on cycle ergometers and treadmills

> $\dot{Q} = SV \times HR$
> Cardiac output is the stroke volume times the heart rate during a one minute interval.

The **Rockport one-mile walk test** is often used. See Table 5.2 for the Rockport Fitness test. In order to correctly evaluate cardiovascular fitness, walk the distance in the shortest amount of time, find your age and corresponding time and estimated fitness level.

Recovery heart rate is taken after an exercise session is completed, typically for thirty seconds, and multiplied by two for a per minute count. The higher a person's level of cardiovascular fitness, the less time it will take after exercise for the heart rate to return to a pre-exercise level. One minute after the cessation of exercise, a conditioned male heart rate should have returned to below 90, and a female heart rate to below 100 beats per minute. Five minutes post exercise, both male and female heart rates should be below 80 beats per minute. This is an indication not only of one's fitness level, but also of the adequacy of a cooldown period.

TABLE 5.2 ♦ Rockport One-Mile Walk Test

	Age (years)			
Fitness Category	13–19	20–29	30–39	40+
Men				
Very Poor	>17:30	>18:00	>19:00	>21:30
Poor	16:01–17:30	16:31–18:00	17:31–19:00	18:31–21:30
Average	14:01–16:00	14:31–16:30	15:31–17:30	16:01–18:30
Good	12:31–14:00	13:01–14:30	13:31–15:30	14:01–16:00
Excellent	<12:30	<13:00	<13:30	<14:00
Women				
Very Poor	>18:01	>18:31	>19:31	>22:01
Poor	16:31–18:00	17:01–18:30	18:01–19:30	19:01–22:00
Average	14:31–16:30	15:01–17:00	16:01–18:00	16:31–19:00
Good	13:01–14:30	13:31–15:00	14:01–16:00	14:31–16:30
Excellent	<13:00	<13:30	<14:00	<14:30

Because the one-mile walk test is designed primarily for older or less conditioned individuals, the fitness categories listed here do not include a "superior" category.

Source: Modified from Rockport Fitness Walking Test.

> "Fitness isn't just for highly skilled athletes. It is for all of us. It's our natural state of being, particularly when we are young. Being out of shape is really being out of sorts with ourselves."
> —Kenneth H. Cooper, M.D.,
> *The Aerobics Way*

Muscular fitness includes two specific components: muscular strength and muscular endurance. **Muscular strength** is the force or tension a muscle or muscle group can exert against a resistance in one maximal effort. **Muscular endurance** is the ability or capacity of a muscle group to perform repeated contractions against a load, or to sustain a contraction for an extended period of time. The American College Of Sports Medicine's (ACSM) position stand on progressive resistance training is that everyone would benefit from resistance training two to three days per week, working eight to ten muscle groups with one to two sets of eight to twelve repetitions. For more information, these statements can be found on the ACSM Web site listed at the end of this chapter.

Several **physiological adaptations** occur as a result of resistance training. *Strength gains can be seen within the first six weeks, with little or no change in muscle size, and are attributed to neural changes.* These changes include decreased activation of antagonistic muscles, learning how to perform the activity, changes in activation of the motor unit, improved recruitment patterns of muscle fibers, change in the gain of the muscle spindle and Golgi tendon organ, and reduction in the sensitivity of force-producing limiting factors.

As strength training activities continue, hypertrophy, or an increase in the size of the muscle fibers, occurs (see Figure 5.6 on page xx). Another result from training is an increase in the amount of energy available for contraction. Carbohydrates are stored in the form of **glycogen** in the muscle and can be used as the primary energy source for contraction. These muscle

Top Ten Reasons to Work Your Muscles

1. *Gain lean body mass and lose body fat.* For each pound of muscle you gain, you'll burn 35 to 50 more calories daily.
2. *Get strong.* Extra strength makes it easier to carry suitcases and accomplish some daily activities, such as lifting children or groceries.
3. *Build denser bones.* Weight training can increase spinal bone mineral density by 13 percent in six months.
4. *Reduce risk of diabetes.* Weight training can boost glucose utilization in the body by 23 percent in four months and lower the likelihood of developing diabetes.
5. *Fight heart disease.* Strength training reduces harmful cholesterol and lowers blood pressure.
6. *Beat back pain.* In a twelve-year study, strengthening the low-back muscles had an 80 percent success rate in eliminating or alleviating low-back pain.
7. *Move easier.* Weight training can ease arthritis pain and strengthen joints, so you feel fewer aches.
8. *Improve athletic ability.* Whatever your sport, strength training may improve proficiency and decrease risk of injury.
9. *Feel younger.* Even men and women in their 80's and 90's can make significant gains in strength and mobility with weight training.
10. *Boost your spirits.* Strength training reduces symptoms of anxiety and depression and instills greater self-confidence.

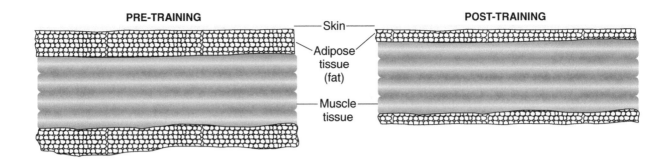

FIGURE 5.6
Changes in Body Composition from Combined Aerobic and Strength Training Program

glycogen stores increase as a result of training. Bone and connective tissue also undergo changes with resistance training, including an increase in bone matrix, an increase in **bone mineral density,** and an increase in mass and tensile strength of ligaments and tendons. These increases help prevent injury and decrease the chance of development of osteoporosis after middle age. More muscle mass increases an individual's basal metabolic rate, which is why weight training is excellent for "dieters," or those wanting to reduce their percentage of body fat. "An increase in one pound of muscle elevates basal metabolic rate by approximately 2–3 percent" (Powers and Dodd, 2009).

Along with the physiological adaptations previously discussed come benefits that improve the quality of one's life. These benefits of muscular fitness include:

- an increase in muscular strength,
- power and endurance,
- a higher percentage of muscle mass,
- improved posture,
- increased metabolic rate,
- improved ease of movement,
- increased resistance to muscle fatigue,
- increased strength of tendons, ligaments, and bones,
- decreased risk of low back pain,
- increased energy and vitality,
- Improved balance and coordination.

Importance of the Core Musculature in Functional Movement

An important new buzzword in the group fitness and personal training field is **functional movement.** Functional movement is exercise based on real-life movement. The actions done by an athlete in his/her sport are functional. Functional movement usually involves gross motor, multi-planar, multi-joint movements which place demand on the body's core—from the hips to the sternum. The core involves the muscles of the abdomen, back, and hips (listed below). Functional exercises (medicine ball warm-up, full squat with military press, wood choppers) attempt to incorporate as many variables as possible (balance, multiple joints, multiple planes of movement). This is in contrast to weight training for a specific muscle group such as the biceps brachai when doing a biceps curl. It is important to stabilize the pelvis, train the core muscles, align the spine, and achieve muscle equity to use functional movement when exercising. *A strong core reduces the risk of injury and increases the efficiency of movement.* Training the core can lead to an increase in balance and coordination, as well as gains in strength, power, and endurance.

- **Pelvic floor muscles**—The pelvic floor muscles run collectively from the pubic bone to the tailbone. Contraction of these muscles contributes to spinal stability, which is the foundation from which we move.
- **Abdominal wall**—The rectus abdominus, the internal and external obliques, and the transverse abdominus together are responsible for spinal flexion, extension, and rotation, as well as for assisting in stabilization.
- **Back muscles**—The erector spinae and multifidus produce spine extension, lateral flexion, and rotation. The interconnections of these muscles help contribute to stability of the lower back and pelvis.
- **Hip muscles**—The adductor and abductor muscles of the hip, when in balance, provide optimum stability and mobility to the hip and lumbopelvic area.
- **Lats and glutes**—Both of these muscle groups attach to the spine or pelvis, so each has an important role in the stability and mobility of the trunk.

These muscle groups make up the "core" and play an important role in core stabilization, muscle balance, and proper alignment, as well as strength and flexibility (see Figures 5.7a and 5.7b).

The following are general definitions regarding the use of weight training for developing an exercise protocol to increase muscular endurance (Cissik, 1998).

- **Frequency** refers to how often one should lift, with the recommendation being three nonconsecutive days per week.
- **Load** defines the amount of weight lifted. This will vary with each individual, with the recommended amount being a weight that will allow twelve to fifteen repetitions with good form. If form is compromised, the load should be decreased.
- **Repetition** is simply the performance of a movement from start to finish one time.
- **Set** is the specific number of repetitions performed without resting. Twelve to fifteen repetitions per set are recommended while performing two to three sets of each exercise. Each exercise session should contain eight to ten different exercises, with at least one being a full-body exercise.
- **Recovery** is the amount of time between each set. Thirty seconds is considered the optimum amount; however, taking more time is not detrimental.
- **Repetition-maximum** (also called one rep max) is the maximum amount of weight that can be lifted one time without compromising form.

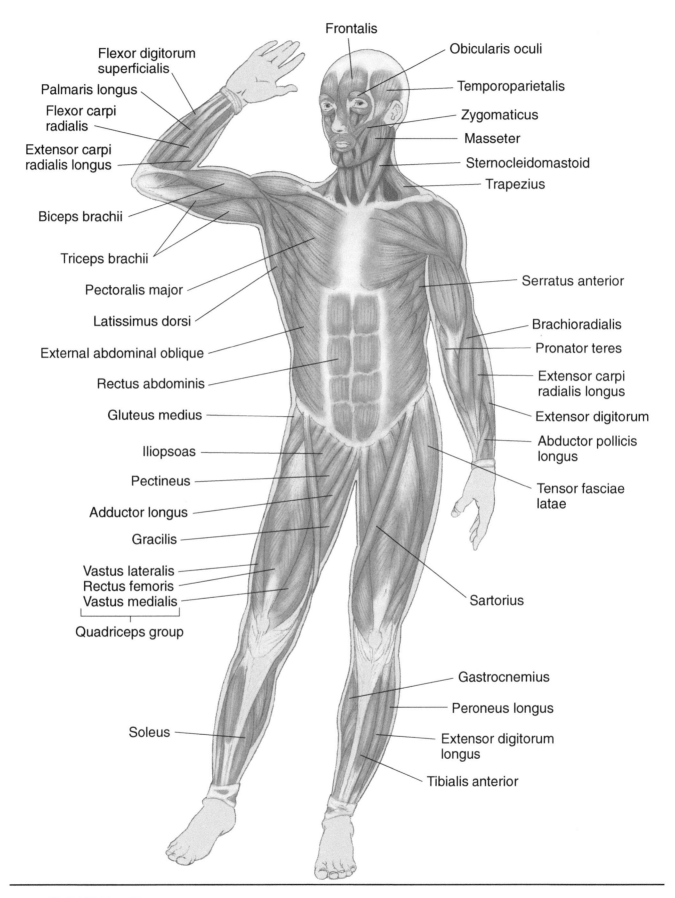

FIGURE 5.7a

Major Anterior Muscles in the Human Body

Source: Kendall/Hunt Publishing Company.

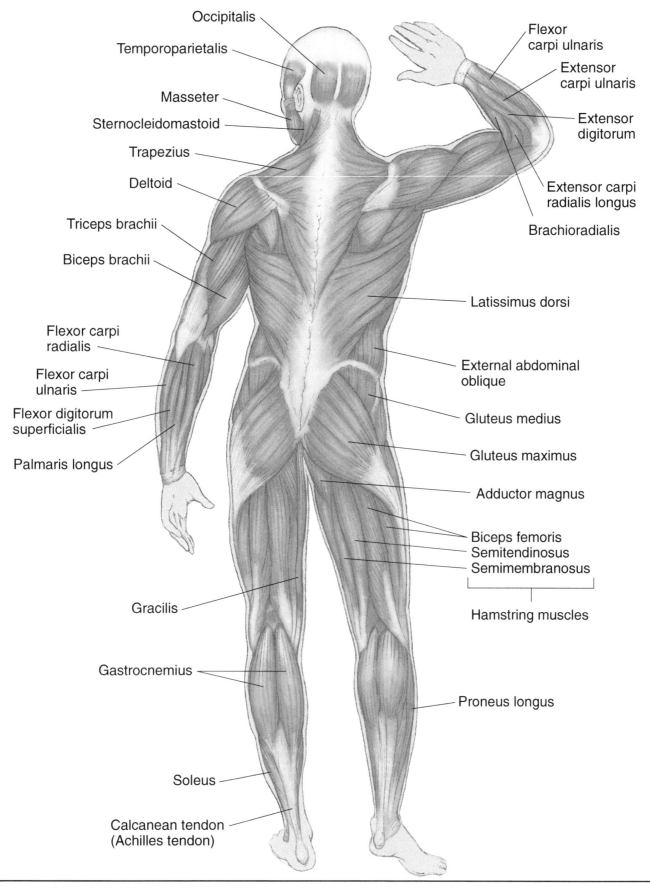

FIGURE 5.7b

Major Posterior Muscles in the Human Body
Source: Kendall/Hunt Publishing Company.

- **Intensity** is the stress level of the exercise and is expressed as a percent of a one repetition-maximum. The recommended intensity level is less than 70 percent of a one repetition-maximum.

Training for the Best Results

The exercise prescription for developing muscular strength is more intense than for endurance. The number of repetitions decreases from eight to one, with the number of sets increasing from three to five. The percent of the one repetition maximum that should be lifted increases to 80 to 100 percent. Due to the increase in intensity, the rest period is extended, lasting three to five minutes. Exercise prescription will vary depending on one's goals and objectives for training (see Figure 5.8 p. xx).

Muscle soreness often accompanies resistance training and will occur at various times during the training process. Muscle soreness that begins late in an exercise session and continues during the immediate recovery period is known as acute muscle soreness. This soreness will last only a brief period of time and is typically gone within twenty-four hours, while **delayed-onset muscle soreness** begins a day or two after the exercise session and can remain for several days. Eccentric contraction seems to be the primary cause of delayed-onset muscle soreness. According to Wilmore and Costill (1999), the causes of delayed-onset muscle soreness include structural damage to muscle cells and inflammatory reactions within the muscles. Muscle soreness can be prevented or minimized by: reducing the eccentric component of muscle action during early training, starting training at a low intensity and gradually increasing it, or beginning with a high-intensity, exhaustive bout, which will cause much soreness initially, but will decrease future pain. During delayed-onset muscle soreness, strength production is reduced by as much as 50 percent during the first five days and a reduction in strength can occur for as long as fourteen days. The best technique for prevention of delayed-onset muscle soreness is to maintain an appropriate training program. Diet supplements of vitamin E, an antioxidant, may also help to reduce damage to muscle fiber membranes (Evans, 2000).

Always focus on correct form when doing resistance exercise.

As with any type of training program, watch for signs of **over-training** called burn-out. With muscular fitness training, these indicators include a decrease in physical performance, weight loss, increase in muscle soreness, increase in resting heart rate, sleeplessness, nausea after a workout, constant fatigue, and decreased interest in exercise.

True or False? Weight Training Myths

1. **Myth:** Weight training causes one to lose flexibility.
   Fact: Resistance training will increase muscle size, but it does not necessarily make one less flexible. In fact, proper strength training can actually increase flexibility when a full range of motion is used.
2. **Myth:** Resistance training or "spot reducing" is beneficial in reducing deposits of fat from specific areas on the body, such as in the hips, thighs, and waist.
   Fact: Resistance training focuses on the muscles used. Fat is not removed from one area of the body by working the muscles

Preparation:
- Establish goals. Why do you want to weight train? Your goals will affect the way you train.
- Don't train on an empty stomach; fuel your body and brain before working out with a healthy snack.
- Consume a balanced diet daily.
- Stay hydrated with water before, during, and after the workout.
- Dress properly in loose clothing and wear non-slip athletic shoes.
- Commit to a regularly scheduled training time weekly for best results.
- Get instruction in a safe, effective, and balanced program.
- Adjust the weight machine to your body height.

Lifting:
- Always warm up; try ten minutes of cardiovascular activity or calisthenics followed by joint-specific movements for the body parts you will target.
- Progress slowly; use common sense when overloading.
- Work all the major muscle groups; avoid focusing on just one or two areas of the body.
- Consider balance—working opposing muscle groups (biceps-triceps, hamstring-quadriceps, lower back-abdominals).
- Work the larger muscle groups first, progressing to the smaller muscles.
- Complete multi-joint exercises before doing single-joint exercises (squats before leg curls).
- Keep the weights as close to the body as required for proper form and technique.
- Use a full range of motion for each exercise.
- Make sure stretching needs are appropriate to the individual; only stretch the muscle, not the ligaments or joint capsule.
- Maintain good posture throughout each movement.
- Movement should be slow and controlled, with a smooth rhythm.
- Breathe, exhaling on the exertion phase.
- Avoid breath holding during heavy lifting.
- Expect some soreness; excessive muscle soreness may be a sign of overuse or injury.
- Allow forty-eight hours of recovery for a particular body part after heavy lifting; light activity is fine during recovery.
- Get plenty of rest, water, and good nutrition to allow for muscle repair and recovery.
- Enlist a workout buddy to help with technique, motivation, and safety.
- Use a workout log to track your progress, recording weight, sets, and repetitions.

Safety:
- Learn the rules at your facility.
- Don't overtrain; listen to your body.
- Never sacrifice form for additional repetitions or weight.
- When squatting, maintain a natural lordotic curve in your spine; descend until the thighs are parallel to the floor.
- Use collars or other locking devices to keep the plates on the bars when using free weights.
- Use experienced spotters for heavy lifting.
- Always lift free weights with a partner.
- Avoid locking out or hyperextending any joints while lifting.
- Don't lift if you are too fatigued to maintain good form.

Etiquette:
- Practice good weight room etiquette; don't drop or bang the weights.
- Always re-rack free weights.
- Use a towel to wipe equipment when you are done at an exercise station.
- Be considerate of others who are waiting to use the equipment.

It is important to learn proper form and technique when weight training.

FIGURE 5.8
Beginning Strength Training Guidelines

in that area. Creating a caloric deficit consistently, through diet, exercise, or a combination of both, loses fat. The location of fat deposits is determined genetically. The majority of women tend to be "pears" with fat deposits collecting on the hip and thigh region. The majority of men tend to be "apples" with fat deposits collecting around the torso. Abdominal fat has been shown to indicate an increased health risk.

3. **Myth:** Fat will be converted to muscle with resistance training.
   Fact: Fat is not converted to muscle with exercise, nor is muscle converted to fat through disuse. Muscle cells and fat cells are different entities. The size of muscle cells can be increased with resistance training. Fat cell size is increased with sedentary living combined with a poor diet.
4. **Myth:** Dietary supplements will make one bigger and stronger.
   Fact: A balanced diet and hard work in the weight room will increase muscle size and strength. Most dietary supplements will only cause the manufacturer's wallet to become bigger. Often when a person spends money on a supplement believing that supplement to work, the placebo effect might result in some apparent short-term improvement.
5. **Myth:** Performance-enhancing drugs such as steroids, growth hormones, diuretics, and metabolism boosters will help make one fit.
   Fact: These drugs are extremely dangerous and potentially fatal. They can contribute to aesthetic changes, but can also have a negative impact on health. Some fitness enthusiasts have lost their lives searching for a short-cut to health by using supplements.
6. **Myth:** Women will become masculine in appearance by participating in resistance training activities.
   Fact: Masculinity and femininity are determined through hormones, not through resistance training. Resistance training will cause an increase in muscle tone, which is perceived to increase the attractiveness of both males and females.
7. **Myth:** Kids should not weight train.
   Fact: Pre-adolescent children can and should use their body as resistance. Swinging on the monkey bars or climbing a tree are good examples of using one's body as resistance; push-ups, sit-ups, and tumbling activities are also great. Teaching 11- to 13-year-olds proper technique and form lifting light weight with proper supervision helps lay a strong foundation for future training. Proper training can improve flexibility, as well as strengthen muscles and the skeletal structure. Body composition and self-esteem are also usually enhanced with a training program, which can be a positive outcome with childhood obesity on the rise.

Flexibility

Can you touch your toes? Think of how much your flexibility has changed in the last ten years. How much more will it change in the next ten years? Truly, if you don't stretch or if you are not active, flexibility will be lost. Why is this a concern? Loss of flexibility with age or injury can greatly affect a person's quality of life. Simple activities such as putting on your socks, or bending to lift a toddler can be painful or worse, impossible. Individuals who are active tend to be more flexible simply because they tend to use a full range of motion in their activity. Active individuals are also more likely to

> **Yoga** is gaining in popularity in the United States. Most forms of yoga encourage the buildup of heat within the body to facilitate movement and internal focus. Relaxation is a common goal of most yoga participants, yet most experience enhanced flexibility and increased strength as a fringe benefit. One style of yoga, Bikram yoga, advocates practicing yoga in rooms with temperatures as high as 105 degrees. The premise is that the heat will allow the tendons, ligaments, and muscles to loosen up more and stretch further. A common yoga truism is that even steel, when heated hot enough, will bend.

engage in health-enhancing behaviors. Several factors can have an impact on the amount of flexibility a person can achieve, including gender, age, genetic composition, activity level, muscle core temperature, and previous or current injury. Old injuries often hamper flexibility for adults later in life, therefore affecting future activity.

Flexibility and balance are a concern for the aging population. Non-impact activities such as tai chi and yoga are gaining in popularity with all ages, both of which are appropriate to the aging population.

Flexibility is defined as the range of motion around a joint. Flexibility is also specific to individual joints. For instance, an individual may have complete range of motion in the wrist but be very limited or stiff in the shoulder. An individual could be very flexible on the right side of the body, and inflexible on the left side. Flexibility exercises should be included in all exercise programs regardless of the objectives. The benefits of maintaining flexibility include having the ability to perform daily activities without developing muscle strains or tears and being able to participate in sports with enhanced performance. Consider a swimmer who increases shoulder flexibility is able to reach further, pull more water, and thus swim faster.

The athlete, whether serious about competition or a weekend recreator, will have a greater ability to perform particular sports skills with an increased range of motion. Consider a football coach encouraging his receivers to bench press as much as possible. That receiver can be very strong in the weight room; however, if he cannot apply that strength on the football field, he will not be an effective player. Athletes should train for **functional strength.** A wide receiver should train to jump high and extend from his shoulders to catch a pass. Strength without flexibility is limiting. It is especially important to include flexibility exercises in a muscular fitness workout. Flexibility also helps to prevent injuries through a reduction in strains and muscle tears.

Stretching exercises are identified through three specific categories: **ballistic, static,** and **PNF.**

Yoga isn't about touching your toes, it is about touching your soul—M.C. Yogi, *Eight Limbs*, Mantra, Beats, and Meditations.

1. **Ballistic** stretching involves dynamic movements, or what is commonly referred to as "bouncing." Ballistic stretching is not recommended for the general population as a means to improve flexibility. An exception is athletes who have ballistic movement in their sport. This type of stretch actually stimulates receptors in the muscle that

> Time, Inc reported (Aug. 5, 2002) that tai chi is the perfect exercise for seniors. **Tai chi** is an ancient martial art involving graceful movement performed slowly with great concentration and focus on breathing. The atmosphere is non-competitive, and participants are encouraged to progress at their own pace. The Oregon Research Institute reports that studies show older men and women who are inactive yet relatively healthy attain many benefits from participation in tai chi.

are designed to help prevent injury due to over-extending the muscle. Thus, the ballistic stretch can cause the muscle to contract rather than relax, and can contribute to muscle soreness. A more appropriate type of muscular stretching for the general population is identified as static stretching.

2. **Static** stretching involves slowly moving the joint to the point of mild discomfort in the muscle and maintaining that angle for approximately thirty seconds before allowing the muscle to relax. The entire procedure should be repeated several times for maximum benefit. As previously noted, a warmup is highly recommended prior to stretching for injury prevention and to facilitate the stretch. A warm environment and a warm muscle will greatly enhance the stretch. If the stretch hurts the muscle or the joint, then stop. Learn to distinguish between the mild tensions needed to overload from pain, indicating a potential injury.

3. A third type of stretching activity is called **proprioceptive neuromuscular facilitation** or **PNF.** This activity requires a partner to provide resistance. The basic formula for this activity is to isometrically resist against a partner using the muscle groups surrounding a particular joint, causing contraction, and then relaxing the same muscle group. For example, in stretching the hamstring, both the hamstrings and the quadriceps will be contracted and then relaxed. This contraction and relaxation process will increase the range of motion in the hamstrings. When stretching with a partner, communication is essential to avoid injury to the joint.

When You Should Not Exercise

Injuries

Although injuries do occur during exercise, the benefits of regular exercise far out-weigh the risk of injury. In most cases, proper training, clothing, and equipment will prevent injuries. Avoiding injury requires common sense and moderation. One should not attempt to self-diagnose, nor try to "train through the pain." Pain is a signal that something is wrong, and activity should be stopped until the source of the pain is identified and a trained medical professional can advise you. Some common injuries resulting from exercise include joint sprains, muscle strains, and other musculoskeletal problems. Knowing how to treat an acute, or immediate, injury is important (see Table 5.3). **RICE** most injuries, such as a twisted ankle: Rest, Ice, Compression, and Elevation.

> **RICE**
> **R—Rest** the injured limb, preventing further injury.
> **I—Ice** will help reduce swelling by reducing circulation and easing pain. Apply ice in thirty-minute periods several times per day. A Styrofoam cup with frozen water can be used as an ice rub. A bag of frozen peas also works well!
> **C—Compression** will help reduce swelling and fluid collection at the injury site. An elastic bandage works well to wrap the injured limb.
> **E—Elevating** the injured limb will reduce swelling. Ideally, raise the injured area above the heart. Placing the injured area on pillows on a stool is helpful.

Table 5.3 ♦ Reference Guide for Exercise-Related Problems

Injury	Signs/Symptoms	Treatment*
Bruise (contusion)	Pain, swelling, discoloration	Cold application, compression, rest
Dislocations, fractures	Pain, swelling, deformity	Splinting, cold application, seek medical attention
Heat cramps	Cramps, spasms and muscle twitching in the legs, arms, and abdomen	Stop activity, get out of the heat, stretch, massage the painful area, drink plenty of fluids
Heat exhaustion	Fainting, profuse sweating, cold/clammy skin, weak/rapid pulse, weakness, headache	Stop activity, rest in a cool place, loosen clothing, rub body with cool/wet towel, drink plenty of fluids, stay out of heat for two to three days
Heat stroke	Hot/dry skin, no sweating, serious disorientation, rapid/full pulse, vomiting, diarrhea, unconsciousness, high body temperature	Seek immediate medical attention, request help and get out of the sun, bathe in cold water/spray with cold water/rub body with cold towels, drink plenty of cold fluids
Joint sprains	Pain, tenderness, swelling, loss of use, discoloration	Cold application, compression, elevation, rest, heat after thirty-six to forty-eight hours (if no further swelling)
Muscle cramps	Pain, spasms	Stretch muscle(s), use mild exercises for involved area
Muscle soreness and stiffness	Tenderness, pain	Mild stretching, low-intensity exercise, warm bath
Muscle strains	Pain, tenderness, swelling, loss of use	Cold application, compression, elevation, rest, heat after thirty-six to forty-eight hours (if no further swelling)
Shin splints	Pain, tenderness	Cold application prior to and following any physical activity, rest, heat (if no activity is carried out)
Side stitch	Pain on the side of the abdomen below the rib cage	Decrease level of physical activity or stop altogether, gradually increase level of fitness
Tendinitis	Pain, tenderness, loss of use	Rest, cold application, heat after forty-eight hours

*Cold should be applied three or four times a day for fifteen to twenty minutes. Heat should be applied three times a day for fifteen to twenty minutes.

Of course, using proper equipment, wearing proper clothing and shoes, and practicing correct technique are essential for injury prevention. Weight-bearing forms of exercise will obviously cause more stress on the joints, but also have benefits that non-weight-bearing activities do not have, such as increasing strength of the bones and other connective tissues.

It many seem trivial, but proper footwear is critical to success in weight bearing exercise. Shoe technology has come a long way in the past decade. Sport-specific shoes are highly recommended to avoid injury and to enhance performance. Unfortunately, the consumer pays for the research and technology, as well as the logo on the shoes. A good cross trainer shoe is the way to go if you like to do a variety of activities. Cross trainers are not, however, recommended for aerobic dance or running. Running shoes are also not appropriate

How to Buy Athletic Shoes

For many aerobic activities, good shoes are the most important purchase you'll make. Take the time to choose well. Here are some basic guidelines:

- Shop for shoes in the late afternoon, when your feet are most likely to be somewhat swollen—just as they will be after a workout.
- For walking shoes, look for a shoe that's lightweight, flexible, and roomy enough for your toes to wiggle, with a well-cushioned, curved sole; good support at the heel; and an upper made of a material that breathes (allows air in and out).
- For running shoes (see the figure), look for good cushioning, support, and stability. You should be able to wiggle your toes easily, but the front of your foot shouldn't slide from side to side, which could cause blisters. Your toes should not touch the end of the shoes because your feet will swell with activity. Allow about half an inch from the longest toe to the tip of the shoe.
- For racquetball shoes, look for reinforcement at the toe for protection during foot drag. The sole should allow minimal slippage. There should be some heel elevation to lessen strain on the back of the leg and Achilles tendon. The shoe should have a long throat to ensure greater control by the laces.
- For tennis shoes, look for reinforcement at the toe. The sole at the ball of the foot should be well padded because that's where most pressure is exerted. The sides of the shoe should be sturdy, for stability during continuous lateral movements. The toe box should allow ample room and some cushioning at the tips. A long throat ensures greater control by the laces.

Don't wear wet shoes for training. Let wet shoes air dry, because a heater will cause them to stiffen or shrink. Use powder in your shoes to absorb moisture, lessen friction, and prevent fungal infections. Break in new shoes for several days before wearing them for a long-distance run or during competition.

Source: Canadian Podiatric Sports Medicine Academy.

What to Look for When You Buy Running Shoes

- Well-molded Achilles pad prevents irritation of Achilles tendon
- Well-padded tongue prevents extensor tendinitis and irritation of dorsum of foot
- Laces not too long so they stay tied longer
- High, rounded toe box (at least 1½" high) prevents subungual hematomas ("black toes")
- Studded sole absorbs shock and provides traction in mud and snow
- Firm heel counter for hindfoot stability
- Flared heel for stability and beveled or rounded heel for quick roll-off
- Soft, raised heel wedge to absorb impact at heel strike
- Flexible midsole helps prevent Achilles tendon problems

---

for "studio activities" such as aerobic dance, step aerobics, BOSU activities, as well as court activities like tennis or racquetball. Running shoes have little lateral support and the higher flared heel can actually cause a person participating in step aerobics to be more prone to twisting an ankle or knee joint. Some steps also have a rubber top which can grip the waffle sole of the running shoe and increase the risk of injury.

In any athletic shoe, fit and comfort are of the utmost importance. It is worth going to a store staffed by knowledgeable personnel. Often they can give you good insight into the type of shoe that is most appropriate for your foot and your gait.

Christopher McDougall's 2009 publication *Born to Run* started a discourse between runners regarding what is the best way to run—wearing $160.00 state-of-the-art running shoes or running barefoot? The barefoot crowd claims that a barefoot runner has a more natural gait, striking with the mid or forefoot first. Shoes with cushioning and a high heel cause the runner to strike the ground

© EdBockStock, 2012, Shutterstock, Inc.

Proper footwear is critical to success in weight bearing exercise. Sport-specific shoes help prevent injury and enhance performance.

with the heel first. Vibram's successful FiveFingers brand popularized the minimalist shoe movement in California and first-place runners in marathons have been seen wearing them. Barefoot runners using the FiveFingers shoe can "feel" the road and may run with a more natural gate and yet still have protection from irritants on the ground like small shards of glass. There are 2 categories of minimalist shoes—the FiveFingers type that fits snug on your foot, and a minimalist running shoe (a cross between a traditional running shoe and being barefoot) which has just a bit of structure and support. If you decide to try the minimalist route, progress slowly and use caution. (Ellingson, Linda, *The Basics of Barefoot/Minimalist Running,* Jan. 2012, REI expert advice online.)

Environmental Conditions

Take into consideration environmental conditions such as temperature, air pollution, wind-chill, altitude, and humidity that can affect one's health and safety. Dressing appropriately is important when exercising in extreme weather conditions.

When exercising in the cold weather, layering of clothes is advised. There are new fabrics that can wick away moisture from the body better than fabrics such as wool, polypropylene, and cotton. Avoid cotton as a base layer in cold weather because if it gets wet with perspiration it will stay wet and make you colder. The Dupont company pioneered such a fabric called ComfortMax (Powers and Dodd, 2009). This is an advantage because moisture from perspiration can be transferred away (called "wicking") from the body, allowing evaporation to occur. This type of fabric is excellent as a first layer when skiing, jogging, or hiking in cold weather. Outer layers are ideally a waterproof shell that has mesh or zippered compartments to "breathe" and can be peeled off as needed. It is advisable to limit exercise time in extremely cold weather to avoid hypothermia.

Exercising in the heat can be a challenge (see Figure 5.9). It is important to acclimate to the heat and humidity, especially when moving from an area that is cool and arid. Gradually increase duration and intensity when exercising in a new type of environment. Especially in the Southern states, heat injuries are a real concern. **Heat cramps**, **heat exhaustion**, and **heat stroke** can all occur with prolonged exposure to the heat. Heat stroke is a life-threatening condition, and necessitates hospitalization. Heat exhaustion is more common, with individuals typically suffering from dehydration. In order for the body to effectively cool, evaporation of sweat needs to occur. *In a humid environment when perspiration drips off of a person, evaporation is not occurring, and therefore cooling is not taking place.* A hot and humid environment is especially risky to the very young, the old, and those with low cardiovascular fitness levels.

Heat injuries are much less likely to occur if a person is adequately hydrated. Proper hydration is necessary for the body to function properly. Water aids in

Guidelines for Exercising in the Heat
1. Stay hydrated with cool water (cool water is absorbed best in the gut) before, during and after activity.
2. Dress appropriately in clothes that can wick moisture away from the body.
3. Limit exposure time.
4. Exercise with a buddy, or let someone know your plan and stick to it.
5. Wear lightweight sunglasses for eye protection against sun glare, dust and debris in the air. (Important if you exercise near a construction site or near traffic.)
6. Exercise in the coolest time of the day if possible.
7. Stop activity if you experience nausea, dizziness, or extreme headache.
8. Monitor your heart rate, staying within your target heart rate zone.
9. Check the heat index to make sure it is safe to exercise.

FIGURE 5.9
Heat and Humidity Chart

Apparent temperature (what it feels like)

Air temperature (F°)	70°	75°	80°	85°	90°	95°	100°	105°	110°	115°
Relative Humidity 0%	64°	69°	73°	78°	83°	87°	91°	95°	99°	103°
10%	65°	70°	75°	80°	85°	90°	95°	100°	105°	111°
20%	66°	72°	77°	82°	87°	93°	99°	105°	112°	120°
30%	67°	73°	78°	84°	90°	96°	104°	113°	123°	135°
40%	68°	74°	79°	86°	93°	101°	110°	123°	137°	151°
50%	69°	75°	81°	88°	96°	107°	120°	135°	150°	
60%	70°	76°	82°	90°	100°	114°	132°	149°		
70%	70°	77°	85°	93°	106°	124°	144°			
80%	71°	78°	86°	97°	113°	136°				
90%	71°	79°	88°	102°	122°					
100%	72°	80°	91°	108°						

Apparent temperature:	Heat stress risk with exertion:
90°–105°	Heat cramps and heat exhaustion possible.
105°–130°	Heat cramps or heat exhaustion likely; heat stroke possible.
130° and above	Heat stroke highly likely with continued exposure.

To determine the risk of exercising in the heat, locate the outside air temperature on the top horizontal scale and the relative humidity on the left vertical scale. Where these two values intersect is the apparent temperature. For example, on a 90°F day with 70 percent humidity, the apparent temperature is 106°F. Heat cramps or heat exhaustion are likely to occur, and heat stroke is possible during exercise under these conditions.

Source: Adapted from U.S. Department of Commerce, National Oceanic and Atmospheric Administration, Heat index chart, in *Heat wave: A major summer killer.* Washington, D.C.: Government Printing Office, 1992.

controlling body temperature, contributes to the structure and form of the body, and provides the liquid environment for cell processes. When the thirst mechanism is activated, dehydration has already begun. It is important to pre-hydrate, drink before thirst occurs, and especially drink before exercising. The standard recommendation is to drink at least eight eight-ounce glasses of water a day. Exercise increases the body's demand for water due to an increase in metabolic rate and body temperature. Therefore, this amount should be increased. Drinking water every waking hour is a good habit for individuals who exercise on a regular basis. Hydration is very critical in a humid environment. Before, during, and after aerobic exercise, increase the amount of water consumed. *Water is necessary for the efficient functioning of the body; thus, the importance of hydration cannot be overstated.* Electrolyte levels, especially calcium, sodium, and potassium, are critically important in muscle contraction and should also be carefully maintained. This may be accomplished through re-hydrating with sports drinks. Sports drinks are useful for glycogen replacement when the duration of an activity is sixty to ninety minutes or longer or if the athlete is in a tournament with multiple events.

Hyponatremia

Avoiding dehydration is critical when exercising, especially in the heat and high humidity. There is, however, a possibility of over-hydration, which can be just as critical. Due to the popularity of marathon and triathlon training programs, more people are participating in longer road races. There are more marathon walkers than ever before. A walker can be on the course for a much longer time than a runner—perhaps six or seven hours. If the walker is hydrating the entire time, it is possible to over-hydrate. This over-hydration can lead to a condition called hyponatremia, also called **water intoxication**. Hyponatremia is characterized by a low sodium concentration in the blood. Hyponatremia is seen in some medical conditions such as certain forms of lung

Adverse Effects of Dehydration

Exercise in the heat can be extremely dangerous, depending on exercise intensity, ambient temperature, relative humidity, clothing, and state of hydration (water content of the body). Although some forms of heat injury can occur prior to significant weight loss due to sweating, the table in this box shows how weight loss during exercise can be a predictor of some of the dangers associated with exercise in the heat. The loss of body weight during exercise in the heat is simply due to water loss through sweating. Thus, prolonged, profuse sweating is the first warning signal of impending dehydration.

% Body Weight Loss	Symptoms	% Body Weight Loss	Symptoms
0.5	Thirst	6.0	Impaired temperature regulation, increased heart rate
2.0	Stronger thirst, vague discomfort, loss of appetite	8.0	Dizziness, labored breathing during exercise, confusion
3.0	Concentrated blood, dry mouth, reduced urine output	10.0	Spastic muscles, loss of balance, delirium
4.0	Increased effort required during exercise, flushed skin, apathy	11.0	Circulatory insufficiency, decreased blood volume, kidney failure
5.0	Difficulty in concentrating		

Source: From *Total Fitness and Wellness*, 5th Edition, by Powers and Dodd, Benjamin Cummings Publishers.

cancer. Exercise-associated hyponatremia involves excess ADH (antidiuretic hormone) being secreted from the pituitary gland. The longer a person sweats, the higher the risk of hyponatremia due to lost electrolytes. Hyponatremia can be life-threatening, and unfortunately the hyponatremia symptoms mimic the symptoms of heat illness (fatigue, light headedness, nausea, cramping, headache, dizziness). If you treat a hyponatremia victim the same way you would a heat illness victim, you could accelerate their decline. The best way to avoid hyponatremia, or dehydration for that matter, is to be aware of fluid loss and fluid intake. After approximately sixty minutes of activity, it is best to rehydrate in part with sports drinks that contain electrolytes such as sodium and chloride. It is also prudent to eat a normal diet including salt-containing foods unless you are restricted by your physician from sodium in your diet. When competing in races, avoid ingesting aspirin, ibuprofen, or acetaminophen, which can interfere with kidney functioning. As with other things in life, balance is the key.

Use common sense when ill. If you have cold symptoms with no fever, then possibly a light workout might make you feel better. If fever is present, you have a headache, extreme fatigue, muscle aches, swollen lymph glands, or if you have flu-like symptoms, then bed rest is recommended. Marathon efforts of high intensity and long duration have been shown to temporarily suppress the immune system. Mild to moderate exercise has been shown to enhance the immune system and to reduce risk of respiratory infections (ACSM, 1989).

Allergies and asthma can make exercising a challenge. If you suffer from either or both conditions, take precaution when exercising. Dealing with asthma and allergies does not preclude exercising outdoors-it means you have to be prepared and manage your symptoms. People who suffer from allergies have an immune system that responds to triggers in the environment.

Common Sense Concerns for Outside Activity

- *Lightning*—DO NOT exercise if there is lightning in the area. Stay indoors.
- *Air pollution*—When the air quality is poor, exercise early in the morning, later in the evening, or preferably indoors, especially for those with lung or heart disease. Pay attention to air pollution alerts. Avoid high traffic areas.
- *Allergens*—Check weather reports for pollen counts, and avoid outdoor vigorous exercise when the pollen count is high.
- *Night exercise*—It is common for some to walk, jog, or bike on the shoulder of roads. Drivers need to use caution; sometimes the glare from oncoming traffic can obscure visibility. However, night exercisers must be responsible and make themselves more visible at night:

1. Use a flashlight.
2. Dress in *white* clothing—there is an amazing difference in visibility between gray and white shirts at night.
3. Wear a reflective vest or reflective arm bands.
4. Walk in a well-lit, safe area if possible.
5. Be safe, be aware of your surroundings, and use common sense.
6. Don't go alone; go with friends or borrow a dog if you don't have your own.
7. Carry ID with you.
8. Do NOT let headphones or texting distract you from traffic or safety concerns.
9. Use flashing lights that can be attached to a belt or arm band.
10. Remember when walking to face the oncoming traffic if possible.
11. Let someone know your route and expected time back.
12. Be aware that drivers may have difficulty seeing you at twilight.

Use common sense when environmental conditions are significant!

My Workout Considerations

Goals: What is my purpose in exercising? There is a big difference between getting up at 5:00 am to run alone, body-building for 3 hours at a time in the gym, rock-climbing, or joining a Zumba class.

Activity: do something you like to do-enjoyment is key to sticking with the activity.

Equipment: What do I need to be successful in my chosen activity? Examine the quality of your equipment to ensure safety.

Time: Workout when it is convenient. If you don't have time, make time. You are worth the effort. Regular exercise often enhances sleep quality-so getting up 30 minutes earlier can actually help you get more from your sleep.

Intensity: If you have only a short amount of time, increase exercise intensity. This is the basis of interval type classes like Tabata and Cross-Fit. It is good mix up longer endurance activities along with shorter higher intensity activities.

Warm up: Warm up prepares you both physically and mentally. Rehearse movements to be used-like a swimmer will swing his arms and

Be safe: Work hard, train safe. If it hurts, likely your body is signaling you to stop. Consult a personal trainer if you are unsure about your exercise.

Focus: Consider leaving your phone in your locker. Focus on what you are doing at the moment.

Afterwards: Let your heart rate drop gradually. When your muscles are warm is the best time to stretch for increased flexibility. Stretch, breathe and relax

Eat: Fuel your body to get the most out of your exercise, eating a light snack or meal that has protein and carbohydrate. After activity replace muscle glycogen with again, complex carbohydrates and protein and plenty of fluids.

Drink: Water! Before, during and after.

Rest: Rest is the most significant part of an exercise program, allowing the body to adapt and recover from the stress of exercise.

Variety: Try new things to challenge yourself and avoid boredom.

Commit: Put your workout time on the calendar. Inform family and friends of your intention and ask for their support to help you follow through.

The Biggest Risk to Exercise Is Not Starting!

> If you're interested in training for a 5K or 10K fun run, visit Smart Coach at runnersworld.com. You will find a beginning runner's training guide that includes free advice according to your personal fitness and training level.

The internal conditions of the body before and during exercise are even more crucial than exercising with the proper external conditions. Eating a regular meal immediately before exercising will usually result in poor performance, stomach cramps, and sometimes even vomiting. The days of a steak and potato pre-game meal are gone. It is important to fuel your body with high quality protein, low fat, and high complex carbohydrate foods prior to competition; however, even more important is fueling your body on a daily basis. *Everyday good nutrition will cause an athlete to perform better in practice, thereby optimizing training that may result in a better performance in competition.* This is also sound advice for non-athletes trying to stay active. The recommendations for individuals involved in a regular exercise program are: 55 to 60 percent of total calories consumed should be from carbohydrates, 25 to 30 percent from fat, and 12 to 15 percent from protein. Individuals who are involved in a high-intensity muscular training program should consume a higher amount of protein and less fat for muscle growth and maintenance. Adequate hydration is also critical and can make a difference in the quality of exercise and performance.

Staying active has clearly been shown to enhance a person's quality of life. **Exercise is for everyone; it is never too early or too late to start.** Most people know that they would benefit from participating in an exercise program, but for many it is difficult to get started. Find an activity you enjoy. Make a plan. Write it down. Get a workout buddy. Start slowly, and listen to your body. Pain is usually a signal that something is wrong. The old adage 'no pain, no gain' can cause beginners to become frustrated. Balance activity, leisure time, and rest each week. With consistency, activity can have a positive impact on reducing risk for many conditions associated with too little activity, called hypokinetic conditions. And most importantly, you should experience increased stamina, enthusiasm, and enhanced mental well-being in your daily life.

In 2008, the U.S. Department of Health and Human Services (DHHS) released new physical activity guidelines for Americans (see Figure 5.1).

FIGURE 5.10

At-A-Glance: A Fact Sheet for Professionals

Physical Activity Guidelines for Americans

These guidelines are needed because of the importance of physical activity to the health of Americans, whose current inactivity puts them at unnecessary risk. The latest information shows that inactivity among American children, adolescents, and adults remains relatively high, and little progress has been made in increasing levels of physical activity among Americans.

Adults (aged 18–64)
- Adults should do two hours and thirty minutes a week of moderate-intensity, or one hour and fifteen minutes (seventy-five minutes) a week of vigorous-intensity aerobic physical activity, or an equivalent combination of moderate- and vigorous-intensity aerobic physical activity. Aerobic activity should be performed in episodes of at least ten minutes, preferably spread throughout the week.
- Additional health benefits are provided by increasing to five hours (three hundred minutes) a week of moderate-intensity aerobic physical activity, or two hours and thirty minutes a week of vigorous-intensity physical activity, or an equivalent combination of both.
- Adults should also do muscle-strengthening activities that involve all major muscle groups performed on two or more days per week.

Source: U.S. Department of Health & Human Services

References

American College of Sports Medicine (ACSM). *Exercise and the Common Cold.* 1989.

American College of Sports Medicine. *Perceived Exertion* Current Comment, 2014.

ACSM/AHA Joint Position Stand "Exercise and Acute Cardiovascular Events: Placing Risks into Perspective." *Medicine and Science in Sports and Exercise.* 2007.

American Heart Association. *Heart and Stroke Statistical Update.* Dallas: American Heart Association. 2012.

Bishop, J. G. and Aldana, S. G. *Step Up to Wellness.* Needham Heights, MA: Allyn & Bacon. 1999.

Cissik, J. M. *The Basics of Strength Training.* New York: McGraw-Hill Companies, Inc. 1998.

Corbin, C. B. and Lindsey, R. *Concepts of Fitness and Wellness: Active Lifestyles for Wellness* (15th ed). McGraw Hill. 2009.

Corbin, C. et al. Physical Activity for Children: A Statement of Guidelines for Children Age 5–12, NASPE. Dec. 2003.

Ellingsen, Jan, The Basics of Barefoot/Minimalist Running Jan. 2012, REI expert advice online. www.physicalactivityplan.org

Evans, W. J. Vitamin E, Vitamin C, and Exercise. *American Journal of Clinical Nutrition,* Vol. 72, 647s-652s. August 2000.

Fox, E., Bowers, R., and Merle, F. *The Physiological Basis for Exercise and Sport* (5th ed). Madison, WI: WCB Brown & Benchmark Publishers. 1989.

Haskell, W. L. et al. Physical Activity and Public Health: Updated Recommendations for Adults from the American College of Sports Medicine. *Medicine and Science in Sports and Exercise* 39 (8):1424–1434 Belmont, CA: Wadsworth/Thompson Learning. 2007.

Healthier U.S. Initiative; www.whitehouse.gov

Journal of Obesity Vol 2012, Article ID 480467. doi:10.1155/2012/480467

McArdle, W. D., Katch, F. I., and Katch, V. L. *Exercise Physiology: Energy, Nutrition, and Human Performance.* Baltimore: Williams and Wilkins. 1999.

Pate, R., Pratt, M., Blair, S., Haskell, W., Macera, C., et al. Physical Activity and Public Health: A Recommendation from the Centers for Disease Control and Prevention and the American College of Sports Medicine. *Journal of the American Medical Association,* 273: 402–407. 1995.

Payne, W. A. and Hahn, D. B. *Understanding Your Health* (6th ed). St. Louis, MO: Mosby. 2000.

Physical Activity and Health: A Report of the Surgeon General. Atlanta: U.S. Department of Health and Human Services, Centers for Disease Control and Prevention, National Center for Chronic Disease Prevention and Health Promotion. 1996.

Pollock, M. L., Gaesser, G. A., Butcher, J. D., Despres, J-P., Dishman, R. K., et al. ACSM Position Stand on the Recommended Quantity and Quality of Exercise for Developing and Maintaining Cardiorespiratory and Muscular Fitness, and Flexibility in Adults. *Medicine & Science in Sports & Exercise, 30:* 975–991. 1998.

Powers, S. K., and Dodd, S. L. *Total Fitness and Wellness* (5th ed). San Francisco: Pearson Benjamin Cummings. 2009.

Powers, S. K., and Howley, E. T. *Exercise Physiology: Theory and Application to Fitness and Performance* (6th ed). New York: McGraw-Hill Companies, Inc. 2006.

Rosato, F. *Fitness to Wellness: The Physical Connection* (3rd ed.) Minneapolis: West. 1994.

Sabo, E. *Good Exercises for Bad Knees.* www.healthology.com; Retrieved June 14, 2005.

Sharkey, B. J. *Fitness and Health.* Champaign, IL: Human Kinetics Publishing. 1997.

Sieg, K. W., and Adams, S. P. *Illustrated Essentials of Musculoskeletal Anatomy.* Gainesville, FL: Megabooks Inc. 1985.

2005 Dietary Guidelines, www.health.gov.

Why We Are Losing the War on Obesity; Health Annual Editions 05/06, 26th edition, McGraw-Hill/Dushkin.

Wilmore, J. H. and Costill, D. L. *Physiology of Sport and Exercise.* Champaign, IL: Human Kinetics Publishing Company. 1999.

Contacts

American College of Sports Medicine (ACSM)
http://www.acsm.org/index.asp

American Council of Exercise
Cardiovascular Fitness Facts
www.acefitness.org/fitfacts/fitfacts_list.cfm#1

American Running Association
http://www.americanrunning.org

American Heart Association
www.americanheart.org/statistics/

American Heart Association Web site for tips, health facts, a personal trainer, and more
www.justmove.org/fitnessnews

American Medical Association
http://www.ama-assn.org

Centers for Disease Control and Prevention
http://www.cdc.gov/nccdphp/sgr/mm.thm

National Institute of Arthritis and Musculoskeletal and Skin Diseases
http://www.healthfinder.gov/

National Institute for Health Web site for lowering blood pressure.
http://www.nhlbi.nih.gov/hbp/

President's Council on Physical Fitness and Sports
Fitness Fundamentals
http://www.hoptechno.com/book11.htm

President's Council on Physical Fitness and Sports
The Link Between Physical Activity and Morbidity and Mortality
http://www.cdc.gov/nccdphp/sgr/mm.htm

Tucker Center—Women in Sport
http://www.kls.coled.umn.edu/crgws/

Results from the President's Council on Physical Fitness
http://www.girsite.org/Html/nike2.htm

Shape Up America!
http://www.shapeup.org

Excellent current information regarding osteoporosis treatment and prevention.
www.osteo.org/osteo.html

www.nhlbi.nih.gov
www.healthfinder.gov
www.medlineplus.gov
www.nutrition.gov
www.fitness.gov

UCBerkelyWellnessLetter.com

Women's Health womenshealth.gov
exercisemedicine.com

Activities

Notebook Activities

Calculating Your Activity Index
Karvonen Formula
Developing an Exercise Program for Cardiorespiratory Endurance
Assessing Your Current Level of Muscular Endurance
Check Your Physical Activity and Heart Disease I.Q.
Assessing Cardiovascular Fitness: Cooper's 1.5-Mile Run

Name _____ Section _____ Date _____

NOTEBOOK ACTIVITY

Calculating Your Activity Index

Frequency: How often do you exercise?

If you exercise:	Your frequency score is:
Less than 1 time a week	0
1 time a week	1
2 times a week	2
3 times a week	3
4 times a week	4
5 or more times a week	5

Duration: How long do you exercise?

If your total duration of exercise is:	Your duration score is:
Less than 5 minutes	0
5–14 minutes	1
15–29 minutes	2
30–44 minutes	3
45–59 minutes	4
60 minutes or more	5

Intensity: How hard do you exercise?

If exercise results in:	Your intensity score is:
No change in pulse from resting level	0
Slight increase in pulse from resting level	1
Slight increase in pulse and breathing	2
Moderate increase in pulse and breathing	3
Intermittent heavy breathing and sweating	4
Sustained heavy breathing and sweating	5

Multiply your three scores:

Frequency _____ × Duration _____ × Intensity _____ = Activity index _____

To determine your activity index, refer to the following table:

If your activity index is:	Your estimated level of activity is:
Less than 15	Sedentary
15–24	Low active
25–40	Moderate active
41–60	Active
Over 60	High active

Name _____ Section _____ Date _____

NOTEBOOK ACTIVITY

Karvonen Formula

Determining Target Heart Rate Zone (THRZ)

Take your resting heart rate early in the morning before you rise, counting for sixty seconds. Use your index and middle finger to palpate either your carotid (neck) or radial (wrist) artery. It is best to do this three different times and then average the three resting heart rates.

Finding your target heart rate zone is beneficial so that you can determine at any given time during a workout how hard your heart is working. This gives you feedback that helps you construct a proper workout that matches your goals. Working out with intensity high enough to bring the heart rate above the minimum threshold is important to attain cardiovascular benefits of exercise.

EXAMPLE:

AGE: 20 yr. old RESTING HEART RATE: 68 bpm

Formula for calculating Maximum Heart Rate (Max HR)
220 − age (in years) = Maximum Heart Rate

Example
220 − 20 = 200 beats per minute (bpm)

Formula for calculating Heart Rate Reserve
Max HR − Resting HR = Heart Rate Reserve

Example
200 − 68 = 132 beats per minute

Formula for calculating Threshold of Training HR
HR Reserve × 60%
Plus Resting HR = Threshold of Training HR

Example
132 × .60 = 80 bpm
80 + 68 = 148 bpm

Formula for calculating the Upper Limit of the THRZ
HR Reserve × 85%
Plus Resting HR = Upper Limit for the THRZ

Example
132 × .85 = 112
112 + 68 = 180 bpm

The target zone for this 20-year-old with a resting HR of 68 bpm is 148 − 180 bpm.

Divide these numbers to get a 10-second working heart rate 24–30 bpm/10 sec

Your age: _____ Your Resting HR _____ bpm

Max HR = 220 − _____ = _____

Max HR − RHR = HR Reserve _____ − _____ = _____

HR Reserve × 60% + RHR = Minimum Threshold
_____ × .60 = _____ + _____ = _____

HR Reserve × 85% + RHR = Upper Limit
_____ × .85 = _____ + _____ = _____

Your target heart rate zone is _____ bpm (60%) to _____ bpm (85%)

Now, divide these numbers by 6 to determine your working heart rate for 10 seconds _____ bpm/10 sec to _____ bpm/10 sec

Name _____ Section _____ Date _____

NOTEBOOK ACTIVITY

Developing an Exercise Program for Cardiorespiratory Endurance

Goals: Identify three goals you want to accomplish as a result of this program. Goals should be accomplished by the end of the semester.

1. _____

2. _____

3. _____

Activities: Identify three different activities you will perform.

1. _____

2. _____

3. _____

Duration: Fill in an amount of time for each exercise session and activity.
Activity   Duration

1. _____

2. _____

3. _____

Name _____ Section _____ Date _____

NOTEBOOK ACTIVITY

Assessing Your Current Level of Muscular Endurance

Push-up Test:
Men should use the standard push-up position with hands shoulder-width apart and feet on the floor. Women may modify the standard push-up position by putting their knees on the floor. Complete as many push-ups as possible without stopping, and evaluate your performance according to the following.

MEN				
Age	20s	30s	40s	50s
Good	40	36	30	27
Fair	35	30	25	22
Poor	30	25	21	18

WOMEN				
Age	20s	30s	40s	50s
Good	38	33	27	22
Fair	32	27	22	18
Poor	27	22	18	15

Curl-up Test:
Begin by lying on your back, arms by your sides with palms down and on the floor and fingers straight. Your knees should be bent at about ninety degrees, with your feet twelve inches away from your buttocks. To perform a curl up, curl your head and upper back upward, keeping your arms straight. Slide your fingers forward along the floor until you touch the back of your heels. Then curl back down until your back and head reach the floor. Palms, feet, and buttocks remain on the floor the entire time. Perform as many curlups as you can in one minute without stopping to rest, and evaluate your performance according to the following.

MEN				
Age	20s	30s	40s	50s
Good	25	25	25	25
Fair	22	22	21	19
Poor	13	13	11	09

WOMEN				
Age	20s	30s	40s	50s
Good	25	25	25	25
Fair	22	21	20	15
Poor	13	11	06	04

Source: Department of Health and Human Services, National Institutes of Health.

Name _____ Section _____ Date _____

NOTEBOOK ACTIVITY

Check Your Physical Activity and Heart Disease I.Q.

Test how much you know about how physical activity affects your heart. Mark each statement true or false. See how you did by checking the answers on the back of this sheet.

1. Regular physical activity can reduce your chances of getting heart disease. T F

2. Most people get enough physical activity from their normal daily routine. T F

3. You don't have to train like a marathon runner to become more physically fit. T F

4. Exercise programs do not require a lot of time to be very effective. T F

5. People who need to lose some weight are the only ones who will benefit from regular physical activity. T F

6. All exercises give you the same benefits. T F

7. The older you are, the less active you need to be. T F

8. It doesn't take a lot of money or expensive equipment to become physically fit. T F

9. There are many risks and injuries that can occur with exercise. T F

10. You should consult a doctor before starting a physical activity program. T F

11. People who have had a heart attack should not start any physical activity program. T F

12. To help stay physically active, include a variety of activities. T F

Answers to the Check Your Physical Activity and Heart Disease I.Q. Quiz

1. **True.** Heart disease is almost twice as likely to develop in inactive people. Being physically inactive is a risk factor for heart disease along with cigarette smoking, high blood pressure, high blood cholesterol, and being overweight. The more risk factors you have, the greater your chance for heart disease. Regular physical activity (even mild to moderate exercise) can reduce this risk.

2. **False.** Most Americans are very busy but not very active. Every American adult should make a habit of getting thirty minutes of low to moderate levels of physical activity daily. This includes walking, gardening, and walking up stairs. If you are inactive now, begin by doing a few minutes of activity each day. If you only do some activity every once in a while, try to work something into your routine everyday.

3. **True.** Low- to moderate-intensity activities, such as pleasure walking, stair climbing, yardwork, housework, dancing, and home exercises can have both short- and long-term benefits. If you are inactive, the key is to get started. One great way is to take a walk for ten to fifteen minutes during your lunch break, or take your dog for a walk every day. At least thirty minutes of physical activity everyday can help improve your heart health.

4. **True.** It takes only a few minutes a day to become more physically active. If you don't have thirty minutes in your schedule for an exercise break, try to find two fifteen-minute periods or even three ten-minute periods. These exercise breaks will soon become a habit you can't live without.

5. **False.** People who are physically active experience many positive benefits. Regular physical activity gives you more energy, reduces stress, and helps you to sleep better. It helps to lower high blood pressure and improves blood cholesterol levels. Physical activity helps to tone your muscles, burns off calories to help you lose extra pounds or stay at your desirable weight, and helps control your appetite. It can also increase muscle strength, help your heart and lungs work more efficiently, and let you enjoy your life more fully.

6. **False.** Low-intensity activities—if performed daily—can have some long-term health benefits and can lower your risk of heart disease. Regular, brisk, and sustained exercise for at least thirty minutes, three or four times a week, such as brisk walking, jogging, or swimming, is necessary to improve the efficiency of your heart and lungs and burn off extra calories. These activities are called aerobic—meaning the body uses oxygen to produce the energy needed for the activity. Other activities, depending on the type, may give you other benefits such as increased flexibility or muscle strength.

7. **False.** Although we tend to become less active with age, physical activity is still important. In fact, regular physical activity in older persons increases their capacity to do everyday activities. In general, middle-aged and older people benefit from regular physical activity just as young people do. What is important, at any age, is tailoring the activity program to your own fitness level.

8. **True.** Many activities require little or no equipment. For example, brisk walking only requires a comfortable pair of walking shoes. Many communities offer free or inexpensive recreation facilities and physical activity classes. Check your shopping malls, as many of them are open early and late for people who do not wish to walk alone, in the dark, or in bad weather.

9. **False.** The most common risk in exercising is injury to the muscles and joints. Such injuries are usually caused by exercising too hard for too long, particularly if a person has been inactive. To avoid injuries, try to build up your level of activity gradually, listen to your body for warning pains, be aware of possible signs of heart problems (such as pain or pressure in the left or mid-chest area, left neck, shoulder, or arm during or just after exercising, or sudden light-headedness, cold sweat, pallor, or fainting), and be prepared for special weather conditions.

10. **True.** You should ask your doctor before you start (or greatly increase) your physical activity if you have a medical condition such as high blood pressure, have pains or pressure in the chest and shoulder, feel dizzy or faint, get breathless after mild exertion, are middle-aged or older and have not been physically active, or plan a vigorous activity program. If none of these apply, start slow and get moving.

11. **False.** Regular physical activity can help reduce your risk of having another heart attack. People who include regular physical activity in their lives after a heart attack improve their chances of survival and can improve how they feel and look. If you have had a heart attack, consult your doctor to be sure you are following a safe and effective exercise program that will help prevent heart pain and further damage from overexertion.

12. **True.** Pick several different activities that you like doing. You will be more likely to stay with it. Plan short-term and long-term goals. Keep a record of your progress, and check it regularly to see the progress you have made. Get your family and friends to join in. They can help keep you going.

Name _____ Section _____ Date _____

NOTEBOOK ACTIVITY

Assessing Cardiovascular Fitness: Cooper's 1.5-Mile Run

(Please note: If you are not comfortable with the run, use the Rockport Fitness test found in Table 5.2 on page xx.)

This test is optimally done on a track for six laps.

Prior to testing, get a good night's rest, drink water, and try to choose a time of day when the weather is agreeable. If you cannot run the entire test, you may walk-run as best you can. It may be advisable to practice running a 1.5-mile distance prior to testing to determine a reasonable pace for you. Use a stopwatch for accuracy in timing. The objective of this test is to complete the six laps as quickly as possible.

Warm up first.
Run; note time.
Recover, stretch, drink water.

Fitness Categories for Cooper's 1.5-Mile Run Test to Determine Cardiorespiratory Fitness

Fitness Category	Age (years)					
	13–19	20–29	30–39	40–49	50–59	60+
Men						
Very poor	>15:30	>16:00	>16:30	>17:30	>19:00	>20:00
Poor	12:11–15:30	14:01–16:00	14:46–16:30	15:36–17:30	17:01–19:00	19:01–20:00
Average	10:49–12:10	12:01–14:00	12:31–14:45	13:01–15:35	14:31–17:00	16:16–19:00
Good	9:41–10:48	10:46–12:00	11:01–12:30	11:31–13:00	12:31–14:30	14:00–16:15
Excellent	8:37–9:40	9:45–10:45	10:00–11:00	10:30–11:30	11:00–12:30	11:15–13:59
Superior	<8:37	<9:45	<10:00	<10:30	<11:00	<11:15
Women						
Very poor	>18:30	>19:00	>19:30	>20:00	>20:30	>21:00
Poor	16:55–18:30	18:31–19:00	19:01–19:30	19:31–20:00	20:01–20:30	20:31–21:31
Average	14:31–16:54	15:55–18:30	16:31–19:00	17:31–19:30	19:01–20:00	19:31–20:30
Good	12:30–14:30	13:31–1554	14:31–16:30	15:56–17:30	16:31–19:00	17:31–19:30
Excellent	11:50–12:29	12:30–13:30	13:00–14:30	13:45–15:55	14:30–16:30	16:30–18:00
Superior	<11:50	<12:30	<13:00	<13:45	<14:30	<16:30

Times are given in minutes and seconds. (> = greater than; < = less than)
From Cooper, K. *The aerobics program for total well-being.* Bantam Books, New York, 1982.

Date _____ Temperature _____ Relative Humidity _____

Location of test _____

Finish time _____

Fitness category _____

Chapter 6
Nutrition and Metabolism

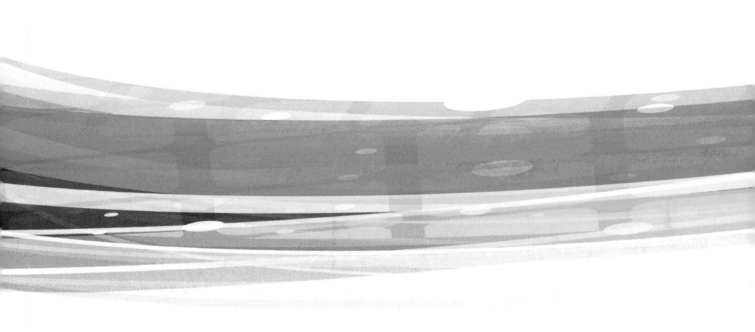

OBJECTIVES

After reading this chapter, you will be able to:
- Describe how the body metabolizes and provides sources of different nutrients.
- Identify and describe different types of metabolism.
- Describe body temperature, its source, and its regulation.
- Explain and the role of a balanced diet and physical activity in a healthy lifestyle and regulation of body weight.

From *An Introduction to Anatomy and Physiology, 1st Edition* by Coast Learning Systems, John L. Erickson. Copyright © 2010 by Kendall Hunt Publishing Company. Reprinted by permission.

AT A GLANCE

Nutrients and Water
- Carbohydrates
- Proteins
- Lipids
- Vitamins
- Minerals
- Water

The Difference between Food and Nutrients
- Food Groups
- Food Pyramid
- Metabolism
- Types of Metabolic Reactions
- Carbohydrate Metabolism
- Lipid Metabolism
- Protein Metabolism
- Body Temperature

Summary
Key Terms
Test Yourself
Thought Questions
Medical Questions

Nutrition
The study of the substances ingested and their effect on the function of the body.

Nutrients
Chemicals from which we obtain the essential ingredients to perform all of the chemical and electrical processes the body requires to maintain homeostasis.

If we define nutrients as substances we need to perform chemical and electrical processes within our body, do we all need the same nutrients? What happens if those nutrients are in short supply? In our brain, the hypothalamus makes us feel hungry when our nutrient levels are low. Why do some people prefer sweets over meat and potatoes? Why might a pregnant woman crave unusual combinations of food? The study of the substances ingested and their effect on the function of the body is known as **nutrition**.

The chemical processes that occur in the body, especially concerning the conversion of available energy into adenosine triphosphate (ATP), are known as *metabolism*. Virtually every living cell in the human body performs a form of metabolism, although some very active cells, such as muscle fibers or neurons, require immensely higher levels of ATP than passive adipocytes or bone cells. In this chapter we first consider the chemicals we must ingest as nutrients and then discuss the processes by which energy is extracted from nutrients and used to produce ATP.

Nutrients and Water

Nutrients are chemicals from which we obtain the essential ingredients to perform all of the chemical and electrical processes the body requires to maintain homeostasis. Nutrients provide the following functions:

- They are a source of energy. The quantity of energy available from a nutrient source is measured in terms of calories. Lipids are the greatest source of energy, containing 9 calories per gram while carbohydrates and proteins each have 4 calories per gram.

- They are the components, such as amino acids, necessary to produce molecules involved in chemical reactions, synthesis of enzymes, and the building of cellular structures, such as muscle or cellular membrane proteins.
- They are cellular signaling molecules that initiate the functions of each cell.

Six substances are essential to the normal functioning of the human body:

- Carbohydrates
- Proteins
- Lipids
- Vitamins
- Minerals
- Water

Each of these components must be taken in from external sources on a regular basis.

Carbohydrates

Carbohydrates are sugars. Recall from Chapter 2 that carbohydrates can be found in three forms: (1) single sugars—monosaccharides, (2) two single sugars bonded together—disaccharides, or (3) many single sugars chained together—polysaccharides. Single sugars can be absorbed through the wall of the jejunum and ileum into the bloodstream without any alteration. Disaccharides and polysaccharides must be broken apart by enzymes into single sugars in order to pass through the columnar cells of the small intestine and be absorbed into the bloodstream.

Proteins

Proteins are chains of amino acids. Proteins are too large to be absorbed directly into the bloodstream from the small intestine. Enzymes break the protein chains into individual amino acids so that they can pass through the intestinal wall and diffuse into the bloodstream. Cells can bond the amino acids together to form appropriate proteins to meet their needs. For example, cells can form amino acid chains into membrane proteins such as ion channels or receptor sites then insert them into its cell membrane. Outside the cell, collagen fibers, made of protein, may be knit together in the matrix around cells such as osteoblasts or fibroblasts to form additional bone or scar tissue. The human body uses 20 amino acids to form proteins. Of these, nine are considered *essential amino acids* because the body cannot synthesize them and they must be taken in from other sources. The remaining amino acids are designated as *nonessential*, not because they are not important, but because they can be synthesized from the essential amino acids when needed. Proteins containing the nine essential amino acids are referred to as *complete proteins*; these include milk, cheese, eggs, meat, fish, and poultry. *Incomplete proteins* do not possess all the essential amino acids. Some incomplete proteins are legumes (beans or peas), grains, and leafy green vegetables.

Lipids

Lipids, such as fats and oils, are handled differently than carbohydrates and proteins. In Chapter 2 we discussed the fact that water is polar, possessing an electrical charge, and oils are nonpolar, or uncharged. Because of this electrical difference, oil and water do not mix. Carbohydrates and proteins are polar and can dissolve in water and pass directly into the bloodstream. Lipids, because they cannot mix with water, require another form of transport throughout the

Carbohydrates
Sugars.

Proteins
Chains of amino acids.

Lipids
Fats and oils.

body. Most lipids in the digestive system are in the form of triglycerides and have been coated with bile, charged particles that emulsify fats (see Chapter 15). Triglycerides are too large to pass through the intestinal mucosa, so they are partially disassembled by pancreatic lipase, an enzyme that allows the parts to pass through the lining of the jejunum and ileum. Inside the cells of the mucosa the triglycerides are reassembled and then coated with charged proteins. These coated molecules are too large to diffuse into capillaries as single sugars and amino acids do when they exit the other side of the mucosa, so they pass instead into lymphatic vessels. They circulate through the lymphatic system and into the left subclavian vein to enter the bloodstream indirectly.

> **COMPREHENSION CHECKUP**
>
> 1. The six types of nutrients are _____ , _____ , _____ , _____ , _____ , and _____ .
> 2. Most lipids in the digestive system are in the form of _____ .
>
> 1. carbohydrates, proteins, lipids, vitamins, minerals, water 2. triglycerides

Vitamins
Substances that cannot be manufactured in the body in sufficient quantity yet are essential to the performance of various chemical processes throughout the body.

Vitamins are cofactors of enzymes that cannot be manufactured in the human body in sufficient quantity, except for vitamin D and must be ingested on a regular basis in order for homeostasis to be maintained. They are essential to the performance of various chemical processes throughout the body. They can be found in food and vitamin supplements. These chemicals must be taken in regularly for homeostasis to be maintained. Vitamins can be divided into two general categories: water-soluble (polar) and lipid-soluble (nonpolar) (Table 6.1 on pages 158–159). Water-soluble vitamins are the B complex vitamins and vitamin C. The level of the water-soluble vitamins can be controlled by the kidneys. Excess water-soluble vitamins are normally excreted in the urine. Lipid-soluble vitamins include vitamins A, D, E, and K and are not as easily regulated. They are stored and maintained by the digestive system, primarily the liver. Because they are not controlled by the kidneys, it is easier to overdose on lipid-soluble vitamins than on those that are water-soluble.

TABLE 6.1 ♦ Vitamins

Vitamin	Significance	Sources	Daily Requirement	Effects of Deficiency	Effects of Excess
Fat-Soluble Vitamins					
A	Maintains epithelia; required for synthesis of visual pigments	Leafy green and yellow vegetables	1 mg	Retarded growth, night blindness, deterioration of epithelial membranes	Liver damage, skin peeling, central nervous system effects of nausea, anorexia
D	Required for normal bone growth, calcium and phosphorus absorption at gut, and retention at kidneys	Synthesized in skin exposed to sunlight	None[a]	Rickets, skeletal deterioration	Calcium deposits in many tissues disrupting functions

TABLE 6.1 ♦ Vitamins (Continued)

Vitamin	Significance	Sources	Daily Requirement	Effects of Deficiency	Effects of Excess
E (tocopherols)	Prevents breakdown of vitamin A and fatty acids	Meat, milk, vegetables	12 mg	Anemia; other problems suspected	None reported
K	Essential for liver synthesis of prothrombin and other clotting factors	Vegetables; production by intestinal bacteria	0.7–0.14 mg	Bleeding disorders	Liver dysfunction, jaundice
Water-Soluble Vitamins					
B_1 (thiamine)	Coenzyme in decarboxylation reactions	Milk, meat, bread	1.9 mg	Muscle weakness, central nervous system and cardiovascular problems, including heart disease; called *beriberi*	Hypotension
B_2 (riboflavin)	Part of FMN and FAD	Milk, meat	1.5 mg	Epithelial and mucosal deterioration	Itching, tingling sensations
Niacin (nicotinic acid)	Part of NAD	Meat, bread, potatoes	14.6 mg	Central nervous system, gastrointestinal, epithelial, and mucosal deterioration; called *pellagra*	Itching, burning sensations, vasodilation, death after large dose
B_6 (pyridoxine)	Coenzyme in amino acid and lipid metabolism	Meat	1.42 mg	Retarded growth, anemia, convulsions, epithelial changes	Central nervous system alterations, perhaps fatal
Folacin (folic acid)	Coenzyme in amino acid and nucleic acid metabolism	Vegetables, cereal, bread	0.1 mg	Retarded growth, anemia, gastrointestinal disorders	Few noted except in massive doses
B_{12} (cobalamin)	Coenzyme in nucleic acid metabolism	Milk, meat	4.5 mg	Impaired iron absorption causing *pernicious anemia*	Polycythemia
Biotin	Coenzyme in decarboxylation reactions	Eggs, meat, vegetables	0.1–0.2 mg	Fatigue, muscular pain, nausea, dermatitis	None reported
Pantothenic acid	Part of acetyl-CoA	Milk, meat	4.7 mg	Retarded growth, central nervous system disturbances	None reported
C (ascorbic acid)	Coenzyme; delivers hydrogen ions	Citrus fruits	60 mg	Epithelial and mucosal deterioration; called *scurvy*	Kidney stones

aUnless there is poor exposure to sunlight for extended periods; alternative sources are provided in fortified milk
FMN = Flavin mononucleotide
NAD = Nicotinamide adenine dinucleotide
FAD = Flavin adenine dinucleotide

> **Minerals**
> Metals and nonmetals required by the body primarily in the form of ions to perform chemical reactions or create action potentials.

Minerals

Minerals, metals and nonmetals, required by the body are in the form of ions (Table 6.2 on page 160). They can be absorbed directly into the bloodstream from the small intestine, where they remain in ionized form. We have previously discussed the importance of sodium and potassium ions for depolarization and repolarization of cells. Calcium was also mentioned as being essential for activation of the thin filament in each sarcomere in muscle. Phosphorus is ionized to phosphate and used to make ATP. Calcium and phosphorus are major components in the matrix of our bones. Like vitamins, minerals also serve as cofactors in the function of many enzymes.

TABLE 6.2 ♦ Minerals

Mineral	Significance	Total Body Content	Primary Route	Recommended Daily Intake
Bulk Minerals				
Sodium	Major cation in body fluids; essential for normal membrane function	110 g, primarily in body fluids	Urine, sweat, feces	1.1–3.3 g
Potassium	Major cation in cytoplasm; essential for normal membrane function	140 g, primarily in cytoplasm	Urine	1.9–5.6 g
Chloride	Major anion in body fluids	89 g, primarily in body fluids	Urine, sweat	1.7–5.1 g
Calcium	Essential for normal muscle and nerve function, structural support of bones	1.36 kg, primarily in skeleton	Urine, feces	0.8–1.2 g
Phosphorus	As phosphate in high-energy compounds, nucleic acids, and structural support of bones	744 g, primarily in skeleton	Urine, feces	0.8–1.2 g
Magnesium	Cofactor of enzymes; required for normal membrane functions	29 g, 17 g in skeleton and the rest in cytoplasm and body fluids	Urine	0.3–0.4 g
Trace Minerals				
Iron	Component of hemoglobin, myoglobin, and cytochromes	3.9 g, 1.6 stored (ferritin or hemosiderin)	Urine (traces)	10–18 mg
Zinc	Cofactor of enzymes systems, notably carbonic anhydrase	2 g	Urine, hair (traces)	15 mg
Copper	Required for hemoglobin synthesis, as cofactor	127 mg	Urine, feces (traces)	2–3 mg
Manganese	Cofactor for some enzymes	11 mg	Feces, urine (traces)	2.5–5 mg

Water

Water is found in most areas of the body and is used for the transport of substances to and from each cell. It is also necessary for any metabolic reaction within the cell and is a major component of saliva and secretions from the stomach, pancreas, liver, and small intestine. The average adult typically takes in about 2,500 ml of water in food and beverages. Additional water is added to the digestive tract from saliva (1,500 ml), from the stomach (2,000 ml), from the pancreas (1,500 ml), from the liver (500 ml), and from the small intestine (1,500 ml). All of the secretions from the digestive organs obtain their water from storage in body tissue. In a 24-hour period, approximately 9,000 to 9,500 ml of water passes through the digestive system (Figure 6.1 on page 161). The intestines reabsorb all but about 500 ml of this water back into the bloodstream. If a significant amount of the water in the digestive system is not reabsorbed, as occurs with diarrhea, dehydration may occur.

The kidneys, along with input from the brain, can then determine how much of the water in the bloodstream is essential to maintain homeostasis. The excess is excreted in the urine. The 500 ml remaining in our feces simply provides some softness. Water is also lost from the body through sweating or exhaled during respiration.

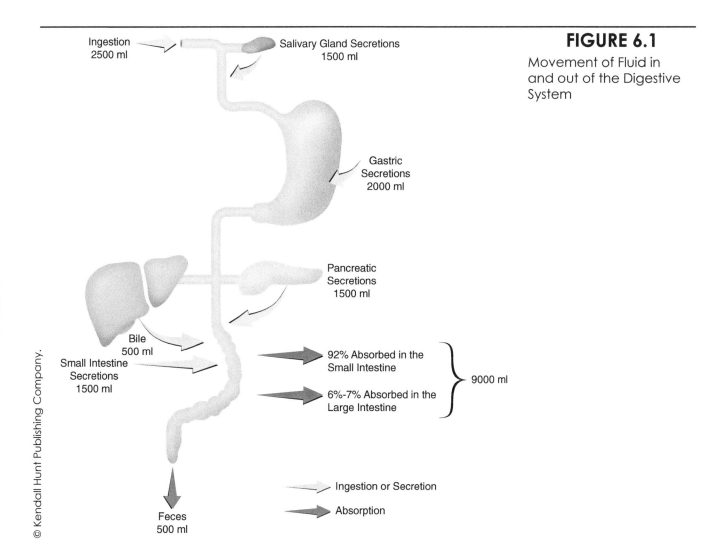

FIGURE 6.1

Movement of Fluid in and out of the Digestive System

> **COMPREHENSION CHECKUP**
>
> 1. The two general categories of vitamins are _____ and _____.
> 2. Sodium, potassium, and calcium are _____, a type of nutrient.
>
> 1. water soluble, lipid soluble
> 2. minerals

The Difference between Food and Nutrients

Not everything we eat as food is nutritious. There are some substances in our food for which we do not possess enzymes to break their chemical bonds to reduce them to their simplest components, for example, cellulose. Commonly known as *plant fiber*, cellulose is composed of chains of glucose. Although it would seem to be a great source of nutrition, animals do not produce the enzymes necessary to break these bonds. Instead, fiber primarily found in fruits and vegetables adds bulk to our digestive system and helps maintain intestinal health by keeping the intestinal wall clean. It can also bind cholesterol to assist in removing excess levels from the body.

Diarrhea and Its Effect on the Body

Almost everyone has experienced diarrhea at some point. Typically the cause is some type of irritation to the intestinal mucosa, as is commonly caused by bacteria or some type of virus. The response of the digestive system is to empty its contents as quickly as possible. When gastric motility, that is, the speed of the digestive process, increases substantially, there is little time to reabsorb water and electrolytes and water is lost from the body. This excessive loss may lead to serious health concerns, especially in children. The digestive organs obtain water and nutrients that are essential to the production of their secretions from the bloodstream and from internal tissue fluid. Typically 9000 ml of water is reabsorbed by the intestine back into the cardiovascular system. If, however, a large volume of that normally reabsorbed water is lost as a result of diarrhea, the cardiovascular system may have a significant loss of plasma volume, making it difficult for the heart to maintain blood pressure. Most systems in the body can be affected by inadequate blood flow. Diarrhea is of particular concern in small children because they do not have the large reservoir of fluid in their tissue typically found in adults. As a result, diarrhea can become serious much more quickly in the small child.

Food Groups

The food we eat can be divided into five distinct groups depending on their source and content: fruits, vegetables, grains, milk products, and meats.

Food Pyramid

The U.S. Department of Agriculture developed a food pyramid to assist individuals in determining the quantity of each food group to take in each day (Figure 6.2).

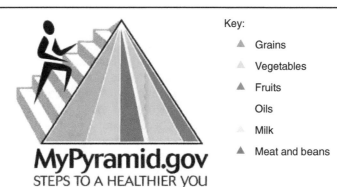

FIGURE 6.2

Food Pyramid
(From: http://www.righthealth.com, Food Pyramid.)

Each color in Figure 6.2 represents a different food group and the quantity to be consumed:

▲ Grains such as rice, bread, cereals, and pasta—6 to 11 servings per day
▲ Vegetables—3 to 5 servings per day
▲ Fruits—2 to 4 servings per day
  Fats, oils, and sweets—occasional use
▲ Dairy products, including cheese and yogurt—2 to 3 servings in 2 days
▲ Meat, fish, eggs, nuts, and beans or legumes—2 to 3 servings per day

Grains are to be consumed in the largest quantities because they are composed primarily of complex carbohydrates (polysaccharides as well as protein). Chains of sugars take more time to digest than disaccharides (two single sugars bonded together), which are typically found in sweets. As a result, complex carbohydrates provide a source for energy over a much longer period than do disaccharides.

Fruits and vegetables also contain carbohydrates. They are an excellent source of many of the vitamins and minerals needed for normal body function. Undigestible plant fiber helps clean the walls of the intestine and also provides bulk to keep the movement of food through the digestive system efficient.

Dairy products contain protein as well as vitamins and minerals. They are, however, also high in fat, which is why they are recommended to be consumed in lower quantities in adults than the previously mentioned food groups.

Meats, fish, eggs, nuts, beans, and legumes are an excellent source of protein. They may also contain vitamins, minerals, and carbohydrates. The proteins in this food group are broken down into amino acids for reassembly into needed proteins throughout the body.

For maximum health and wellness, it is recommended that fats, oils, and sweets be limited to only occasional servings. We can obtain the lipids and sugars we need from the other groups. It is not necessary to take them in separately. Although fat is an excellent energy source, the consumption of excessive fat, especially for individuals with low activity levels, causes the body to store more than is used. There are two types of fats found in our diet: saturated and unsaturated. Saturated fats have been linked to a buildup of plaque in blood vessels, resulting in heart disease and stroke. Polyunsaturated fats may actually decrease the inflammation of blood vessels that leads to the buildup of plaque. Sweets (disaccharides) are absorbed into the bloodstream too fast, causing a rapid rise in blood glucose followed by a large increase in insulin. The purpose of insulin is to lower blood sugar back to homeostatic levels, but it also causes additional storage of other nutrients and increases the content of adipose tissue. The effects of increased insulin levels are discussed in greater detail later in this chapter.

The concept of the food pyramid is that appropriate consumption of each food group, combined with daily physical activity, provides the level of nutrients needed to perform all of the necessary chemical reactions and processes in the body while helping the individual maintain his or her current weight and increase overall health. It is important to understand that physical activity is an essential factor in maintaining health and weight. There is no substitute for exercise.

The hypothalamus controls our food intake. It contains both a hunger center, which increases our desire to eat, and a satiety center, which determines our nutrient level is sufficient and decreases our hunger. If we have low levels of a specific nutrient we may have cravings for foods containing that substance. This often affects women during pregnancy when cravings for specific foods become very strong. We do not just eat because our hypothalamus stimulates our hunger center. We may consume food because we are bored, tired, or stressed. Perhaps you have heard the term, "comfort foods." There may also be foods we associate with unpleasant experiences which we refuse to eat. We may even be conditioned to like certain foods, such as sweets or meat, over others, like fruits and vegetables.

It is a concern that there is an increase in obesity in the United States, especially in children. Food portions are becoming larger and physical activity is decreasing. It is an issue that needs to be addressed today because the patterns learned as children are very likely to continue throughout life. Perhaps the best alternative is to not only provide appropriate diets for children but also to encourage them to be physically active.

Advantages of Physical Activity

Studies have shown that obese individuals typically do not overeat; they underexercise. There are numerous advantages to getting off the couch and taking a walk.

Some of the benefits of exercise are that it:

- Increases metabolism, which uses excess fat, resulting in weight loss or allowing the individual to stay trim.
- Improves circulation as blood flow increases to muscles performing their work.
- Increases assistance of venous return of blood to the heart.
- Increases stroke volume because the heart pumps higher volumes of blood during exercise, so during relaxation the larger stroke volume allows the heart more time to rest while it maintains normal cardiac output.
- Improves movement of extracellular fluid out of the tissue to decrease potential swelling.
- Strengthens bones.
- Keeps joints lubricated and moving freely.
- Relieves stress, which allows your defensive system to work more efficiently.
- Improves muscle tone, which reduces the likelihood of injury.
- Releases neurotransmitters in the brain that improve mental health.
- Allows you to think more clearly.
- Increases stamina.
- Improves the body's ability to maintain homeostasis.

None of these advantages can be achieved only by diet. If you provide the body with the appropriate level of nutrients combined with physical activity you increase probability of improved health and well-being.

COMPREHENSION CHECKUP

1. _____ and _____ are two food groups that are a good source for vitamins, minerals, and fiber.
2. The food group that is an outstanding source of protein is _____.

1. Fruits, vegetables
2. meats, fish, eggs, nuts, beans, and legumes

Metabolism

Metabolism is the series of chemical reactions that occur in the body that result in the conversion of energy in the form of nutrients into a form our cells can use to perform whatever work is needed at the time. For example, energy is needed to cause muscle contraction, to move nutrients through the digestive mucosa, for the kidneys to separate excess substances in the plasma and excrete them as urine, for the maintenance of ion balance to allow nerve impulse conduction, and for contraction of the heart to pump blood. Recall from Chapter 2 that all energy on the surface of the Earth comes from the sun. Plants capture the solar energy by a process known as photosynthesis. During photosynthesis the plant uses the energy from the sun to cause the single electron around hydrogen atoms to spin faster. Once the electron possesses the additional energy, it continues to retain that energy until a later time when the energy can be released and used by the plant. The hydrogen atoms come from water in the plant. The plant then attaches those energized hydrogen atoms to carbon dioxide molecules to form glucose ($C_6H_{12}O_6$) and oxygen. The chemical equation for photosynthesis is:

$$6\ H_2O + 6\ CO_2 + \text{solar energy} \rightarrow C_6H_{12}O_6 + 6\ O_2$$

We, as animals, cannot use light as a source of energy in our bodies. We eat plants or we eat animals that eat plants and obtain energy-rich molecules from them for metabolism. We can reverse the process by breaking glucose down, when oxygen is available, to water and carbon dioxide, and we can also release the energy contained in the glucose molecule.

$$C_6H_{12}O_6 + 6\ O_2 \rightarrow 6\ H_2O + 6\ CO_2 + \text{energy}$$

It is possible to break down glucose without oxygen to release a small amount of energy, but the process is temporary and not available to all cells.

Animals store the released energy from the breakdown of nutrients in the high-energy bonds of the ATP molecule. The energy is then transferred by ATP to the appropriate locations in the cell for work to be accomplished. This process that occurs within the cell is known as *cellular respiration*.

Types of Metabolic Reactions

Throughout the process of metabolism, two types of chemical reactions occur. The combining of small molecules that results in the synthesis of larger ones is referred to as **anabolism**. The breakdown of large molecules into smaller ones is known as **catabolism**. For example, in Chapter 2 we discussed that ATP (Adenosine-Pi~Pi~Pi where ~ represents a high energy bond) was constructed by adding a third phosphate molecule to adenosine diphosphate (ADP; 2 phosphates) but energy was required to add the third phosphate onto ADP. Later, the third phosphate can be broken off ATP to release that energy for use elsewhere. Adding the third phosphate to ADP builds the larger molecule ATP by anabolism. To obtain the energy necessary to add that third phosphate, through catabolism we must break glucose down into smaller molecules to cause its energy to be released.

The general purposes of metabolism occur in the following order:

1. Remove the energized hydrogen atoms containing additional energy from their source.
2. Cause the hydrogen electron to release its additional energy.
3. Use that energy to produce ATP in order to transfer the energy to the appropriate location in the cell.

Energy is available from carbohydrates, lipids, and proteins. The process of removing this energy from nutrients is not very efficient. Only 25% of the

---

Metabolism
The series of chemical reactions that occur in the body that result in the conversion of energy in nutrients into a form our cells can use to perform work.

Anabolism
The combining of small molecules that results in the synthesis of larger ones.

Catabolism
The breakdown of large molecules into smaller ones.

available energy actually is used for work such as causing muscle contraction or the pumping of ions. Seventy-five percent of the energy in our nutrients becomes heat, raising the temperature of the body. The processes involved with extracting energy from these nutrients are addressed individually.

> ### COMPREHENSION CHECKUP
>
> 1. The source of energy on the surface of the earth is the _____.
> 2. Of the energy in nutrients, ___% of that energy actually is used to perform work and the remaining ____% turns into heat.
>
> 1. sun
> 2. 25, 75

Anaerobic metabolism
The production of a small amount of ATP per glucose molecule without the use of oxygen.

Aerobic metabolism
A series of oxygen dependent chemical reactions by which a relatively large amount of ATP is produced.

Carbohydrate Metabolism

Glucose is a major source of available energy. The catabolism of glucose to extract excess energy to make ATP occurs in two pathways. The first series of chemical reactions does not require oxygen to complete the process and is referred to as **anaerobic metabolism**. Essentially, only carbohydrates are catabolized during anaerobic metabolism. The second pathway is oxygen dependent and is known as **aerobic metabolism**. Lipids and proteins can also be catabolized during aerobic metabolism. Anaerobic metabolism involves only a few chemical reactions to complete its process, so it is a rapid source of ATP but its yield per glucose is small. Aerobic metabolism, on the other hand, involves many chemical reactions, but the yield of ATP per glucose is much greater. The process involving oxygen during the production of ATP is referred to as *oxidative phosphorylation*. During skeletal muscle contractions, we often use anaerobic metabolism as a quick source of ATP production and then later, when oxygen is available, we continue to aerobic metabolism for increased ATP production.

Anaerobic Metabolism

The catabolism of glucose is accomplished by a series of chemical reactions that breaks glucose in half and rearranges it into two molecules known as *pyruvic acid*. In the process of breaking glucose apart, called *glycolysis*, a small amount of ATP is produced (Figure 6.3-1). In most cases, the pyruvic acid moves on to aerobic metabolism. If oxygen is in short supply, however, it becomes essential to do something to remove the pyruvic acid; otherwise, the catabolism of glucose will back up and come to a halt. The production of pyruvic acid is somewhat like an object being made and then placed on a conveyor belt. At the end of the belt someone must remove the product, or it will stack up so high that work cannot continue. In the same way, pyruvic acid needs to be moved out of the way and a coenzyme, NAD, which is a carrier of energized electrons during the process, needs to be regenerated so that the production of ATP through anaerobic metabolism can continue. Alternatively, muscle can convert pyruvic acid into lactic acid. The second phase of anaerobic metabolism results in the production of lactic acid which regenerates NAD, allowing the breakdown of glucose (glycolysis) to continue. Lactic acid is somewhat like taking out a loan until oxygen is available, sometimes referred to as *oxygen debt*. This allows muscle to continue working even though the bloodstream is unable to transport sufficient oxygen at the moment to keep up with the demand.

If there is excess pyruvic acid and the demand for ATP has decreased, pyruvic acid can be converted back into glucose as an anabolic reaction. This process occurs in the liver and is known as *gluconeogenesis* (the creation of new glucose).

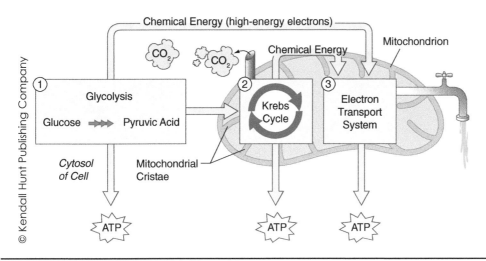

FIGURE 6.3

Glucose Metabolism. (1) In the cytsol, glycolysis breaks glucose into 2 pyruvic acids and produces a small amount of ATP. (2) Pyruvic acid enters the Krebs cycle in the mitochondrion, where hydrogen atoms with their high-energy electrons are removed. In the process, a small amount of ATP and carbon dioxide is produced. (3) The chemical energy from glycolysis and the Krebs cycle is transferred to the electron transport system, where the energy contained in the high-energy hydrogen electron is used to form ATP. Hydrogen atoms that have given up their excess energy are bonded with oxygen to form water

Aerobic Metabolism

When oxygen is available, pyruvic acid is fed into a cycle of chemical reactions, known as the *Krebs cycle*, designed to extract substantially more energy. The remaining hydrogen atoms are removed from what had been pyruvic acid and are sent to the electron transport system, where the electron gives up its excess energy, allowing large amounts of ATP to be produced. The carbon and oxygen atoms, which, along with hydrogen atoms, composed pyruvic acid, are released as carbon dioxide. The hydrogen atoms, which have given up their excess energy, are bonded with oxygen to produce water. The carbon dioxide and water are transported to the lungs from the bloodstream so that they can be exhaled out of the body into the atmosphere (Figure 6.3-2 and 6.3-3).

Lipid Metabolism

Fats are carbon chains surrounded by hydrogen atoms. Those hydrogen atoms contain excess energy, making lipids an excellent option for storing large amounts of energy. In fact, there is more than twice the amount of energy in fat than in either carbohydrates or protein, so the resulting energy from fat allows us to be active for a longer period than that provided by the other two nutrients. When oxygen is available, the liver can catabolize fatty acids by a process known as *lipolysis* and, like pyruvic acid, send them into the Krebs cycle and electron transport system to produce a great deal of ATP (Figure 6.4). The process of lipolysis produces some by-products known as *ketone bodies* that have a fruity odor. If a diabetic individual has taken too much insulin for the amount of food eaten and is unconscious, often the person's breath will have a fruity odor, indicating that the body is rapidly catabolizing lipids because their available glucose is too low.

If there are excess carbohydrates, they can be converted into fatty acids and stored in liver or adipose tissue. The creation of lipids by the body is known as *lipogenesis*.

Protein Metabolism

Proteins are chains of amino acids. Amino acids may be used as a source for ATP as well, although they contain a nitrogen atom, which is unusable by the

FIGURE 6.4

Carbohydrates, Lipids, and Proteins. These can be sources of ATP through aerobic metabolism in mitochondria

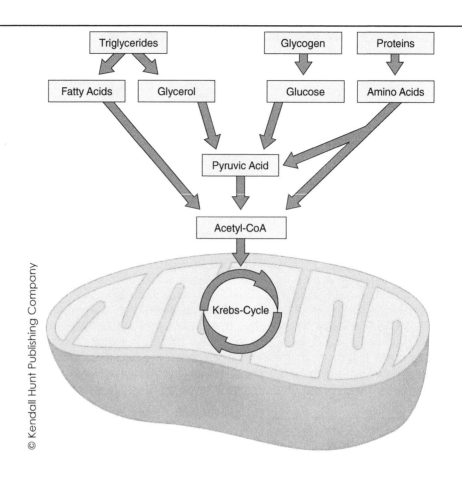

> **COMPREHENSION CHECKUP**
>
> 1. The primary nutrient that has some energy extracted to make ATP without the use of oxygen is _____.
> 2. The nutrient that contains more energy than the other two combined is _____.
> 3. To use amino acids for an energy source, our cells must first remove _____.
>
> 1. carbohydrates or glucose   2. lipids   3. nitrogen

Krebs cycle. As a result, during amino acid catabolism, the nitrogen atom is removed and discarded as urea. Urea is eliminated in the urine. Once urea has been exposed to the air, it incorporates another hydrogen atom, converting it to ammonia. It is the excretion of urea that causes a diaper pail to smell of ammonia.

Excess pyruvic acid can also be converted into amino acids when a nitrogen atom from some of the urea remaining in the body is added to the molecule. Those amino acids can then be bonded together in a specific order inside cells to form protein through protein synthesis.

Body temperature is regulated at an average of 98.6°F (37°C). The heat for maintaining our temperature comes from metabolism. As we convert the energy in nutrients into ATP, 50% of that energy escapes as heat. When transferring the energy in ATP to its intended process, another 25% of that energy becomes heat as well. That heat energy is available, somewhat like a constantly running furnace, to increase body temperature.

Regulation of Body Temperature

Our body temperature is derived from metabolism. We can regulate our temperature by (1) retaining or releasing heat into the environment through our skin and (2) altering the rate of metabolism to adjust the generation of heat.

The hypothalamus controls body temperature to maintain it within homeostatic range. If we become too warm—for example, if we are exercising—the hypothalamus causes an increase in blood flow to the skin and increases sweating to allow evaporation to draw additional heat from the skin and cool the body. If we become cold, the hypothalamus causes a decrease in blood flow to the skin to minimize heat loss. We also increase the need for more ATP by shivering, a rapid contraction of muscle that produces no coordinated movement but uses up ATP, which results in an increase in metabolism to generate more heat. Another method for increased heat production is to stimulate our sympathetic nervous system, our "fight-or-flight" system. The process of getting our body ready for action increases the production of ATP, which can also warm us when we are cold.

Thermal receptors in our skin can inform the brain about the temperature of our environment. For example, stepping into a sauna causes the sensing of heat. This results in the appropriate response by our temperature-regulating system. If we jump into a cold swimming pool, the sensing of the temperature of the water causes responses to cold.

Regulation of Body Weight

Scientists determined a measure of heat energy in terms of calories. A **calorie** is the amount of energy required to raise 1 g of water 1°C. The levels of calories in food, and also those required by our body, have been determined and are in the millions. When dealing with relatively large numbers, it is easier to discuss heat energy in terms of kilocalories (1000 calories), also written as Calories. Scientists then determined the minimum level of Calories (kilocalories) required by a resting, alert individual to perform essential activities, such as breathing and circulating blood. They called this the **basal metabolic rate (BMR)**. Any activity performed during the day requires additional energy above BMR.

If the Calories taken in equals the energy used, then weight remains unchanged. When more energy is taken in than Calories used, the body stores the excess primarily as fat because it is the most efficient means to do so. As a result, there is weight gain. The reverse is also true; when the individual uses more energy than is taken in, the body draws from its reserves and weight is lost.

For the body to perform all of the functions necessary to maintain homeostasis and well-being throughout life, it is important that it receive the essential nutrients (9 amino acids, most vitamins, and a few fatty acids) on a regular basis; most nutrients can be synthesized in the body from other nutrients. Each system is more capable of performing its tasks when it possesses all of the chemicals it requires than when those chemicals are in short supply. As a result, the interaction of each system improves because they are more capable of meeting the other's needs. Not only are our defenses more capable of reacting to invasion, but repair of tissue damage is also more rapid. Exercise relieves stress, improves circulation, strengthens bones, and provides numerous health benefits. Our bodies become trim and resilient, so injury is minimized. It has even been found that running, for example, can be as effective an antidepressant as drugs. Adequate nutrition combined with physical activity improves the quality of life because it provides the body with what it needs and creates the opportunity for our body to function in the way it was designed.

Calorie
A measure of the amount of energy in a substance, usually determined in food.

Basal metabolic rate (BMR)
The minimum level of Calories (kilocalories) required by a resting, alert individual to perform essential activities such as breathing and circulating blood.

Body mass index (BMI)
Comparison of body weight to height as a rough estimate of the condition of the body.

Body Mass Index versus Percentage of Fat Content

A rough estimate of the condition of the body can be made by comparing weight to height. This is known as **body mass index (BMI)**. BMI is calculated by the following equation:

$$\text{BMI} = \frac{\text{Wt (kg)}}{(\text{Height [meters]})^2} \quad \text{or} \quad \frac{\text{Wt (lb.)} \times 703}{(\text{Height [inches]})^2}$$

Normal BMI is between 18 and 25. A person with a BMI of less than 16.5 is considered severely underweight. A person with a BMI between 25 and 30 is overweight, and someone with a BMI greater than 30 is defined as obese.

For example, if a 5'5" woman weighed 135 pounds, her BMI would be calculated as follows:

$$\text{BMI} = \frac{135 \times 703}{(65)^2} = \frac{94{,}905}{4225} = 22.46$$

The BMI for this woman is within normal range.

There are times when BMI can be misleading. For example, a professional football player may have a high quantity of muscle mass even though he is relatively lean, causing his BMI to indicate an unhealthy weight. Conversely, an inactive person may not be overweight but his proportion of fat to muscle may be at an unhealthy level even though his BMI would calculate him to be within normal range.

Although BMI is a method for calculating the general relationship between weight and height, it is not taking into account the ratio of fat to muscle. It is possible, through measurement of the thickness of specific areas of the pinched skin, to determine the fat content of the body, giving a more accurate health assessment than BMI. Normal body fat content should be around 8% to 22% for men and 22% to 33% for women. There is a slight normal increase with age but not more than a few percentage points every 20 years. An individual who is greater than 20% over the normal fat content for age and gender is considered obese.

COMPREHENSION CHECKUP

1. The control center for body temperature is the _____.
2. The minimum amount of energy in Calories that a resting, alert individual needs to perform minimal activities is known as _____.

1. hypothalamus   2. basal metabolic rate

Factors Involved with Weight Control

Many factors are involved in weight control, including genetic predisposition, hormone levels, and disease processes. For the average person, though, four major factors come into play when trying to lose, gain, or maintain weight. They are:

- Avoid foods that increase your insulin level. Insulin is a hormone that tells your body you have more than enough glucose for normal function. Not only does it cause you to store excess glucose, it also causes some of that excess to be converted into fat. In addition, it prevents the removal of fat from adipose tissue. The primary source of insulin that raises glucose is sucrose (table sugar). We need a daily supply of glucose, so it is unwise to stay away from all carbohydrates, however, taking in large quantities of sucrose by eating sweets or consuming soft drinks increases insulin to high levels. A side effect of increased insulin is that it may actually cause the storage of so much glucose that it results in low blood sugar, causing a person to feel irritable, confused, and fatigued. For the person trying to lose weight, insulin keeps fat in storage. For the person trying to gain weight, it is a short-lived benefit. It still may result in low blood glucose about 30 to 40 minutes after taking in high levels of sugar.

- Do not allow your body to go into "starvation mode." If the glucose level becomes low, the body switches into starvation mode. It assumes no additional food will be coming any time soon, so it slows metabolism to conserve the supplies it has and it saves as many additional nutrients as it can. Skipping meals is counterproductive as a weight loss method. For example, a person who skips breakfast will typically gain 6 to 8 pounds a year. A significantly better alternative is to eat a relatively small quantity of food every 2 or 3 hours. Avoid foods high in sugar.

- Follow the suggested amounts for each food group in the food pyramid to obtain the nutrients your body requires. Do everything in moderation. Just because a certain amount is good, it does not mean doubling it is better. If you give your body what it needs, it will be able to maintain itself in most cases.

- Get regular exercise, which increases your basal metabolic rate. Much of your use of nutrients occurs after the exercise is over, as you replenish the energy used during the activity. Vigorous exercise 6 days a week is optimal. Three days should involve aerobic exercise to help increase blood flow and allow your heart to work more efficiently. Three days should involve anaerobic exercise to build muscle through strength training, which strengthens and tones muscle and increases basal metabolic rate. Note: Strength training for women does not result in bulk like in men because they have lower levels of testosterone, which is needed to produce large muscle mass. Women can become trim and shapely.

Apples and Pears

Obesity has reached epidemic proportions in the United States. There are two types of obesity compared with shapes of fruit—used to describe the general distribution of fat in the obese person. Although both types can be found in both genders, there is a tendency for one shape to be more commonly found in each sex. Individuals whose adipose tissue is distributed primarily on the thighs and hips are considered pear-shaped, or gynoid. This form is most commonly seen in females and is correlated with their reproductive hormones. Fortunately, this form of obesity carries much fewer health risks than the second type—apple-shaped, or android. In the case of android obesity, most commonly seen in males, the distribution of adipose tissue is on the trunk. In this condition, however, the excess fat is not only in the hypodermis but is also found surrounding the abdominal organs. This visceral fat releases chemical signals that lead to diseases such as insulin resistance, which can result in insulin-independent diabetes mellitus, heart disease, and stroke.

Homeostasis—Holding in Balance

Although we are not discussing a body system specifically, nutrition and metabolism play an active role in the homeostasis of body temperature. The generation of heat by the process of metabolism provides the source that allows the hypothalamus to control body temperature. By adjusting blood flow to the skin, we can increase or slow the release of heat into the environment. If we become overheated, metabolism slows. If we become cold, metabolic rate is increased through shivering or other activities to increase heat generation to maintain body temperature.

There Is No Substitute for Exercise

The more we exercise, the more efficient the body becomes at converting nutrients into ATP. We increase blood flow to provide nutrients and oxygen and remove carbon dioxide and metabolic wastes more efficiently. We produce additional mitochondria in our muscle to increase aerobic metabolism. We also increase myoglobin production, a chemical in skeletal muscle that is able to hold additional oxygen, making it more readily available for aerobic metabolism.

Summary

I. There are six basic nutrients needed for normal function of the body that must be regularly ingested.
   A. Carbohydrates—sugars, our primary source for energy
   B. Proteins—chains of amino acids, primarily reused to make new proteins
   C. Lipids—fats and oils, a source of high levels of energy and also an excellent method for storing energy in our body
   D. Vitamins—chemicals essential for other chemical reactions to occur in the body; not manufactured by the body or produced in insufficient quantities
   E. Minerals—ions essential for chemical reactions and the creation of action potentials in the body
   F. Water—the nutrient in greatest quantity in the body; used primarily as a means of transporting other substances throughout the body

II. Food is the total of all the material ingested, including both nutritive and nonnutritive substances. Nutritive substances are those from which we obtain energy and chemicals essential to perform normal body functions.

III. Food can be divided into groups depending on the nutrients they contain and the effect they have on the normal function of the body.

IV. The Food Pyramid graphically represents the quantities of nutrients that should be ingested daily.
   A. Grains composed primarily of complex carbohydrates and protein supply the greatest quantity of our energy source.
   B. Fruits and vegetables are excellent sources of vitamins and minerals. They also contain fiber to help maintain the general condition of the digestive system.
   C. Fats, oils, and sweets are usually excessive substances normally obtained in sufficient quantities from the other food groups and are recommended to be consumed only sparingly.
   D. Dairy products are high in protein, vitamins, and minerals, but they are also high in fat content.
   E. Meat, fish, eggs, nuts, beans, and legumes are an excellent source of protein, as well as some vitamins, minerals, and carbohydrates.

V. Metabolism includes the chemical processes by which we obtain energy from nutrients.
   A. Anabolism is the process of combining chemicals to form larger molecules.
   B. Catabolism is the breaking down of large molecules into smaller ones.

VI. Energy is primarily transferred in the human body in the form of ATP.

VII. There are two methods by which energy is transferred from nutrients into ATP.
   A. Anaerobic metabolism does not require oxygen but produces only a small amount of ATP. Only carbohydrates are metabolized by anaerobic metabolism. Glucose is catabolized into two pyruvic acids.
   B. Aerobic metabolism requires oxygen and produces large amounts of ATP. Pyruvic acids, lipids, and, if necessary, amino acids can be metabolized to produce ATP.
   C. Nutrients not needed at the moment can be bonded into larger molecules by anabolism and stored for later use.

VIII. Heat is produced during metabolism and can be used to maintain body temperature by two methods.
   A. Controlling the release of heat through the skin into the environment.
   B. Controlling the rate of metabolism to regulate heat production.

IX. Regulation of body weight is dependent on the intake or storage of Calories—a measure of available energy in nutrients. The rate at which Calories are used at rest is known as our *basal metabolic rate*. Weight gain or loss depends on the ratio between Calories taken in and Calories used.

KEY TERMS

aerobic metabolism (166)
anaerobic metabolism (166)
anabolism (165)
basal metabolic rate (BMR) (169)
body mass index (BMI) (170)
calories (169)
carbohydrates (157)
catabolism (165)

lipids (157)
metabolism (165)
minerals (160)
nutrients (156)
nutrition (156)
proteins (157)
vitamins (158)

TEST YOURSELF

Match the following terms with their definition:

1. carbohydrates
2. catabolism
3. aerobic metabolism
4. lipids
5. calories
6. vitamins
7. anaerobic metabolism
8. proteins
9. minerals
10. anabolism

A. chains of amino acids
B. metal and nonmetal ions
C. building larger molecules from smaller ones
D. chains of sugars
E. production of ATP without oxygen
F. a measure of energy in food
G. breaking down large molecules into smaller ones
H. fats and oils
I. cofactors of enzymes the body cannot produce
J. production of ATP by the use of oxygen

Choose the best answer to the following multiple choice questions:

1. Although dairy products contain protein, vitamins, and minerals, they should not be eaten in large quantities because they also contain
   a. contaminants.
   b. too many sugars.
   c. toxins.
   d. fat.

2. Why is it recommended that we take in few fats, oils, and sweets?
   a. They are toxic to us.
   b. We take in sufficient quantities in other food groups.
   c. They will give us too much energy.
   d. We cannot break them down.

3. The process by which plants convert solar energy into glucose and oxygen is known as
   a. photosynthesis.
   b. parthenogenesis.
   c. solar metabolism.
   d. light chain reaction.

4. Lipid-soluble vitamins are vitamins
   a. A, B, C, and D.
   b. A, D, E, and K.
   c. D, E, and K.
   d. B and C.

5. When we are cold, we shiver. Why?
   a. It helps shake nutrients through our digestive system faster.
   b. We release so much adrenaline when cold that our fight-or-flight system makes us jittery.
   c. It is rapid muscle contraction that increases the need for ATP to speed up metabolism and generate heat.
   d. The cold makes us so weak that we shake while trying to remain upright.

6. In the small intestine, lipids are absorbed
   a. into the urinary system before being recycled into the bloodstream.
   b. directly into capillaries within the intestinal lining.
   c. only in the colon, where bacteria can break them into small enough droplets to be reabsorbed.
   d. into lymphatic vessels and passed through the lymphatic system into the bloodstream.

7. The average adult takes in about _____ ml of water in food a beverages per day.
   a. 1,000
   b. 350
   c. 6,000
   d. 2,500

8. During anaerobic metabolism, glucose is converted into pyruvic acid to produce 2 ATP molecules. If oxygen is still not available, muscle will temporarily convert pyruvic acid into _____ to allow the conversion to continue. Excessive levels of this product causes muscle soreness.
   a. lactic acid
   b. uric acid
   c. hydrochloric acid
   d. nicotinic acid

9. When cold, the body can increase the generation of heat through metabolism and can
   a. increase sweating.
   b. increase absorption of nutrients.
   c. decrease loss of heat to the environment.
   d. decrease heart rate so that less heat is transported through the body.

10. A graphic method for demonstrating the quantities of food groups we should take in daily is known as the
    a. Venn diagram.
    b. Food Pyramid.
    c. Nutrient Circle.
    d. Square Meal Chart.

THOUGHT QUESTIONS

1. Why do complex carbohydrates keep us from becoming hungry much longer than sweets?
2. If I do not eat carbohydrates, my body will be forced to use up lipids. Is this a wise method to diet?
3. What can a vegetarian eat to maintain sufficient levels of protein in the body?
4. If I am exercising by lifting weights and am not sore the next day, does it mean I have not worked hard enough?
5. If I do not like vegetables and do not eat much fruit, will it really hurt me if I do not eat them?

MEDICAL QUESTIONS

1. Why is diarrhea so much more serious for a small child than for an adult?
2. As a stunt, a college student, wearing only swimming trunks, spray painted his entire body with gold spray paint. Several hours later he was hospitalized with heat stroke. Why?
3. If an individual had a stroke that damaged the hypothalamus, how would body temperature regulation and weight control be affected?
4. If I come across an unconscious individual who is a known diabetic and she has a fruity smell on her breath, what needs to be done to treat her?
5. If I have a body mass index of 38 and carry my obesity on my trunk, creating an apple shape, should I be concerned about my overall health or do I just need to lose weight?

Chapter 7
Scientific Principles of Weight Management

OBJECTIVES

Students will be able to:
- Identify problems associated with fast food dining.
- Discuss diet supplements.
- Present guidelines for a successful weight-loss program.
- Identify a healthy Body Mass Index.
- Identify causes of obesity and complications associated with obesity.
- Discuss the importance of activity for weight management.
- Recognize the pitfalls of Fad Dieting.

Adaptation of *Health and Fitness: A Guide to a Healthly Lifestyle, 5/e* by Laura Bounds, Gayden Darnell, Kirstin Brekken Shea and Dottiede Agnor. Copyright © 2012 by Kendall Hunt Publishing Company.

"Thou shouldst eat to live; not live to eat."

—Socrates, 469 bc–399 bc

Why have Americans gained so much weight over the last fifty years? Approximately two-thirds of adult Americans are overweight, and typically so are their children. The causes of obesity are numerous and complex. The bottom line is nutritional balance (see Figure 7.2): calories eaten versus calories expended. However, even if some folks exercise and attempt to eat healthy, it is more complicated than that. Because our society has become increasingly automated, Americans are saving lots of energy in the form of stored fat. "Easier," "automated," "instant," and "remote control" are all terms which describe using less energy. Saving energy relates to moving less, which means saving calories. A person who moves less typically stores more energy in the form of fat. Chapter 5 discusses lifestyle activity, which means looking for opportunities to expend more energy—like taking the stairs instead of the elevator. Every time you drive through for fast food, think of all of the energy you are saving. Another culprit in contributing to larger Americans is the Internet, games, texting, and so on. Balancing sedentary activities with playing baseball, tag, and climbing trees is important.

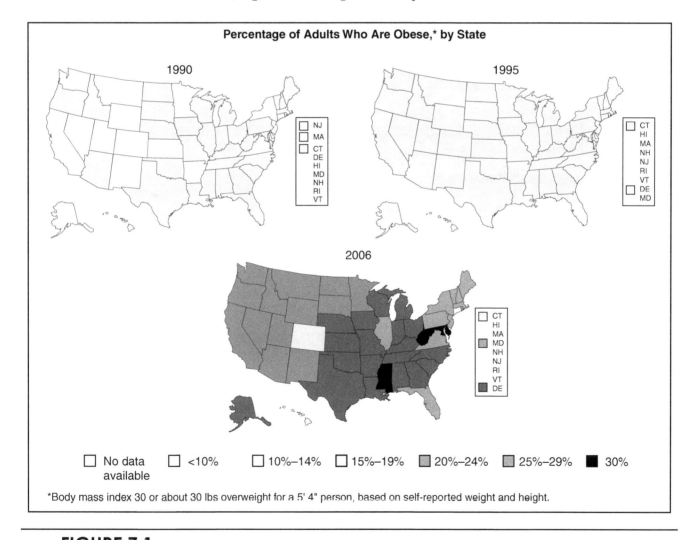

FIGURE 7.1

Percentage of Adults Who Are Obese,* by State from 1990 to 2006

Source: Behavioral Risk Factors Surveillance System, CDC.

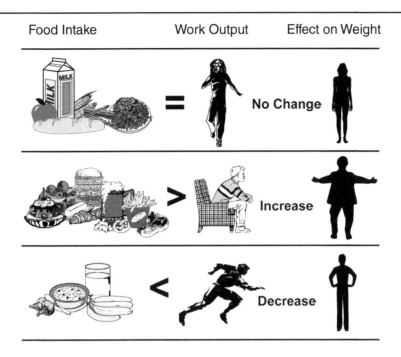

FIGURE 7.2

Caloric Balance: Caloric intake should equal caloric expenditure in order to maintain weight

Causes of Obesity and Being Overweight

Patrick O'Neil, director of the Weight Management Center at the Medical University of South Carolina in Charleston, determined that 40 percent of people's weight problems are due to whom they choose for their parents. Gene pool plays a big role in body size. Through evolution we are programmed to hunt and gather, then eat plenty as the food is available to store energy. Dr. O'Neil quotes the old adage, "Genes load the gun, but environment pulls the trigger." We typically learn eating patterns and develop a familiarity with certain types of foods that we eat when we are younger. Starting with what we eat as infants through adulthood, we learn particular eating styles.

Foods have changed. Fifty years ago foods were less processed. There was less convenience food. Foods were eaten more in their wholesome state. Portion sizes were smaller than they are today. Today's small McDonald's Happy Meal French fries were the "regular" size when McDonald's opened. A McDonald's Super Size serving of French fries contains 540 calories with 230 calories coming from fat. It is only in the last 25 years that convenience stores and fast-food eateries have offered the 42-ounce size of sweetened soft drinks. A 42-ounce Dr. Pepper has 525 calories. Many in the health industry think that the increase in corn syrup sweetened soft drink consumption has contributed to America's obesity problem, especially childhood obesity. Portion control is a critical component in weight management. Check product labels to determine how much food is considered a serving, as well as how many calories, grams of fat, and so on, are in a serving. Many prepackaged foods contain two or more servings—always read the label! See What Counts as a Serving on page 179. Trans-fatty acids have entered the diet via hydrogenation, a process by which liquid oils (which are unsaturated and healthy) are reconstituted to a solid convenient form. As you recall from the nutrition chapter, the problem with ingesting trans-fatty acid is that it raises LDL (the bad cholesterol) and lowers HDL (the good cholesterol). Results from Nurses' Health Study determined that a diet high in trans-fatty acids is highly associated with an increase in cardiovascular risk. A diet high in trans fats also may be linked to an increased abdominal fat measurement, as well as to an increase in the risk of Type 2 diabetes (IUFOST, 2006). What you eat, when

Calories Count
Between 1971 and 2004, the average American woman increased her caloric consumption by 22%. Between 1971 and 2004, the average American man increased his caloric consumption by 10%. *Men, women, and children are eating more carbohydrates in the form of starches, more refined grains and sugars, larger portion sizes, more fast food, and more sugar sweetened beverages (AHA, 2012).* To combat this, consider eating more foods the way nature made them. Try adding more raw foods to your diet and minimizing processed foods. Limit high sugar drinks, including juices, and drink more water.

you eat, and how much you eat make a difference. How you choose to eat makes a difference as well. Do you sit down and enjoy your food, or do you eat on the run?

Although the causes of overweight and obesity are complex and numerous, the risks of being overweight or obese are very real and tangible. An individual who is overweight is at increased risk for conditions such as high blood pressure, high cholesterol, stroke, heart disease, diabetes, and certain types of cancers. For most Americans, food is plentiful. Americans are typically not malnourished due to a lack of food, but the World Health Organization (WHO) predicts that there will soon be a world epidemic of overweight and malnourished people resulting from the unhealthy types of foods that are being eaten. A person living on only fast food that is high in sodium, fat, and cholesterol and low in vitamins and fiber can experience a lack of some essential nutrients. *Eating a variety of foods such as vegetables, fruits, whole grains, low-fat dairy, fish, lean cuts of meat (or other quality protein), and beans are the building blocks of a solid nutritional practice.*

Chapters 5 and 3 emphasized the importance of regular activity for overall health. Exercise can help you get fit and stay fit and is a critical aid in efforts to lose weight. The three largest contributors to overweight are lack of exercise, eating choices and behaviors, and genetics. Exercise may be the most significant. Sobering studies show that most people who successfully lose weight will have gained it back within two years time. The 5 percent that successfully maintain weight loss have exercise in their life—it is part of their lifestyle (see Figure 7.3). It isn't important to take up running or any activity you are uncomfortable doing. The important thing is to just move!

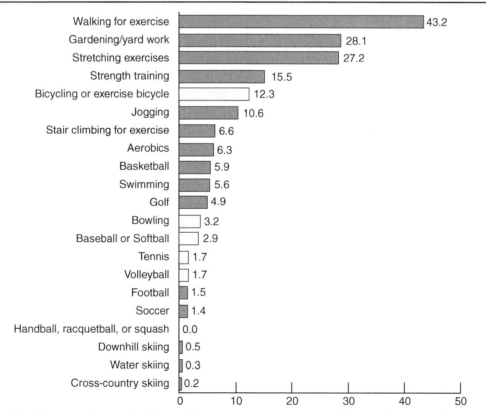

FIGURE 7.3
Most Popular Adult Physical Activities in the United States

Note: Data are weighted to the U.S. population and age-adjusted to the year 2000 population standard. "Participation" in activity reported as being done at least once during the past two weeks.

Source: Centers for Disease Control and Prevention, National Health Interview Survey (NHIS), 1998, Atlanta.

The Weight Loss 'Halo' Effect

Bariatrics is the field of medicine that specializes in treating morbid obesity. A surgeon may perform bariatric surgery to assist in weight loss only after lifestyle changes and other conventional methods are unsuccessful. There is associated short term and long term risks with this type of surgery and this option must be considered carefully. Patti Neighmond reported for NPR on a study directed by Dr. John Morton, Stanford Bariatric Surgery Director. Dr. Morton considers obesity a "family disease." We typically sit around the table together, sharing and learning eating patterns. He noticed that after his patients' surgeries, family members seemed to mimic the patient's new weight loss habits such as eating smaller portions and being more physically active. Dr. Morton calls this a 'halo' effect; where good habits rub off on family members so much so that in a study of 35 of his bariatric patients one year post-surgery, family members had lost an average of 5% of their body weight. Five percent is enough to significantly enhance health and reduce risk of heart disease and diabetes. Being supportive is positive for the weight loss patient, but due to the 'halo' effect supportive family members often benefit as well (NPR, 2012).

Being supportive is positive for the weight loss patient, but due to the 'halo' effect supportive family members often benefit as well.

What Is a Healthy Body Weight?

There is no single ideal body weight; rather there is a range of healthy body weights that are acceptable for a certain height. Activity level, age, eating patterns, body composition, pregnancy or lactation, and gender, as well as genetic predisposition, can determine weight. You may find that your body seems to change, but it hovers around the same weight. The **set-point theory** postulates that the body regulates metabolism in order to maintain a certain weight, much the same as a thermostat regulates temperature. If fat stores fall below the "set" point, then the body responds by increasing appetite. If we overeat, then appetite may be reduced or all of the calories available may not be stored. Set-point gradually creeps up with poor health habits. Lowering your set-point takes patience and regular physical activity.

Body Mass Index or **BMI** is a simple way of using a ratio between height and weight to determine if your body weight falls within a healthy range. BMI is not meant to be used alone as a sole parameter of healthy body weight. Used with other measures like percentage of body fat, diet history, and exercise patterns, BMI is an easy, economical, and reproducible value to measure health risk. BMI does not take into account body type, age, sex, gender, bone density, or muscle mass. For that reason, BMI results are not good indicators for athletes, children, pregnant or lactating women, or the elderly. Fat distribution is not considered with a BMI measure. A person with abdominal obesity (an apple-shaped person) has a higher health risk than a person with fat accumulation around the hips and thighs (pear shape). Two people with different shapes could have the same BMI but not the same health risk. This is why an athlete with lean muscle and very little fat could have a high or "unhealthy" BMI. Even so, BMI is considered superior to traditional height-weight charts. See Table 7.1 to determine your BMI.

1 kg = 2.2 lbs
1 meter = 39.37 inches

A BMI over 25 is considered overweight and a BMI over 30 is considered obese. A higher BMI may indicate you are at an elevated risk for heart disease, Type 2 diabetes, and most of the conditions related to obesity. Underweight is a BMI of 18.5 or below. It is interesting to note that on the catwalks of New York and Paris, models now have to weigh in and are unable to participate if they are below a BMI of 18. It is encouraging that the fashion world is participating in increasing awareness regarding the dangers of being too thin.

$$BMI = \frac{\text{weight in kilograms}}{\text{height in meters squared}}$$

Body composition is one of the five health-related fitness components discussed in Chapter 5. A person's body composition is a measure of health, estimating the amount of fat mass relative to the lean body mass. Lean body mass is comprised of muscle, bone, and internal organs. Body composition is a more accurate indicator of overall fitness than using a person's body weight.

The ideal range for college-aged females is 18–23 percent body fat and 12–18 percent body fat for college-aged men. Essential fat is that fat which is necessary for normal physiological functioning. If a female gets below 11–13 percent of essential body fat, she typically experiences hormonal disturbances and may have menstruation cessation. Essential fat for men is around 3 percent body fat.

There are numerous different methods to determine an estimate of percent body fat. Skinfold calipers are commonly used in schools. At health fairs, bioelectrical impedance is a simple and inexpensive test to administer. The accuracy of this method is highly questionable due to variations in hydration levels in people throughout the day. The air displacement method uses pressure sensors inside an airtight

Body composition is a more accurate indicator of overall fitness than is a person's body weight.

TABLE 7.1

Body Mass Index Table

To use the table, find the appropriate height in the left-hand column labeled Height. Move across to a given weight (in pounds). The number at the top of the column is the BMI at that height and weight. Pounds have been rounded off.

BMI	19	20	21	22	23	24	25	26	27	28	29	30	31	32	33	34	35
Height (inches)								Body Weight (pounds)									
58	91	96	100	105	110	115	119	124	129	134	138	143	148	153	158	162	167
59	94	99	104	109	114	119	124	128	133	138	143	148	153	158	163	168	173
60	97	102	107	112	118	123	128	133	138	143	148	153	158	163	168	174	179
61	100	106	111	116	122	127	132	137	143	148	153	158	164	169	174	180	185
62	104	109	115	120	126	131	136	142	147	153	158	164	169	175	180	186	191
63	107	113	118	124	130	135	141	146	152	158	163	169	175	180	186	191	197
64	110	116	122	128	134	140	145	151	157	163	169	174	180	186	192	197	204
65	114	120	126	132	138	144	150	156	162	168	174	180	186	192	198	204	210
66	118	124	130	136	142	148	155	161	167	173	179	186	192	198	204	210	216
67	121	127	134	140	146	153	159	166	172	178	185	191	198	204	211	217	223
68	125	131	138	144	151	158	164	171	177	184	190	197	203	210	216	223	230
69	128	135	142	149	155	162	169	176	182	189	196	203	209	216	223	230	236
70	132	139	146	153	160	167	174	181	188	195	202	209	216	222	229	236	243
71	136	143	150	157	165	172	179	186	193	200	208	215	222	229	236	243	250
72	140	147	154	162	169	177	184	191	199	206	213	221	228	235	242	250	258
73	144	151	159	166	174	182	189	197	204	212	219	227	235	242	250	257	265
74	148	155	163	171	179	186	194	202	210	218	225	233	241	249	256	264	272
75	152	160	168	176	184	192	200	208	216	224	232	240	248	256	264	272	279
76	156	164	172	180	189	197	205	213	221	230	238	246	254	263	271	279	287

Source: www.nhlbi.gov

chamber to measure the amount of air displaced by the person inside the chamber. This is a bulky and expensive container. Hydrostatic weighing is popular with laboratories and athletic centers. The clinicians determine how much a person weighs under water. Dual energy X-ray absorptiometry (DEXA) is the preferred method in research facilities. Each method has pros and cons. If measuring a percentage of body fat in a pre- and post-comparison, it is important to replicate the same environment and to use the same technique in the post-test that was used for the pre-test.

Positive Dietary and Eating Guidelines

What is "Normal" Eating?

Normal eating is eating when you are hungry and stopping when you are "full." It is choosing foods you enjoy and eating them until your hunger is satisfied—you don't have to 'clean your plate' nor do you have stop eating simply because your original portion has been consumed—it is balance. Normal eating is giving some thought to variety and nutritional content of food but not obsessing about your next meal or the calories and fat it contains.

A person with "normal" eating habits might eat three meals a day or they may eat 4 or 5 smaller meal or choose to snack throughout the day. It is finding out what pattern works best for their lifestyle and schedule. It is knowing that it is ok to eat ice cream or cookies and it is also ok to pass them up or not eat all that is available because you know there will be more later.

Normal eating can be overeating and feeling too full sometimes, for example, after a holiday meal or a special fest and it can be under eating and wishing you had more! Normal eating is listening to your body—it avoids being overly rigid in what or how much is consumed. It is flexible.

I. Eating regularly throughout the day is an important component of keeping basal body metabolism high. Eating within the first hour after waking up and every 2 to 5 hours thereafter, depending on how hungry a person is and when food is available, will keep metabolism at its peak.

II. Carry food with you. Put a box of breakfast bars and water in your car for a quick and easy breakfast. Keep things that won't spoil like trail mix, nuts, chesses and crackers, peanut butter in your book bag.

III. It's ok to eat at night. Closer to bedtime carbohydrates are a good option because they are easy to digest and won't keep most people awake.

IV. Carbohydrates should make up about 55–60% of ingested calories. Fruits, grains, pasta, rice, and milk are examples of complex carbohydrates. About 2/3 of your plate should be filled with these types of foods.

V. Proteins represent 12–15% of daily calories. Requirements for protein are based primarily on body weight and somewhat on activity type and level. Protein is in every food except fruits and fats so, ingesting adequate protein should be easy to do.

VI. Fat should make up 25–30% of your total calories. For most college age students that is about 45 to 60 grams of fat per day. Everything you buy or consume doesn't need to be low fat or not fat. Fats are required nutrients!

VII. Listening to your body and being aware of your activity level is the best way to balance calories in with calories out. Knowing when you are hungry rather than bored and recognizing how hungry you are and when you are full and not overeating are key to this balance.

VIII. You should not have "forbidden foods." No food is a good food or a bad food. People don't gain weight because they ate a certain food, they gain weight because they regularly over eat and or don't exercise regularly. If you want a cookie, eat a cookie. Otherwise, you will obsess about it and may end up over indulging when you finally give in.

IX. Have a realistic goal about weight and focus on what your body can do rather than what it looks like. Genetics play a huge role in what a healthy weight is for you and in how you will "carry" your weight. You want to be at your body's natural "set point." This is the weight you attain when you eat and exercise and your weight doesn't fluctuate. Don't focus on a specific weight as a goal. Instead focus on developing positive eating and exercise habits.

It's All about Balance and Portion Size!

- One small chocolate chip cookie is equivalent to walking briskly for ten minutes.
- The difference between a large gourmet chocolate chip cookie and a small chocolate chip cookie could be about forty minutes of raking leaves (200 calories).
- One hour of walking at a moderate pace (twenty min/mile) uses about the same amount of energy that is in one jelly-filled doughnut (300 calories).
- A fast-food "meal" containing a double-patty cheeseburger, extra-large fries, and a twenty-four-ounce soft drink is equal to running two and one-half hours at a ten min/mile pace (1,500 calories).

(Surgeon General, 2005).

Determining Caloric Needs

Caloric needs are different for every individual. To a large degree, each person's need is determined by their current body weight and by the level of physical activity they choose to engage in. See Table 7.2 to determine your own daily caloric needs. Notice the different caloric requirements for active individuals compared to sedentary individuals.

Obesity

Overweight is defined as an excess of body weight to some height standard, or a BMI between 25 and 30. **Obesity** is a term that refers to excess fat with an accompanying loss of function and an increase in health problems, or a BMI of 30 or more. **Creeping obesity** is a gradual increase of percent body fat as activity decreases with age. This typically results in a one-half to one pound fat gain per year, with an approximate simultaneous loss of one-half pound of fat-free mass or muscle. Consider that if you overeat just 100 calories per day, you will gain one pound in a month. An extra ten pounds can sneak up on you in one year.

> Some health profesionals think **waist measurement alone** is a valuable indicator of future health risk. High risk for women is a waist measurement over 35", and over 40" for men (National Institute for Health).

TABLE 7.2

Estimated Calorie Requirements (in Kilocalories) for Each Gender and Age Group at Three Levels of Physical Activity[a]

Estimated amounts of calories needed to maintain energy balance for various gender and age groups at three different levels of physical activity. The estimates are rounded up to the nearest 200 calories and were determined using the Institute of Medicine equation.

Gender	Age (years)	Sedentary[b]	Moderately Active[c]	Active[d]
Child	2–3	1,000	1,000–1,400e	1,000–1,400e
Female	4–8	1,200	1,400–1,600	1,400–1,800
	9–13	1,600	1,600–2,000	1,800–2,200
	14–18	1,800	2,000	2,400
	19–30	2,000	2,000–2,200	2,400
	31–50	1,800	2,000	2,200
	51+	1,600	1,800	2,000–2,200
Male	4–8	1,400	1,400–1,600	1,600–2,000
	9–13	1,800	1,800–2,200	2,000–2,600
	14–18	2,200	2,400–2,800	2,800–3,200
	19–30	2,400	2,600–2,800	3,000
	31–50	2,200	2,400–2,600	2,800–3,000
	51+	2,000	2,200–2,400	2,400–2,800

Column header: Activity Level[b,c,d]

[a]These levels are based on Estimated Energy Requirements (EER) from the Institute of Medicine Dietary References Intakes macro-nutrients report, 2002, calculated by gender, age, and activity level for reference-sized individuals. "Reference-size," as determined by IOM, is based on median height and weight for ages up to age 18 years of age and median height and weight for that height to give a BMI of 21.5 for adult females and 22.5 for adult males.

[b]Sedentary means a lifestyle that includes only the light physical activity associated with typical day-to-day life.

[c]Moderately active means a lifestyle that includes physical activity equivalent to walking about 1.5 to 3 miles per day at 3 to 4 miles per hour, in addition to the light physical activity associated with typical day-to-day life.

[d]Active means a lifestyle that includes physical activity equivalent to walking more than 3 miles per day at 3 to 4 miles per hour, in addition to the light physical activity associated with typical day-to-day life.

[e]The calorie ranges shown are to accommodate needs of different ages within the group. For children and adolescents, more calories are needed at older ages. For adults, fewer calories are needed at older ages.

Source: USDA.

The bottom line is that in order to lose weight, the calories you take in must be less than the calories you expend.

Activity is the optimal way to manage current weight or successfully lose weight. The key is to exercise, maintain a healthy diet throughout your life, and avoid gaining excess weight. Participate in planned exercise as well as increased lifestyle activity. Establish support systems to help you with exercise adherence and healthy lifestyle habits.

There is only a 2 to 3 percent success rate for people who lose weight to actually maintain weight loss (Texas A&M University Human Nutrition Conference, 1998). Those who are successful are usually committed to a regular exercise routine. Exercise greatly increases the likelihood of success with a maintenance program after weight loss. Weight gain occurs with inactivity; activity is the best way to reduce the size of fat stores. Even a small weight loss, 10 percent of your weight, helps boost the basal metabolic rate, which is often suppressed when a person diets without exercise. Activity, specifically weight training, has been shown to increase a person's confidence and self-esteem, regardless of actual weight loss.

The bottom line is that in order to lose weight, the calories you take in must be less than the calories that you expend. Once you have determined what your daily caloric needs are, then estimate how many calories you are actually eating on a regular basis. The recommended weight loss is one-half to one pound per week. Losing weight faster often signifies a short-term fix, indicating that the weight loss will be difficult to maintain. Fast weight loss is often followed by fast weight gain. This is called **yo-yo dieting,** and often the weight gain is a bit more than what was initially lost. Repeated bouts of weight loss/gain like this can gradually add unwanted excess weight. Studies have shown that yo-yo dieting through the years can make it more difficult to lose weight in the future. This also adds stress to your cardiovascular system because the workload on your heart constantly fluctuates.

In order to change your lifestyle habits to lose weight, you must change your behaviors. Determine what behaviors and everyday patterns seem to sabotage your efforts. Is it the donut cart mid-morning at the office? Is it going through the drivethrough at 2 a.m. after going out with friends? Anticipate these challenges and plan ahead. Use the behavior change activity at the end of Chapter 1 in order to focus on your goal. Most likely your goal will include examining your current diet. Use your results from the diet analysis activity at the end of Chapter 5 (Nutrition). In order to lose a pound, there needs to be an approximate 3,500 caloric deficit. Eating 500 calories less per day over the week may cause that to occur, or expending an extra 500 calories per day may cause that to occur. The best option is to include both. Exercise is the key to losing weight, and is also the key to maintaining weight loss. The importance of movement illustrates how critical it is to find an activity that you enjoy so that you can embrace it for the rest of your life. Walking is the most popular activity in the United States (refer back to Figure 7.3). See Table 7.3 to determine a reasonable schedule for you to lose weight with a walking and caloric restriction program.

If you eat less food than you regularly eat as you might on a calorie restricted diet, the body's natural tendency is to slow the basal metabolic rate up to 30 percent. Exercise does the opposite—it will increase your metabolism. Staying active when trying to lose weight is critical because it burns calories, increases metabolism, and preserves muscle (see Table 7.4). Fad diets accompanied by no exercise often result in weight lost with an actual increase in percent body fat due to the loss of lean body mass. See the example below of Sally and the school dance to see the effects of fad diets versus a change in lifestyle.

TABLE 7.3 ♦ Countdown to Weight Loss

The combination of walking and cutting calories results in greater weight loss than either alone.

If you walk (minutes)	&	If you cut daily calories by	Days to Lose Weight				
			5 lb.	10 lb.	15 lb.	20 lb.	25 lb.
30		400	27	54	81	108	135
30		800	16	32	48	64	80
45		400	23	46	69	92	115
45		800	14	28	42	56	70
60		400	21	42	63	84	105
60		800	13	26	39	54	65

Source: cdc.gov

TABLE 7.4 ♦ How Much Physical Activity Do I Need?

It really depends on what your health goals are. Here are some guidelines to follow:

Goal	Physical Activity Level for Adults
Reduce the risk of chronic disease	At least 30 minutes of a moderate intensity physical activity, above usual activity, most days of the week.
Manage body weight and prevent gradual unhealthy body weight gain	Approximately 60 minutes of moderate intensity physical activity most days of the week while not exceeding calorie needs.
Maintain weight loss	At least 60 to 90 minutes of moderate intensity physical activity most days of the week while not exceeding calorie needs. Some people may need to talk to their healthcare provider before participating in this level of physical activity.

Source: cdc.gov

Sally was invited to the annual spring dance with her sweetheart. She wanted desperately to lose the "freshman 15"—the extra pounds she had put on in the previous year in order to fit into her favorite little black dress.

Option 1: Sally has two weeks before the event. She decides to crash diet. Sally drinks water and two ounces of fruit juice a day. She succeeds in losing thirteen pounds of water, fat, and muscle weight. Along the way, Sally is hungry, which leads to headaches and severe moodiness. She has little energy and sleeps through class. At the dance, Sally fits into her dress, but she is extremely fatigued. Because severe dieting can also cause an individual's blood pressure to plummet, resulting in dizziness, light-headedness, and fatigue, Sally has no energy. After the dance, Sally engages in some binge-eating behaviors. In several days' time, she gains back all the weight she lost. Unfortunately she does not gain back the muscle she lost, so her body fat increases.

Option 2: Sally can think now of next year's dance. She begins a program of walkjogging for thirty minutes five days a week. With each workout, she burns approximately 150–200 calories. Those 150 calories burned five days a week equals 750 calories each week. It will take Sally about a month to lose one pound if she does not limit her eating. If she sticks to this rather conservative program, Sally will have lost about twelve pounds in a year's time. Not only will Sally fit into her dress, but she will have established a habit and she has positively changed her lifestyle, her attitude, her body composition, her measurements, her lean body mass, and she has:

- increased her energy
- increased her muscular endurance
- increased her cardiovascular endurance
- reduced her risk of mental anxiety and depression
- improved her sleep patterns
- dealt with her stress in a positive manner
- increased her cognitive abilities
- reduced her risk of dying prematurely
- reduced her risk of heart disease, diabetes, high blood pressure
- reduced the risk of some cancers
- helped reduce risk of osteoporosis
- most likely decreased her mile time

Dieting or cutting back on calories is considered "severe" when an individual ingests fewer than 800 calories in a day. It is impossible to get all the nutrients you need with less than 1,000–1,200 calories daily. Physiological and psychological problems can result from chronic caloric restriction. Much of the weight lost with severe caloric restriction is in the form of muscle. Cardiac muscle can be weakened to the point that it is no longer able to pump blood through the body—resulting in death.

> Australian researchers have published a study that indicates our hormones can work against us to make us hungrier after weight loss, making it all the harder to maintain the new weight. The hormone **leptin** is in charge of letting the brain know how much body fat is present. With weight loss, leptin levels decrease which causes the appetite to increase and the metabolism to decrease. Even a year after the weight loss, the leptin levels were still lower. The study, although small, may shed some light on why it is so hard to lose weight. (*New York Times*, 2011)

Eating disorders are medically identifiable, potentially life-threatening mental health conditions related to obsessive eating patterns. Eating disorders are not new—descriptions of self-starvation have been found as far back as medieval times.

Even though more young men are succumbing to eating disorders each year, the mental health condition is typically thought of as a woman's disease. Unfortunately, even grade school girls can feel pressure to "fit in" or look thin. This can be very troubling and disruptive to young girls struggling to build a positive body image.

Typically, a person with an eating disorder seeks perfection and control over their life. Both anorexics and bulimics tend to suffer from low self-esteem and depression. They often have a conflict between a desire for perfection and feelings of personal inadequacy. Such persons typically have a distorted view of themselves, in that when they look into a mirror, they see themselves differently than others see them. Narcissism, or excessive vanity, can be linked to both anorexia and bulimia. (see Figure 7.4)

Eating disorders are often accompanied by other psychiatric disorders, such as depression, substance abuse, or anxiety disorders. Eating disorders are very serious and may be life-threatening due to the fact that individuals suffering from these diseases can experience serious heart conditions and/or kidney failure—both of which can result in death. Therefore, it is critically important that eating disorders are recognized as real and treatable diseases.

Body Image

The media, advertising, and the fashion industry portray women, as well as men, as thin (or fit), beautiful, and youthful. This is an ideal that is difficult to attain and nearly impossible to maintain. It is important to note that glamorous magazine cover models' pictures are airbrushed with imperfections deleted, so that the final product is an almost perfect unachievable image. It wasn't too long ago that the ideal for a woman's body was "fat is where it's at" instead of the current preoccupation with "thin is in." Many of the great masters' paintings portray the female image as having desirable traits such as soft, round, and fleshy bodies. The 50's had curvy movie stars such as Marilyn Monroe. In the 60's, the Twiggy look was in. In the 70's, it was Farrah Fawcett Fit. In the 80's, it was the fit look with Cindy Crawford. In the 90's, Kate Moss exemplified the gaunt heroin look. As Americans grow in size, the current unobtainable look is very thin. The Body Mass Index of Playboy models and Miss America contestants lowered significantly from the 1970's to the 1990's. Even wealthy celebrities who can afford personal chefs, trainers, registered dieticians, and top consultants sometimes battle with their weight.

Anorexics and bulimics tend to suffer from low self-esteem and depression and typically have a distorted view of themselves.

A person's body image is how a person sees him or herself in his or her mind. Body image is affected by a person's attitudes and beliefs, as well as by outside influences such as family, social pressures, and the media. It is important to have good perspective. If you are an "apple" and naturally carry excess fat in your abdominal area, when you gain weight you will become a larger "apple" or you may lose weight and become a smaller "apple." The same is true for "pear" shaped individuals who carry their excess fat in

Biological
Dieting
Obesity/overweight/pubertal weight gain

Psychological
Body image/dissatisfaction/distortions
Low self-esteem
Obsessive-compulsive symptoms
Childhood sexual abuse

Family
Parental attitudes and behaviors
Parental comments regarding appearance
Eating-disordered mothers
Misinformation about ideal weight

Sociocultural
Peer pressure regarding weight/eating
Media: TV, magazines
Distorted images: toys
Elite athletes as at-risk groups

FIGURE 7.4
Major Risk Factors for Eating Disorders

Source: White, Jane. "The Prevention of Eating Disorders: A Review of the Research on Risk Factors with Implications for Practice." *Journal of Child and Adolescent Psychiatric Nursing,* Vol. 13, No. 2, April 2000.

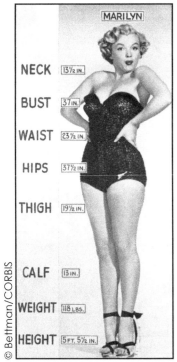 

Body image is affected by a person's attitudes and beliefs, as well as outside influences such as family, social pressures, and the media.

their hips and buttocks. You cannot become a different shape, nor can you diet down to a thin waif. Accept yourself for who you are; then work on behavior changes to become healthier. The fringe benefit to becoming healthier is that you will most likely fit better in your jeans. You will also feel better while you reduce your risk of heart disease and diabetes.

Anorexia Nervosa

Anorexia nervosa is a state of starvation and emaciation, usually resulting from severe dieting and excessive exercise. An anorexic will literally stop eating in an effort to control body size.

Most, if not all, anorexic individuals suffer from an extremely distorted body image. People with this disease look in a mirror and see themselves as overweight or fat even when they have become dangerously thin.

Major weight loss is the most visible and the most common symptom of anorexia. Anorexic individuals often develop unusual eating habits, such as avoiding food or meals, picking out a few "acceptable" foods and eating them in small quantities, or carefully weighing and portioning foods. Other common symptoms of this disease include absent menstruation, dry skin, excessive hair on the skin, and thinning of scalp hair. Gastrointestinal problems and orthopedic problems resulting from excessive exercise are also specific to this illness.

Ways to Love Your Body

- Become aware of what your body does each day, as the instrument of your life, not just an ornament for others.
- Think of your body as a tool. Create a list of all the things you can do with this body.
- Walk with your head held high, supported by pride and confidence in yourself as a person.
- Do something that will let you enjoy your body. Stretch, dance, walk, sing, take a bubble bath, get a massage.
- Wear comfortable styles that you really like and feel good in.
- Decide what you would rather do with the hours you waste every day criticizing your body.
- Describe ten positive things about yourself without mentioning your appearance.
- Say to yourself "Life is too short to waste my time hating my body this way."
- Don't let your weight or shape keep you from doing things you enjoy.
- Create a list of people who have contributed to your life, your community, the world. Was their appearance important to their success and accomplishment? If not, why should yours be?
- If you had only one year to live, how important would your body image and appearance be?

By Margo Maine, Ph.D. and Eating Disorders' Awareness and Prevention

Anorexic individuals can lose between 15 and 60 percent of their normal body weight, putting their body and their health in severe jeopardy. The medical problems associated with anorexia are numerous and serious. Starvation damages bones, organs, muscles, the immune system, the digestive system, and the nervous system.

Between 5 and 20 percent of anorexics die due to suicide or other medical complications. Heart disease is the most common medical cause of death for people with severe anorexia.

Long-term irregular or absent menstruation can cause sterility or bone loss. Severe anorexics also suffer nerve damage and may experience seizures. Anemia and gastrointestinal problems are also common to individuals suffering from this illness.

The most severe complication and the most devastating result of anorexia is death.

Bulimia Nervosa

Bulimia nervosa is a process of bingeing and purging. This disorder is more common than anorexia nervosa. The purging is an attempt to control body weight, though bulimics seldom starve themselves as anorexics do. They have an intense fear of becoming overweight, and usually have episodes of secretive binge eating, followed by purging, frequent weight variations, and the inability to stop eating voluntarily. Bulimics often feel hunger, overeat, and then purge to rid themselves of the guilt of overeating.

Bulimic individuals are often secretive and discreet and are, therefore, often hard to identify. Typically, they have a preoccupation with food, fluctuating between fantasies of food and guilt due to overeating. Symptoms of bulimia can include cuts and calluses on the finger joints from a person sticking their fingers or hand down their throat to induce vomiting, broken blood vessels around the eyes from the strain of vomiting, and damage to tooth enamel from stomach acid.

Because purging through vomiting, the abuse of laxatives, or some other compensatory behavior typically follows a binge, bulimics usually weigh within the normal range for their weight and height. However, like individuals with anorexia, they often have a distorted body image and fear gaining weight, want to lose weight, and are intensely dissatisfied with their bodies.

While it is commonly thought that the medical problems resulting from bulimia are not as severe as those resulting from anorexia, the complications are numerous and serious. The medical problems associated with bulimia include tooth erosion, cavities, and gum problems due to the acid in vomit. Abdominal bloating is common in bulimic individuals. The purging process can leave a person dehydrated and with very low potassium levels, which can cause weakness and paralysis. Some of the more severe problems a bulimic can suffer are reproductive problems and heart damage, due to the lack of minerals in the body.

Binge-Eating Disorder

People with binge-eating disorder typically experience frequent (at least two days a week) episodes of out-of-control eating. Binge-eating episodes are associated with at least three of the following characteristics: eating much more rapidly than normal; eating until an individual is uncomfortably full; eating large quantities of food even when not hungry; eating alone to hide the quantity of food being ingested; feeling disgusted, depressed, or guilty after overeating. Not purging their bodies of the excessive calories they have consumed is the characteristic that separates individuals with binge-eating disorder from those with bulimia. Therefore, individuals suffering from this disease are typically overweight for their height and weight.

Fear of Obesity

Fear of obesity is an over-concern with thinness. It is less severe than anorexia, but can also have negative health consequences. This condition is often seen in achievement-oriented teenagers who seek to restrict their weight due to a fear of becoming obese. This condition can be a precursor to anorexia or bulimia if it is not detected and treated early.

Activity Nervosa

Activity nervosa is a condition in which the individual suffers from the ever-present compulsion to exercise, regardless of illness or injury. The desire to exercise excessively may result in poor performance in other areas of that individual's life due to the resulting fatigue, weakness, and unhealthy body weight.

Female Athlete Triad In 1991, a team was formed by the American College of Sports Medicine to educate, initiate a change, and focus on the medical management of a triad of female disorders that included disordered eating, amenorrhea, and osteoporosis (ACSM, 1991). A triangle is used to depict these disorders because the three are interlinked. Disordered eating behaviors result in weight loss and subsequent loss of body fat that halts menstruation. When amenorrhea occurs, calcium is lost and a decline in bone mass occurs. This in turn causes osteoporosis and can easily result in stress fractures. Many times the inactivity necessary to allow stress fractures to heal causes depression that often leads an individual back into disordered eating behaviors, and the cycle continues (see Figure 7.5).

Who Is at Risk?

By far, more women than men succumb to eating disorders; however, the incidence of eating disorders in men is believed to be very underreported.

It is estimated that one in every hundred teenage girls is anorexic. Anorexia usually occurs in adolescent women (90 percent of all reported cases), although all age groups can be affected. It is estimated that one in every five college-bound females is bulimic.

Individuals living in economically developed nations, such as the United States, are much more likely to suffer from an eating disorder, due to the dual factors of an abundance of available food and external, societal pressure. College campuses have a higher incidence of people with eating disorders, and upper-middle-class

FIGURE 7.5
The Female Athlete Triad

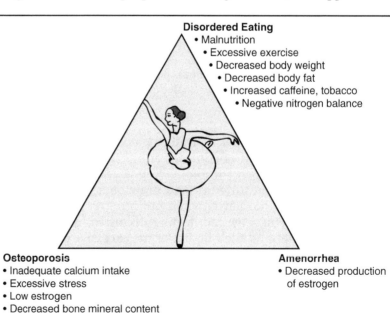

women who are extremely self-critical are also more likely to become anorexic. Being aware of the groups at risk can be a large step toward prevention.

Activities such as dance and dance team, gymnastics, figure skating, track, and cheerleading tend to have higher instances of eating disorders. An estimate of people suffering from anorexia and bulimia within these populations is 15–60 percent. Male wrestlers and body builders are also at risk due to the unsafe practice of attempting to shed pounds quickly in an attempt to "make weight" before a competition.

Causes of Eating Disorders

The causes of anorexia and bulimia are numerous and complex. Cultural factors, family pressure, psychological factors, emotional disorders, and chemical imbalances can all contribute to eating disorders.

Forty to 80 percent of anorexics suffer from depression, as reduced levels of chemical neurotransmitters in the brain have been found in victims suffering from both eating disorders and depression. Links between hunger and depression have been discovered through research, which contributes to the depression a person with an eating disorder may feel.

For some bulimics, seasonality can adversely affect them, causing the disorder to worsen during the dark, winter months. Another startling statistic is that the onset of anorexia appears to peak in May, which is also the peak month for suicides.

Family factors are also critical. One study showed that 40 percent of all 9- to 10-year-old girls were trying to lose weight, many at the encouragement of their mothers. Mothers of anorexics are often overinvolved in their child's life, while mothers of bulimics are many times critical and detached.

It is clear that many people who suffer from eating disorders do not have a healthy body image. From an early age, there is enormous pressure in our culture from society, family, friends, the media, and often from one's self to achieve the unachievable and unnecessary "perfect" body. A woman's self-worth is too often associated with other people's opinions, which in many cases put unrealistic emphasis on physical attractiveness.

Ways to Help

The best course of action for a person who suspects they know someone with an eating disorder is to be patient, supportive, and not judge the individual. Learn what you can about the problem by consulting an eating disorder clinic

Forty to 80 percent of anorexics suffer from depression.

Guidelines for Helping a Friend with an Eating Disorder

DO:
- listen with understanding
- appreciate the openness and trust in sharing with you his/her distress
- share your own struggles, be open and real
- learn more about eating disorders
- give support and be available
- give hope that with help and with patience he/she can free themselves from this disorder
- give your friend a list of resources for help

DO NOT:
- tell your friend he/she is crazy
- blame him/her
- gossip about your friend
- follow him/her around to check their eating or purging behavior
- ignore your friend
- reject him/her
- tell him/her to quit this ridiculous behavior
- feel compelled to solve their problem
- make excess comments about being thin

DO heed the signs. Anorexic behavior includes extreme weight loss (often emaciation), obsessive dieting, and distorted body perception (a thin person thinks he/she is fat when they are not). Clues of bulimia are more subtle. Your friend may eat a great deal of food, then rush to the bathroom. She/he may hide laxatives or speak outright about the "magic method" of having the cake and not gaining weight. Anorexics and bulimics tend to be preoccupied with food and many have specific rituals tied to their eating patterns.

DO approach your friend gently, but persistently. Explain that you're worried; listen sympathetically. Don't expect your friend to admit he/she has a problem right away. The first step is realizing there is a problem; therefore, it is important to help your friend realize this.

DO focus on unhappiness as the reason your friend could benefit from help. Point out how anxious or unhappy he/she has been lately, and emphasize that it does not have to be that way.

DO be supportive, but do not try to analyze or interpret their problem. Being supportive is the most important thing you can do. Show your friend you believe in him/her—it will make a difference in recovery.

DO talk to someone about your own emotions if you feel the need. An objective outsider can emphasize the fact that you are not responsible for your friend; you can only try to help that person help him/herself.

DO be yourself. Be honest in sharing your feelings: i.e., "It's hard for me to watch you destroy yourself."

DO give non-judgmental feedback. For example, "We haven't gone to lunch together in a while, is something wrong?" instead of "You haven't gone to eat with me in a while, do you have a problem?"

DO cooperate with your friend if he/she asks you to keep certain foods out of common storage areas. This may help prevent a binge on such foods.

DON'T keep the "secret" from the family when your friend's health and thinking are impaired.

DON'T forget that denial is a form of selective "deafness."

DON'T be deceived by the excuse: "It's not really bad. I can control myself."

DON'T focus on your friend's weight or appearance. Focus on your concern about his/her health and well-being.

DON'T change your eating habits when you're around your friend. Your "normal" eating is an example to your friend of a more healthy relationship with food.

Source: Student Health Services, Texas A&M University. Guidelines for Helping a Friend with an Eating Disorder, 2002.

or counseling center (common on college campuses), and offer to help the ill person seek professional help.

Often, individuals suffering from an eating disorder do not realize or will not admit that they are ill. For this reason, seeking help or continuing/completing treatment for the disorder is often difficult.

Medical treatment is often necessary for eating disorders. However, it is extremely encouraging to note that eating disorders can be treated and a healthy weight and relationship with food can be restored. Because of the complexity of eating disorders, the best and most successful treatment is usually a combination of counseling, family therapy, cognitive behavior therapy, nutritional therapy, support groups, and drug therapy. Treatment, many times, includes a hospital stay and is usually resisted by the patient. Support for the anorexic or bulimic person by friends and family and the realization of the severity of the problem is critical to successful treatment of the illness.

Eating a healthy diet with fruits, vegetables, whole grains, quality proteins, and unsaturated fats, as well as restricting refined white flour and sugar, is the best method for obtaining an adequate supply of nutrients in your diet. Dietary supplements are popular and provide a means for delivering these nutrients in a more convenient, but often less effective, form. It is a good idea to check with a health professional before beginning supplemention.

Guidelines for a Successful Weight Loss Program

The American College of Sports Medicine has put together the following eleven guidelines in an effort to help individuals recognize potentially successful weight loss programs and avoid unsound or dangerous weight loss programs.

1. Prolonged fasting and diet programs that severely restrict caloric intake are scientifically unsound and can be medically dangerous.
2. Fasting and diet programs that severely restrict caloric intake result in the loss of large amounts of water, electrolytes, minerals, glycogen stores, and other fat-free tissues, but with minimal amounts of fat loss.
3. Mild caloric restriction (500–1,000 calories less than usual per day) results in smaller loss of water, electrolytes, minerals, and other fat-free tissues and is less likely to result in malnutrition.
4. Dynamic exercise of large muscle groups helps to maintain fat-free tissue, including lean muscle mass and bone density, and can result in a loss of body weight (primarily body fat).
5. A nutritionally sound diet resulting in mild caloric intake restrictions, coupled with an endurance exercise program, along with behavior modification of existing eating habits, is recommended for weight reduction. The rate of weight loss should never exceed two pounds per week.
6. To maintain proper weight control and optimal body fat levels, a lifetime commitment to proper eating habits and regular physical activity is required.
7. A successful weight loss plan can be followed anywhere—at home, work, restaurants, parties, and so on.
8. For a plan to be successful, the emphasis must be on portion size.
9. Successful weight loss plans incorporate a wide variety of nutritious foods that are easily accessible in the supermarket.
10. A weight loss plan must not be too costly if it is to be successful.
11. The most essential aspect of a weight loss program is that it can be followed for the rest of an individual's life.

The most essential aspect of a weight loss program is that it can be followed for the rest of an individual's life.

Taken in concentrations higher than the recommended daily allowance, some nutrients have undesirable side effects, and some are even toxic.

Unfortunately some individuals think that if one pill is good, two or more must be better. In most cases this is simply not true. Sometimes, an excess of nutrients can be detrimental as is the case with fat-soluble vitamins A, D, E and K. Megadoses of many vitamins, minerals, and other supplements can cause kidney and liver damage and interact adversely with other supplements, herbs, and drugs. It is possible to interfere with absorption of other nutrients when ingesting megadoses of some supplements. In general, consuming more than the RDA of vitamins and minerals is discouraged. A registered dietician or physician may prescribe supplements for some populations or certain medical conditions (for instance, people who are pregnant, have anemia, are elderly, or are certain types of vegetarians).

Energy drinks and nutrition bars have gained popularity with the public and college students in particular over the past several years. It is easy to see why—they are relatively inexpensive, easy and convenient to obtain and consume, and are reported to increase energy and alertness, boost academic and athletic performance, and offer extra nutrition (supplementation).

The fast-paced lifestyle of many college students makes the convenience of energy-on-the-go by way of drinks and bars especially appealing.

While consuming an occasional energy drink is okay for most individuals and eating a protein/nutrition bar for breakfast is a better option than skipping breakfast, drinking 2 or 3 energy drinks daily and/or consuming a substantial amount of daily calories from bars is poor nutritional planning and will take its toll on your long-term health.

Many of the nutrition bars and caffeine drinks contain excessive amounts of caffeine, sugar, and calories. Excessive sugar and calories will result in unwanted weight gain and poor dental health and large amounts of caffeine can cause an individual to be "jittery," have headaches, an upset stomach, high blood pressure, or even irregular heartbeats.

Nutrition bars and energy drinks, in general, do not make good meal replacements. They lack the variety of nutrients that are needed for brain development and function and good physical performance.

Appetite suppressants claim to help diminish a person's appetite, cut cravings, and increase overall energy and possibly increase metabolism. Over time this may result in weight loss, but consider the negative side effects. Appetite suppressants often contain high levels of caffeine, guarana, or Ma Huang (an herbal form of ephedra) that can cause hypertension, cardiac arrhythmia, myocardial infarction, and/or stroke that can and has led to premature death of the person consuming this type of dietary supplement. Examples of commonly used supplements include Hydroxycut, Xenadrine-EFX, and Trim Spa.

In 1997, fenfluramine and dexfenfluramine (fen/phen) were withdrawn from the market due to a link to the development of a heart valve problem. Serious illness and, in some cases, death occurred. Even FDA-approved drugs need to be used cautiously. In the 1960's and the 1970's amphetamines were often prescribed for weight loss, and in the 1920's weight loss pills were found to contain tapeworm eggs (Hales, 2007). Diet aids and supplements that have

Tips to Weight Loss Success

Be a SMART Planner!

SMART means being **S**pecific, **M**easured, **A**ppropriate, **R**ealistic, and **T**ime-bound about what you plan to achieve. For example, if your goal is to increase your physical activity, then write down the type of activity you plan to do, how many times you can realistically do it each week, and for how long each time. Start with small, short, and easier goals, and work your way up.

Make Yourself an Offer You Can't Refuse

Before starting to reach your next goal, offer yourself a promise like this, "If I reach my goal this (day, week, month), I will treat myself to a well-deserved (fill in a reward here, but not a food reward)." Think of something you want, such as an afternoon off, a massage, a movie, or even a deposit toward a larger reward. Be creative, set up rewards for yourself frequently, and make sure you give them to yourself when you reach your goal.

Balance Your (Food) Checkbook

Keep a diary of what you eat and how much physical activity you get each day. Then, at the end of each week, record your weight in the same diary. You and your healthcare provider can use this information to adjust your eating and physical activity plan to find the best way to reach your goal.

Keep an Eye on the Size!

Did you know that we eat most of what is on our plate, no matter what the size of the plate? When at home, try using smaller plates; they will help you take smaller portions. When eating out, share an entrée! Studies show that portions today are often super-sized—enough for two or more people to share.

How Much Is Enough Activity?

You need to get at least 30 minutes of moderate physical activity per day, most days of the week to help burn up extra calories. But give yourself credit for the activities that you're already doing. Common activities such as climbing stairs, pushing a stroller, gardening, and walking all count as physical activity. Just make sure you do enough of them.

Am I Full Yet?

The question may take longer to answer than you think. It takes 15 minutes or more for the message that we're full to get from our stomachs to our brains. So take a few minutes before digging in for that next helping. Having trouble feeling full? Eight glasses (8 ounces each) or more of water or other non-caloric beverages daily fills you up and keeps you refreshed. Also, vegetables and fruits can help you feel fuller, especially when eaten raw.

Source: www.nhlbi.nih.gov/health/public/heart/obesity/lose_wt/index.htm.

gone through rigorous testing and been approved by the FDA still need to be approached with common sense and caution.

Metabolism boosters are various supplements that speed up or boost an individual's basal metabolism. Most of these types of products claim to act in a way that increases the building of lean muscle mass. Examples of such supplements are creatine phosphate, chromium picolinate, and HMB. Megathin,

Microlean, and Metabolife (now banned by the FDA) have appetite suppressants as well as metabolism boosters. Long-term effectiveness and safety of these types of supplements are unknown.

There are two types of weight loss programs, clinical and nonclinical. A clinical program is typically offered in a healthcare facility with a team of licensed health professionals. The following are examples of nonclinical programs: Jenny Craig, Nutrasystem, Weight Watchers, and Slimfast. Jenny Craig and Nutrasystem sell prepackaged foods which are convenient, but typically are higher in cost. Weight Watchers teaches participants the value of foods on a point system. The positive aspect of Weight Watchers is that participants are taught how to choose, shop, and prepare foods. Group support is essential and highly recommended after the goal weight is attained. Weight loss maintenance is one of the most difficult aspects of dieting, so support from others is often a significant help. Two other programs that offer group support are TOPS (Take Off Pounds Sensibly) and Overeaters Anonymous (OA). Programs such as Slimfast replace one or two meals with shakes or bars. Unless a person learns to make wise choices on their own, it is doubtful that weight loss will be maintained.

Low-carbohydrate high-protein diets are popular. The Zone, Atkins, Sugar-Busters, Protein Power, and the South Beach Diet are all examples of this type of diet. Although there are some variation in the diets, all advocate severely limiting carbohydrate foods such as pasta, potatoes, bread, cereals, juices, sweets, and even fruits and vegetables. Protein-rich foods are plentiful. Steak, ham, bacon, eggs, fish, chicken, and cheese are allowed to be eaten in unlimited quantities for some diets.

The basis for the low-carbohydrate diet is that during digestion, carbohydrates are converted into glucose, which serves as fuel for every cell in a person's body. When blood glucose levels begin to rise, insulin, the hormone that allows the entry of glucose into the cells, is released. This process lowers the level of glucose in the bloodstream. If the available glucose is not rapidly used for normal cellular functions or physical activity, the glucose is converted to and stored as body fat. Individuals who support this type of diet believe that if a person eats fewer carbohydrates and more protein, they will produce less insulin, and as insulin levels drop, the body will look to its own fat stores to meet energy needs.

While research has shown that people participating in low-carbohydrate high-protein diets do initially lose weight more rapidly than an individual who maintains a more nutritionally balanced diet but decreases calorie intake and increases physical activity, these same studies show that at the one-year mark weight loss for many of the dieters in both groups was not significantly different. Low carbohydrate diets also have the potential to result in the loss of B vitamins, calcium, and potassium. This can lead to osteoporosis, constipation, bad breath, and fatigue. Before starting on this type of diet, consider that because the diet is high in protein, and therefore high in fat, it may carry an increased risk for heart disease.

Most days, you're mindful of what you eat. Some days, though, if you're out of your routine, you may be tempted to go overboard on high-calorie, fatty foods. Can one big meal at your favorite restaurant be so bad? Research shows that it may.

Just one bad meal choice can have immediate effects that are most dangerous for people who are at risk for or already have heart disease. After just one large fatty meal, you may have:

- **Stiffer arteries, reduced blood flow.** Blood vessel dilation and expansion is hampered by a big high-fat meal. Large meal digestion forces your heart to work harder, increasing the heart rate, to meet the needs of the digestive tract. Sometimes people who have cardiovascular disease and eat a large meal before exercising will suffer angina or possibly a heart attack.

Fad diets
Fad diets are risky because they...
- tend to be very low in calories
- are limited to a few foods, limiting key nutrients and minerals
- produce only short-term, rapid weight loss—not long-term weight management
- ignore the importance of physical activity in healthy weight loss
- increase risks for certain diseases or health complications
- take the pleasure and fun out of eating
- alter metabolism, making it easier to regain the weight after the diet has ceased

Diets That Don't Work
1. **"Magical"/Same Foods Diets** (i.e., grapefruit, cabbage soup, Subway diet)
   - *Pros*—usually the single food is a nutritious food
   - *Cons*—too few calories, risk of overeating, lacking of specific nutrients, lack variety, do not teach healthy eating habits, do not encourage exercise
2. **High-Protein Diets** (i.e., Atkins)
   - *Pros*—weight loss does occur
   - *Cons*—high in saturated fat and cholesterol increasing risk for heart disease; high protein puts strain on liver and kidneys; lacks vitamins, minerals, complex carbohydrates, and fiber; weight loss is water weight, not fat; lack of carbs causes a condition called ketosis with symptoms of nausea, weakness, and dehydration
3. **Liquid Diets** (i.e., Slimfast)
   - **Pros**—drinks have vitamins, minerals, and high-quality protein
   - **Cons**—do not teach new ways of eating, no long-term weight loss, very low in calories
4. **Gimmicks, Gadgets, and Other "Miracles"**
   - **Pros**—none
   - **Cons**—may be harmful, expensive, do not teach healthy eating, do not encourage exercise

Effective Weight Loss Questionnaire
1. ***Could you follow the diet for the rest of your life?*** Good health and permanent weight loss require a lifestyle change, not just a temporary modification.
2. ***Does the diet promise quick results?*** If so, you're probably losing water and lean muscle tissue. Weight loss of one-half to two pounds a week is safe and will more likely be kept off.
3. ***Does the diet accommodate your lifestyle?*** Any diet that does not allow much freedom or flexibility is less likely to be followed permanently.
4. ***Is the diet very low in calories?*** Any diet that is below 1,200 calories per day could be dangerous. You may not be getting enough energy and nutrients. You may feel deprived and frustrated, both physically and mentally. In the long run, metabolism slows in order to conserve energy with very low calorie diets.
5. ***Does the diet eliminate or restrict certain food groups?*** Many diets leave out one or more food groups. Restricting a type of food may result in elimination of essential nutrients in the diet causing health risks. A balanced diet modeled after the Food Guide Pyramid and including a variety of foods should be followed.
6. ***Does the diet call for unusual items or require you to go to a specialty store?*** Unusual foods or supplements may be very costly and hard to obtain. They may also contain dangerous ingredients that are not regulated by the Food and Drug Administration.
7. ***Will someone make money on the diet?*** If yes, BEWARE! The diet could be a quick way for someone to make a lot of money.
8. ***Is the author or supplier reputable?*** To check for validity and credibility of a book, diet, or supplement, view the list of references provided and check the credentials of the author.

- **Higher blood pressure.** Norepinephrine is a stress hormone that can be released by a big meal. This hormone raises blood pressure and heart rate.
- **High triglycerides.** Triglycerides are fats in the blood, and any meal elevates levels. A super-size meal loaded with fat or refined carbohydrates boosts levels most and keeps them elevated for up to twelve hours.
- **Blood sugar effects.** A diabetic's ability to process glucose can be impaired by a large meal.
- **Heartburn.** The bigger the meal, the more you'll suffer from gastric reflux if you are susceptible to heartburn.

There have been some studies that indicate taking high doses of vitamin C and E before eating a fatty meal help maintain arterial blood flow. However, it is not conclusive that these vitamins—or any others—can protect your heart in either the short- or long-run. Another study determined that young, healthy people who ate a super-size meal (1,000 calories' worth) and then walked quickly for 45 minutes had the benefit of having their arteries' ability to dilate restored. That being said, exercise is not going to wipe out all the bad effects of overeating. It's also worth noting that exercising after eating a big meal may cause more problems for those of us who are older or have less than perfect health.

Overdoing it at the buffet table on occasion shouldn't be a problem if you're healthy. For those who have conditions such as diabetes, high cholesterol or blood pressure, heart disease, or if you smoke and/or are very overweight, ordering your meals super-sized is never a good idea. Plan before you go to a party or out to that favorite restaurant so you're not starving when you get there. Fill your plate with foods high in water content that will serve to fill you up—stick with mainly fruits and vegetables. Take your time eating and enjoy the food so you notice when your body tells your brain that you're full. Once you're finished, take a nice walk.

Fad Diets

Each year billions of dollars are spent in the weight loss industry. Unfortunately, many of these dollars are spent on diet plans that are unhealthy, cannot be maintained long term, or simply do not work. The lure of a quick and easy way to "melt away" the pounds is too tempting for many individuals, and although the diets are more times than not ineffective in the long term, weight loss hopefuls are willing to give almost anything a chance. To avoid the pitfalls of an unsuccessful, unreliable, or even dangerous weight loss plan, one should always take the time to check out as much factual information as possible from a variety of sources. The boxed information on page 200 contains a list of some fad diets that are currently popular and the theories and possible shortcomings within these diets.

Although they are in the minority, many people struggle to gain weight. It is important for individuals who feel that they are "too thin" to recognize that there are healthy and unhealthy ways of accomplishing their goal of weight gain. Overeating all types of foods will simply result in an increase in body fat.

Adding lean body mass through strength training and a slight caloric increase is a much healthier alternative. Chapter 5 outlines the benefits of muscular fitness, as well as presents helpful guidelines for beginning a strength training regimen. Nutrition information on page 201 of Chapter 5 identifies ways to add calories in an appropriate portion and manner to maximize their benefits.

Healthy Food Shopping

The National Heart, Lung, and Blood Institute Obesity Guidelines (www.nhlbi.nih. gov/health/public/heart/obesity/lose_wt/shop.htm) list the following suggestions to help individuals prepare healthier home cooked meals in shorter periods of time. They recommend reading labels while shopping—paying particular attention to serving sizes and the number of servings in the container. Comparing the total number of calories in similar products and choosing the product containing the lower number of total calories will result in a healthier meal. Finally, make cooking at home easier and healthier by shopping for quick, low-fat food items and filling kitchen cabinets with a supply of lower calorie staples such as:

- fat-free or low-fat milk, yogurt, cheese, and cottage cheese
- light or diet margarine
- sandwich breads, bagels, pita bread, English muffins, low-fat tortillas
- plain cereal, dry or cooked
- rice and pastas, dry beans, and peas
- fresh, frozen, canned fruits in light syrup or juice
- fresh, frozen, or no-salt-added canned vegetables
- low-fat or no-fat salad dressings and sandwich spreads
- mustard and ketchup
- jam, jelly, or honey
- salsa, herbs, and spices

Fast Foods/Eating Out

People eat more meals outside the home than ever before. Due to their quick service and relatively low food prices, fast-food chains are the most frequent source for meals eaten away from home. Each day 50 million people line up inside or drive through outside service lanes of one of the over 160,000 fast-food establishments in this country.

When meals are prepared with speed and convenience as the primary focus, good nutrition will, in most cases, suffer. A great majority of fast foods are high in fat, calories, and salt, and low in many of the essential nutrients and dietary fiber.

However, fast food does not have to mean "junk food." While it may take a little more thought and discretion, quick and healthy alternatives do exist.

Everyday choices make a difference!

Depending on what ingredients are used and how the food is prepared, fast foods served in restaurants can be healthy. Most restaurants have nutritional information about the foods they serve posted in the dining area or on their menus. By taking a couple of extra minutes to think about their best and most nutritious options, an individual can make dining out more nutritious, filling, and healthy.

Another pitfall of eating meals prepared away from home is the quality of food an individual is served. In an effort to be competitive, many restaurants serve well beyond an adequate portion size. To control portion sizes when eating out, order from the senior citizens or kids menus, share the entrée with a friend, or take part of the food home for a later meal.

Another way to eat healthy when dining out is to select foods that are steamed, broiled, baked, roasted, or poached rather than foods that are fried or grilled. Asking if the restaurant will trim visible fat off the meat or serve butter, sauces, or dressings "on the side" is another way to ensure a healthy and tasty meal.

Our food choices and habits, our exercise habits, and our genetic make-up all play a role in our ability to maintain a healthy weight. Managing weight is brought about most successfully by a lifestyle choice, not a short-term diet. Even if both of your parents are overweight and you didn't have good nutrition emphasized when you were young, you can make wise choices for yourself today.

Make small positive changes to encourage healthful behaviors. If you are overweight or obese, even a small 5 percent to 10 percent weight loss can have a favorable effect on your overall health risk.

Antioxidants are compounds that aid each cell in the body facing an ongoing barrage of damage resulting from daily oxygen exposure, environmental pollution, chemicals and pesticides, additives in processed foods, stress hormones, and sun radiation. Studies continue to show the ability of antioxidants to

Tips for Healthy Dining at Home and Away

When restaurant eating, ask for a to-go box right away and put half of your order in it as soon as it comes.

Order a dinner salad, but share the entree with a friend when eating out.

Order water with your meal rather than a soda—save money and calories.

Opt for your traditional foods made in a "light" version. Use a smaller plate to encourage smaller portions. Drink a glass of water before your meal.

Eat your salad first.

Eat slowly. Put your eating utensil down and enjoy your meal or converse between bites.

Eat breakfast regularly.

Always have healthy snacks in your backpack or briefcase or car.

Try eating fruits for dessert.

Trim visible fat off meat and take skin off poultry before cooking.

Try to use less refined sugar and processed flour in food preparation. Try whole wheat flour or unbleached white flour.

Wean yourself off sodas—or try to cut way down on your intake.

Read labels: minimize corn syrup, trans fat, coconut oil, palm kernel oil, and cocoa butter.

Use added fats like salad dressings minimally—try dipping your fork into the dressing before skewering the lettuce.

Avoid supersizing your meal.

suppress cell deterioration and to "slow" the aging process. Realizing the potential power of these substances should encourage Americans to take action by eating at least five servings of a wide variety of fruits and vegetables each day (see Table 7.5 on page 203).

There are many proven health benefits of antioxidants. Vitamin C speeds the healing process, helps prevent infection, and prevents scurvy. Vitamin E helps prevent heart disease by stopping the oxidation of low-density lipoprotein (the harmful form of cholesterol); strengthens the immune system; and may play a role in the prevention of Alzheimer's disease, cataracts, and some forms of cancer, providing further proof of the benefits of antioxidants.

Adequate amounts of vitamins, minerals, and antioxidants are crucial to good overall health.

The ability to relate and interact with others is important to a person's overall sense of well-being.

Organic foods are foods that are grown without the use of pesticides. These chemical-free foods are much more difficult to grow because they are more vulnerable to disease and pests. Thus, they are not "high yield" crops. Due to the fact that they are less common, and harder to grow successfully, they are more expensive. Whether the expense is justified by the improved nutritional quality and overall health benefits is yet to be determined.

1 large egg = muffin

Functional foods are foods that have benefits that go above and beyond basic nutrition. A person's overall health can be greatly affected by the food choices they make. Functional benefits of foods that have been consumed for decades are being discovered and new foods are being developed for their helpful dietary components.

Handful of rubber bands = ½ cup pasta

TABLE 7.5 ♦ Antioxidants and Their Primary Food Sources

Vitamin A	Fortified milk; egg yolk; cheese; liver; butter; fish oil; dark green, yellow, and orange vegetables and fruits
Vitamin C	Papaya, cantaloupe, melons, citrus fruits, grapefruit, strawberries, raspberries, kiwi, cauliflower, tomatoes, dark green vegetables, green and red peppers, asparagus, broccoli, cabbage, collard greens, orange juice, and tomato juice
Vitamin E	Vegetable oils, nuts and seeds, dried beans, egg yolk, green leafy vegetables, sweet potatoes, wheat germ, 100 percent whole wheat bread, 100 percent whole grain cereal, oatmeal, mayonnaise
Carotenoids	Sweet potatoes, carrots, squash, tomatoes, asparagus, broccoli, spinach, romaine lettuce, mango, cantaloupe, pumpkin, apricots, peaches, papaya
Flavenoids	Purple grapes, wine, apples, berries, peas, beets, onions, garlic, green tea
Selenium	Lean meat, seafood, kidney, liver, dairy products, 100 percent whole grain cereal, 100 percent whole wheat bread

FIGURE 7.6
A Grain of Wheat

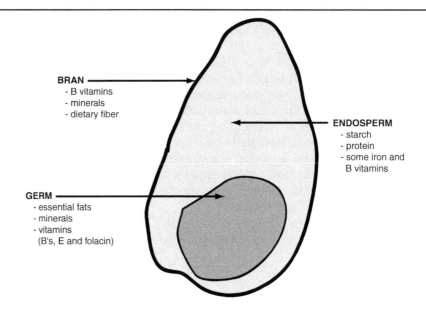

CD = 1 slice bread, waffle, or pancake

6 in. plate = 1 tortilla

tennis ball = 1 serving of vegetable

Baseball = 1 serving of fruit

6 dice = 1½ oz. cheese = 1 serving of dairy

What Counts as a Whole Grain Serving
- Cheerios – 2/3 cup
- Wheat Chex – 2/3 cup
- Oatmeal (hot, cooked) – 1/2 cup
- Quaker Oatmeal Squares or Toasted Oatmeal Cereal – 1/2 cup
- Grape Nuts – 1/5 cup
- Frosted Mini-Wheats (bite-sized) – 9 biscuits
- 100% whole-grain bread – 1 slice
- 100% whole-grain English muffin – 1 half
- Popcorn (popped) – 2 cups
- Sun Chips or baked tortilla chips – 1 oz. (about 15 chips)
- 100% whole-grain crackers (like Triscuits) – 4 crackers
- Whole-wheat pasta – 1/3 cup cooked
- Brown rice, bulgur, sorghum, or barley – 1/3 cup cooked

The concept of building a healthy plate was introduced on page 204 in Chapter 5. This practice is a key healthy habit to form early on and stick with throughout one's lifetime. A "healthy plate" is typically smaller in size, about 9" in diameter, than a typical dinner plate found in most homes and is certainly smaller than the average size plate served in restaurants. Filling half of the plate with fruits and vegetables and making half of the grains consumed whole grains are essential components to building a healthy plate. Eating a variety of small, lean, natural sources of protein and switching to skim or fat free dairy products are also important when building a healthy plate.

MyPlate is an idea based on the 2010 Dietary Guidelines for Americans. The idea behind MyPlate is to simplify the concept of making better/healthier food choices.

MyPlate uses the familiar place setting, using a plate 9 inches in diameter to illustrate the five food groups and the relative proportions in which they should be consumed. When used in conjunction with the ChooseMyPlate.gov website, consumers have access to practical, easy to understand information that will enable them to easily build a healthier diet.

Some select messages ChooseMyPlate uses to help consumers focus in on key behaviors include:

- Balancing Calories
  Eat the right amount of calories for you Everyone has a personal calorie limit. Staying within yours can help you get to or maintain a healthy weight. People who are successful at managing their weight have found ways to keep track of how much they eat in a day, even if they don't count every calorie.

  Enjoy your food, but eat less.
  - Get your personal daily calorie limit at www.ChooseMyPlate.gov and keep that number in mind when deciding what to eat.
  - Think before you eat . . . is it worth the calories?
  - Avoid oversized portions.
  - Use a smaller plate, bowl, and glass.
  - Stop eating when you are satisfied, not full.

  Cook more often at home, where you are in control of what's in your food. When eating out, choose lower calorie menu options.
  - Check posted calorie amounts.
  - Choose dishes that include vegetables, fruits, and/or whole grains.
  - Order a smaller portion or share when eating out.

choosemyplate.gov

Choose MyPlate: 10 Tips to a Great Plate
Making food choices for a healthy lifestyle can be as simple as using these 10 Tips. Use the ideas in this list to *balance your calories,* to choose foods to eat more often, and to cut back on foods to *eat less often.*

1. balance calories
   Find out how many calories YOU need for a day as a first step in managing your weight. Go to www.ChooseMyPlate.gov to find your calorie level. Being physically active also helps you balance calories.
2. enjoy your food, but eat less
   Take the time to fully enjoy your food as you eat it. Eating too fast or when your attention is elsewhere may lead to eating too many calories. Pay attention to hunger and fullness cues before, during, and after meals. Use them to recognize when to eat and when you've had enough.
3. avoid oversized portions
   Use a smaller plate, bowl, and glass. Portion out foods before you eat. When eating out, choose a smaller size option, share a dish, or take home part of your meal.

Deck of cards = 3 oz. meat

Golf ball = 2 Tb. peanut butter

Checkbook = 3 oz. thin fish

1/2 business card = 1 brownie = 1 serving

4. **foods to eat more often**
   Eat more vegetables, fruits, whole grains, and fat-free or 1% milk and dairy products. These foods have the nutrients you need for health—including potassium, calcium, vitamin D, and fiber. Make them the basis for meals and snacks.
5. **make half your plate fruits and vegetables**
   Choose red, orange, and dark-green vegetables like tomatoes, sweet potatoes, and broccoli, along with other vegetables for your meals. Add fruit to meals as part of main or side dishes or as dessert.
6. **switch to fat-free or low-fat (1%) milk**
   They have the same amount of calcium and other essential nutrients as whole milk, but fewer calories and less saturated fat.
7. **make half your grains whole grains**
   To eat more whole grains, substitute a whole-grain product for a refined product—such as eating whole wheat bread instead of white bread or brown rice instead of white rice.
8. **foods to eat less often**
   Cut back on foods high in solid fats, added sugars, and salt. They include cakes, cookies, ice cream, candies, sweetened drinks, pizza, and fatty meats like ribs, sausages, bacon, and hot dogs. Use these foods as occasional treats, not everyday foods.
9. **compare sodium in foods**
   Use the Nutrition Facts label to choose lower sodium versions of foods like soup, bread, and frozen meals. Select canned foods labeled "low sodium," "reduced sodium," or "no salt added."
10. **drink water instead of sugary drinks**
    Cut calories by drinking water or unsweetened beverages. Soda, energy drinks, and sports drinks are a major source of added sugar, and calories, in American diets.

Write down what you eat to keep track of how much you eat. If you drink alcoholic beverages, do so sensibly—limit to 1 drink a day for women or to 2 drinks a day for men.

- Foods to Increase
  Build a healthy plate Before you eat, think about what goes on your plate or in your cup or bowl. Foods like vegetables, fruits, whole grains, low-fat dairy products, and lean protein foods contain the nutrients you need without too many calories. Try some of these options.

  Make half your plate fruits and vegetables.
  - Eat red, orange, and dark-green vegetables, such as tomatoes, sweet potatoes, and broccoli, in main and side dishes.
  - Eat fruit, vegetables, or unsalted nuts as snacks—they are nature's original fast foods.

  Switch to skim or 1% milk.
  - They have the same amount of calcium and other essential nutrients as whole milk, but less fat and calories.
  - Try calcium-fortified soy products as an alternative to dairy foods.

Make at least half your grains whole.
- Choose 100% whole-grain cereals, breads, crackers, rice, and pasta.
- Check the ingredients list on food packages to find whole-grain foods.

Vary your protein food choices.
- Twice a week, make seafood the protein on your plate.
- Eat beans, which are a natural source of fiber and protein.
- Keep meat and poultry portions small and lean.

Keep your food safe to eat—learn more at www.FoodSafety.gov.
- Foods to Reduce

Cut back on foods high in solid fats, added sugars, and salt Many people eat foods with too much solid fats, added sugars, and salt (sodium). Added sugars and fats load foods with extra calories you don't need. Too much sodium may increase your blood pressure.

Choose foods and drinks with little or no added sugars.
- Drink water instead of sugary drinks. There are about 10 packets of sugar in a 12-ounce can of soda.
- Select fruit for dessert. Eat sugary desserts less often.
- Choose 100% fruit juice instead of fruit-flavored drinks.

Look out for salt (sodium) in foods you buy—it all adds up.
- Compare sodium in foods like soup, bread, and frozen meals—and choose the foods with lower numbers.
- Add spices or herbs to season food without adding salt.

Eat fewer foods that are high in solid fats.
- Make major sources of saturated fats—such as cakes, cookies, ice cream, pizza, cheese, sausages, and hot dogs—occasional choices, and not every day foods.
- Select lean cuts of meats or poultry and fat-free or low-fat milk, yogurt, and cheese.
- Switch from solid fats to oils when preparing food.*

*Examples of solid fats and oils	
Solid Fats	**Oils**
Beef, pork, and chicken fat Butter, cream, and milk fat Coconut, palm, and palm kernel oils Hydrogenated oil Partially hydrogenated oil Shortening Stick margarine	Canola oil Corn oil Cottonseed oil Olive oil Peanut oil Safflower oil Sunflower oil Tub (soft) margarine Vegetable oil

• Physical Activity

Be physically active your way Pick activities that you like and start by doing what you can, at least 10 minutes at a time. Every bit adds up, and the health benefits increase as you spend more time being active.

Note to parents
What you eat and drink and your level of physical activity are important for your own health, and also for your children's health.

You are your children's most important role model. Your children pay attention to what you *do* more than what you *say*.

You can do a lot to help your children develop healthy habits for life by providing and eating healthy meals and snacks. For example, don't just *tell* your children to eat their vegetables—*show* them that you eat and enjoy vegetables every day.

Beginning in May 1993, the federal government has required food manufacturers to provide accurate nutritional information about their products on their product labels. Because food labels are standardized, relatively straightforward, and easy to read, much of the guesswork has been taken out of good nutrition.

Ingredients are listed on food labels by percentage of total weight, in order from heaviest or highest to lowest. By reading the listing of ingredients, an individual can determine whether a food is relatively high in fat, sugar, salt, and so on.

Food labels are legally required to include the number of servings per container, serving size, and the number of calories per serving. They must also list the percentage of the daily value of total fat, saturated fat, and, beginning in January 2006, trans fat. Nutrition Facts Labels must also list the percentage of the daily value of cholesterol, sodium, total carbohydrates (including dietary fiber and sugars), proteins, vitamins, and minerals. Figure 7.7 on page 210 provides an example of the required nutrition information found on packaged foods.

The bottom part of Nutrition Facts Labels on larger packages (typically any item that is not packaged for individual sale) contains a footnote

Why Is There No Percentage of Daily Value for Trans Fats?

There have been scientific findings and reports that confirm a link between trans fats and an increased risk of coronary heart disease. However, none of the reports have recommended an amount of trans fat that the Food and Drug Administration could use to establish a daily value, and without a daily value, a percentage of that daily value cannot be calculated.

with Daily Values (DVs) for 2,000- and 2,500-calorie-a-day diets. Because this information is not about a specific food product, it does not change from product to product. It shows recommended advice for all Americans. In the footnote section of the Nutrition Facts Label, the nutrients that have an upper limit or a set amount one wants to stay below are listed first. These nutrients include total fat, saturated fat, trans fat, sodium, and cholesterol. The amount of dietary fiber listed in this section is a minimum amount that should be consumed each day. The daily value for carbohydrates listed is a recommendation based on a 2,000-calorie-a-day diet, but it can vary slightly depending on the amount of fat and protein consumed.

When an individual takes the time to use the main body of the Nutrition Facts Label in conjunction with the footnote section of the label, he or she can get a very accurate picture of not only what source (carbohydrate, fat, or protein) their calories are coming from, but also how close they are coming to meeting the daily requirements necessary to maintain a high level of health.

Although product labels do have accuracy requirements, mistakes can be made and sometimes do occur. For this reason, it is wise to check the accuracy of food labels. One quick and easy way to do this is to divide the number of servings within the container—does it equal the serving size? For example, you have a product that the Nutrition Facts Label shows having a serving size of one-half cup and the number of servings per container is four. If you open the product and check it, does it contain two cups of that food? If so, the label is correct. Another way to check the accuracy of a nutrition label is to calculate the calories (grams of fat times nine, grams of protein and carbohydrates times four). Does the number calculated match the reported calories within 10 to 20 calories? If the numbers are way "off" one should be aware that the label is incorrect. See Figure 7.7 (Nutrition Facts Label) for an example of how to check for label accuracy based on reported calories.

Reported Total Calories per Serving = 250
Reported Calories from Fat = 110

The product contains a total of 12 g of fat. Fat contains 9 calories per gram of fat, so to check for accuracy of reported fat calories, multiply 12 × 9. This equals 108.

To check for accuracy of total number of calories, multiply the total grams of carbohydrates . . . 31 in this product, by 4 (the amount of calories per gram of carbohydrate). This equals 124 calories.

The total grams of protein . . . 5 in this product, by 4 (the amount of calories per gram of protein). This equals 20 calories.

To check for accuracy of the total number of calories per serving, add calories from fat, protein, and carbohydrates. If they are close to the number of calories per serving listed on the Nutrition Fact Label, the label is accurate.

108 + 124 + 20 = 252 actual vs. 250 reported

Reading labels while grocery shopping is important in preparing healthy meals.

Both the total calories and fat calories listed on the Nutrition Facts Label were slightly low. Knowing this, an individual can more accurately determine when he or she has reached their nutritional limit.

FIGURE 7.7 Food Label: Nutrition Facts

Serving Size
Is your serving the same size as the one on the label? If you eat double the serving size listed, you need to double the nutrient and calorie values. If you eat one-half the serving size shown here, cut the nutrient and calorie values in half.

Calories
Are you overweight? Cut back a little on calories! Look here to see how a serving of the food adds to your daily total. A 5'4", 138-lb. active woman needs about 2,200 calories each day. A 5'10", 174-lb. active man needs about 2,900. How about you?

Total Carbohydrate
When you cut down on fat, you can eat more carbohydrates. Carbohydrates are in foods like bread, potatoes, fruits, and vegetables. Choose these often! They give you more nutrients than sugars like soda pop and candy.

Dietary Fiber
Grandmother called it "roughage," but her advice to eat more is still up-to-date! That goes for both soluble and insoluble kinds of dietary fiber. Fruits, vegetables, whole-grain foods, beans, and peas are all good sources and can help reduce the risk of heart disease and cancer.

Protein
Most Americans get more protein than they need. Where there is animal protein, there is also fat and cholesterol. Eat small servings of lean meat, fish, and poultry. Use skim or low-fat milk, yogurt, and cheese. Try vegetable proteins like beans, grains, and cereals.

Vitamins and Minerals
Your goal here is 100 percent of each for the day. Don't count on one food to do it all. Let a combination of foods add up to a winning score.

Nutrition Facts
Serving Size 1 cup (228g)
Servings Per Container 2

Amount Per Serving

Calories 250   Calories from Fat 110

	% Daily Value*
Total Fat 12g	18%
Saturated Fat 3g	15%
Trans Fat 3g	
Cholesterol 30mg	10%
Sodium 470mg	20%
Total Carbohydrate 31g	10%
Dietary Fiber 0g	0%
Sugars 5g	
Protein 5g	
Vitamin A	4%
Vitamin C	2%
Calcium	20%
Iron	4%

*Percent Daily Values are based on a 2,000 calorie diet. Your Daily Values may be higher or lower depending on your calorie needs:

	Calories	2,000	2,500
Total Fat	Less than	65g	80g
Sat Fat	Less than	20g	25g
Cholesterol	Less than	300mg	300mg
Sodium	Less than	2,400mg	2,400mg
Total Carbohydrate		300g	375g
Fiber		25g	30g

Calories per gram:
Fat 9 • Carbohydrates 4 • Protein 4

More nutrients may be listed on some labels.

Total Fat
Aim low. Most people need to cut back on fat! Too much fat may contribute to heart disease and cancer. Try to limit your calories from fat. For a healthy heart, choose foods with a big difference between the total number of calories and the number of calories from fat.

Saturated Fat
A new kind of fat? No—saturated fat is part of the total fat in food. It is listed separately because it's the key player in raising blood cholesterol and your risk of heart disease. Eat less!

Cholesterol
Too much cholesterol—a second cousin to fat—can lead to heart disease. Challenge yourself to eat less than 300 mg each day.

Sodium
You call it "salt," the label calls it "sodium." Either way, it may add up to high blood pressure in some people. So, keep your sodium intake low—2,400 to 3,000 mg or less each day.*

*The AHA recommends no more than 3,000 mg sodium per day for healthy adults.

Daily Value
Feel like you're drowning in numbers? Let the Daily Value be your guide. Daily Values are listed for people who eat 2,000 or 2,500 calories each day. If you eat more, your personal daily value may be higher than what's listed on the label. If you eat less, your personal daily value may be lower.

For fat, saturated fat, cholesterol, and sodium, choose foods with a low percent Daily Value. For total carbohydrate, dietary fiber, vitamins, and minerals, your daily value goal is to reach 100 percent of each.
g = grams (About 28 g = 1 ounce)
mg = milligrams (1,000 mg = 1 g)

You Can Rely on the New Label

Rest assured, when you see key words and health claims on product labels, they mean what they say as defined by the government. For example:

Key Words	*What They Mean*
Fat Free	Less than 0.5 g of fat per serving
Low Fat	3 g of fat (or less) per serving
Lean	Less than 10 g of fat, 4 g of saturated fat, and 95 mg of cholesterol per serving
Light (Lite)	one-third less calories or no more than one-half the fat of the higher-calorie, higher-fat version; or no more than one-half the sodium of the higher-sodium version
Cholesterol Free	Less than 2 mg of cholesterol and 2 g (or less) of saturated fat per serving

To Make	*The Food*
Health Claims About...	Must Be...
Heart Disease and Fats	Low in fat, saturated fat, and cholesterol
Blood Pressure and Sodium	Low in sodium
Heart Disease and Fruits, Vegetables, and Grain Products	A fruit, vegetable, or grain product low in fat, saturated fat, and cholesterol, that contains at least 0.6 g soluble fiber, without fortification, per serving

Other claims may appear on some labels.

Vegetarianism

There have always been people who, for one reason or another (religious, ethical, or philosophical), have chosen to follow a vegetarian diet. However, in recent years, a vegetarian diet has become increasingly popular.

Salmon is a good source of Omega-3 fatty acids.

There are four different types of vegetarian diets. **Vegans** are considered true vegetarians. Their diets are completely void of meat, chicken, fish, eggs, or milk products. A vegan's primary sources of protein are vegetables, fruits, and grains. Because vitamin B12 is normally found only in meat products, many vegans choose to supplement their diet with this vitamin.

Lactovegetarians eat dairy products, fruits, and vegetables but do not consume any other animal products (meat, poultry, fish, or eggs).

Ovolactovegetarians are another type of vegetarians. They eat eggs as well as dairy products, fruits, and vegetables, but still do not consume meat, poultry, or fish.

A person who eats fruits, vegetables, dairy products, eggs, and a small selection of poultry, fish, and other seafood is a partial or **semivegetarian.** These individuals do not consume any beef or pork.

Vegetarians of all four types can meet all their daily dietary needs through the food selections available to them. However, because certain foods or groups of foods that are high in specific nutrients are forbidden, it is critical that a vegetarian is diligent in selecting his or her food combinations so that the nutritional benefits of the foods allowed are maximized. If food combinations from a wide variety of sources are not selected, nutritional deficiencies of proteins, vitamins, and minerals can rapidly occur and proper growth, development, and function may not occur. While a vegetarian diet can certainly be a healthy, low-fat alternative to the typical American diet, without diligent monitoring, it is not a guarantee of good health.

For many individuals who choose a vegetarian diet, it is more than simply omitting certain foods or groups of food, it is a way of living that they have embraced.

The Mediterranean Diet

The Mediterranean Diet is a way of eating that focuses on good heart health. It is based on the eating habits of the people living in countries along the Mediterranean Sea. It focuses on increasing the amount of fish a person eats as well as consuming a lot of whole grains and fresh fruits and vegetables.

Suggestions for following a Mediterranean Style Diet include

- Use MyPlate (p. 204) to plan meals. This plan emphasizes making half your plate fruits and vegetables and this is a key component to the Mediterranean Diet.
- Along with eating fruits and vegetable, consume foods like beans, whole grains, whole grain pastas, nuts, and brown rice that are plant based.
- Limit the amount of red meat consumed. Replace red meat with heart healthy portions of fish that are a great source of Omega -3 oils. Fish should be consumed at least twice a week. Lean poultry is also a good source of nutrition.
- Use olive and canola oils instead of butter, margarine and/or shortening. Olive oil is a monounsaturated fat that is the main fat source of the Mediterranean Diet and it can lower Low Density Lipoprotein (LDL) cholesterol.

- Avoid salt. Use natural spices and herbs such as garlic, rosemary and basil to add flavor to meals without adding excess fat, salt and sugar.
- Processed, pre-packaged foods should be avoided. Many times these processed, pre-packaged foods are high in fat, sugar and salt and can be prepared with fresh ingredients at home in a much healthier manner.

References

American Heart Association. Heart Disease and Stroke Statistics. 2012 Update.

Bishop, A. *Step Up to Wellness: A Stage Based Approach* (1st ed). Needham Heights, MA: Allyn & Bacon. 1999.

The Center for Health and Healthcare in Schools, School of Public Health and Health Services, George Washington University Medical Center. *Childhood Overweight: What the Research Tells Us*. March 2005 Update. www.healthinschools.org

Corbin, C. and Welk, G. *Concepts of Physical Fitness* (15th ed). Dubuque, IA: McGraw-Hill. 2009.

Donatelle, R. J. *Access to Health* (9th ed). Boston: Allyn & Bacon. 2006.

Flegal, K. M., Carrol, M. D., Kuczmarski, R. J., and Johnson, C. L. Overweight and Obesity in the United States: Prevalence and Trends, 1960–1994. *International Journal of Obesity and Related Metabolic Disorders* 22:39–47. 1998.

Floyd, P., Mims, S., and Yelding-Howard, C. *Personal Health: Perspectives and Lifstyles*. Morton Publishing Co. 2007.

Gibbs, W. W. Obesity: An Overblown Epidemic? *Scientific American*, May 23, 2005.

Hahn, D. B. and Payne, W. A. *Understanding Your Health*. McGraw-Hill. 2008.

Hales, D. *An Invitation to Wellness* (Instructor Ed.). Thomson-Wadsworth, 2007.

Hoeger, W. W. K. and Hoeger, S. A. *Lifetime Physical Fitness and Wellness: A Personalized Program* (10th ed). Belmont, CA: Wadsworth. 2009.

http://ahha.org

http://msue.anr.msu.edu/news/mediterranean_diet_have_you_heard_about_it

http://olin.msu.edu/healthed/nutrition/foodguidelines.htm

http://www.cdc.gov/nccdphp/dnpa/healthyweight/physical_activity/index.htm

http://www.nhlbi.nih.gov/health/public/heart/obesity/lose_wt/shop.htm

http://kidshealth.org/pagemanager.jsp?dn-kidsHealthbSlic

http://www.foodinsight.org/Resources

http://www.reachout.com.au/default.asp?ti=2249

http://www.cfsan.fda.gov/~dms/foodlab.html

http://www.win.niddk.nih.gov/publications/tools.htm

http://www.womhealth.org.au/studentfactsheets/bodyimage.htm

Hyman, B., Oden, G., Bacharach, D., and Collins, R. *Fitness for Living*. Dubuque, IA: Kendall-Hunt Publishing Co. 2006.

IUFOST, International Union of Food Science and Technology Bulletin. Trans-fatty Acids, May 2006.

Neighmond, *Patti, Gain together, Lose Together: The Weight-Loss 'Halo' Effect,* Health Blog: NPR http://www.npr.org/blogs/health/2012/03/12, retrieved 3/17/12.

Nordestgaard, B. G., Benn, M., Schnohr, P., Tybjærg-Hansen, A. Nonfasting triglycerides and risk of myocardial infarction, ischemic heart disease, and death in men and women. *JAMA.* 2007; 298(3):299–308, PubMed.

O'Neil, Patrick. Weight Management Center, Medical University of South Carolina. 2009. www.muschealth.com

Powers, S. K., Todd, S. L., and Noland, U. J. *Total Fitness and Wellness* (2nd ed). Boston: Allyn & Bacon. 2005.

Prentice, W. E. *Fitness and Wellness for Life* (6th ed). New York: WCB McGraw-Hill. 1999.

Pruitt, B. E. and Stein, J. *Health Styles.* Boston: Allyn & Bacon. 1999.

Rosato, F. *Fitness for Wellness* (3rd ed). Minneapolis: West. 1994.

Satcher, D. Surgeon General's Report on Physical Activity and Health. Atlanta: U.S. Department of Health and Human Services, CDC. 1996.

Texas A&M University, Student Health Services. Fad Diets: Promise or Profit, 77, 2002.

Texas A&M University Human Nutrition Conference. College Station, TX. 1998.

Wilmore, J. H. Exercise, Obesity, and Weight Control, *Physical Activity and Research Digest.* Washington, DC: President's Council on Physical Fitness and Sports. 1994.

World Health Organization (WHO). Management of Severe Malnutrition: A Manual for Physicians and other Senior Health Workers. Geneva: Author. 1999.

World Health Organization (WHO). Obesity: Preventing and Managing the Global Epidemic — Report of WHO Consultation on Obesity. Geneva, June 1997.

Surgeon General's Call to Action to Prevent and Decrease Overweight and Obesity. 2005. www.surgeongeneral.gov

Diet Books: What the Experts Say, *Consumer Reports,* June 2007, 14–15.

Diet Plans: What the Studies Say, *Consumer Reports,* June 2007, 16–17.

Recommended Reading

Eat, Drink and Be Healthy: The Harvard Medical School Guide to Healthy Eating by Walter C. Willett, M.D. Simon and Schuster Source, 2001.

The Spectrum: A Scientifically Proven Program to Feel Better, Live Longer, Lose Weight, and Gain Health. Ballantine Books, 2007.

Activities

Activities

Body Mass Index Calculator
Facts about My Favorite Fast-Food Meal

Name _____ Section _____ Date_____

HOMEWORK EXPERIENCE

Body Mass Index Calculator

In order to complete this Body Mass Index assignment, go to http://www.cdc.gov/nccdphp/dnpa/bmi/calc-bmi.htm
 Body Mass Index is a mathematical formula that correlates highly with body fat. This weight calculation helps determine whether you are at a healthy weight or have too much fat.

1 kg = 2.2 lbs
1 meter = 39.37 inches

The formula for BMI =

$$\frac{\text{Weight (kg)}}{\text{Height (m)}^2}$$

Note:

If you are under the age of 20 years, you have the option of using the BMI-by-age calculator.

1. Enter your weight and height using English or metric measurements. What is your BMI? _____

2. What is your weight status according to your BMI calculation? _____

3. Click on "What does this all mean?"
   How can two individuals, one fit and one unfit, who weigh the same and are the same height have the same BMI?

4. What does your BMI tell you about your health risk?

   <18   Underweight
   19–26 Healthy Weight (low risk)
   27–29 Overweight (medium risk)
   30–40 Obese (high risk)
   >40   Morbidly obese (very high risk)

A BMI of 25 or higher is associated with an increased health risk of conditions that include coronary heart disease, certain forms of cancer, stroke, high blood pressure, and non insulin-dependent diabetes.

Source: www.cdc.gov

Name _____ Section _____ Date _____

HOMEWORK EXPERIENCE

Facts about My Favorite Fast-Food Meal

1. List your favorite fast-food meal in the space provided below—be specific and detailed. Include anything you consume with the meal and the quantity (i.e., large beverage, four ketchup or salsa packets, etc.).

2. Go to the restaurant and obtain a nutritional analysis of their foods. This information is generally available as a pamphlet. You might also be able to obtain this information from the restaurant's Web site.

3. Determine and list the following for your meal:

    A. total number of calories

    B. grams of total fat

    C. grams of saturated fat

    D. grams of trans fat

4. Use Table 7.2 on page 207 to determine your daily calorie allowance based on your age, gender, and activity level. List that information in the space provided below.

    Age:

    Gender:

    Activity Level:

    Estimated Daily Caloric Allowance:

5. Determine how "healthy" your food choice was in relation to the total number of calories it contains and the number of grams of total fat, saturated fat, and trans fat it contains. When you are looking at these numbers, remember that the calorie allowance and the limit on fat is for all food consumed within a twenty-four-hour period, and the meal listed is probably only about one-third of the calories and fat you will consume during this time period. In the space below, use the numbers you compared and briefly describe how "healthy" your choice was.

6. Other than never eating this meal, what modifications could reasonably be made to keep this meal in your diet but make it a more nutritionally sound choice?

Name _____ Section _____ Date _____

WRITING PROMPTS

- What is your favorite restaurant? What do you typically order? List at least 3 ways, other than not eating the meal, you could realistically improve the nutritional quality of this meal.

- Why are fad diets ineffective for long-term weight loss? Identify successful weight loss programs in your community.

Chapter 8
The Basics of Neuroscience and Drug Addiction

OBJECTIVES

After you have studied this chapter, you should be able to:
- Define and explain what a drug is;
- Describe how drugs are introduced into the body;
- Explain how the method of administration influences rate of action;
- Explain how the rate of action affects the brain;
- Explain what makes a faster-acting psychoactive drug produce more euphoria than a slower-acting one;

From *Drugs: A Healthy Understanding, 1/e* by Larry W. Latterman. Copyright © 2013 by Kendall Hunt Publishing Company. Reprinted by permission.

> Drugs can be administered by oral ingestion, injection, inhalation, and through absorption.

- Describe duration of action;
- Explain how the brain basically communicates;
- Describe how drugs work in the brain;
- List drugs that can cause neurological problems;
- Describe the blood-brain barrier;
- Describe drug dependence;
- Define and explain tolerance;
- Describe withdrawal and how long it lasts;
- Explain how drugs leave the body.

> **Drug**
> Is a chemical compound or substance that can alter the structure and function of the body.

"Drugs are very much a part of professional sports today, but when you think about it, golf is the only sport where the players aren't penalized for being on grass."

—**Bob Hope**

A Drug Defined

A **drug** is a chemical compound or substance that can alter the structure and function of the body. Psychoactive drugs affect the function of the brain, and some of these may be illegal to use and possess.

How Drugs Are Introduced into the Body

Drugs can enter the body in different ways. Drugs can be administered by oral ingestion, injection, inhalation, and through absorption. How the drug is taken will determine how quickly it enters the bloodstream and the drug's rate of effect. Repeated use of any drug by any route of administration can lead to addiction and other adverse health consequences.

Oral ingestion, ingesting a drug by mouth, is the most common and simplest way of taking a drug. Drugs that are ingested include liquids, capsules, and tablets. Drugs that are commonly ingested or swallowed include alcohol, buprenorphine, opium, marijuana, sedatives, and heroin. This method is the most common and simplest, but it is also the slowest method of getting the drug into the bloodstream. Drugs are absorbed into the body when they travel from their point of entry into the blood. When a person takes drugs by mouth, they move through the digestive tract to the liver and enter the bloodstream as the body processes the chemicals.

Oral Ingestion, ingesting a drug by mouth, is the most common and simplest way of taking a drug.

In recent years, smoking or inhalation of drugs has become a popular route of administration among drug users. Various drugs of different classes have been abused by inhalation or smoking, including phencyclidine (PCP), cocaine, heroin, methamphetamine, and marijuana. This increased popularity of smoking drugs has resulted from the fast onset of drug action and fears related to contracting acquired immunodeficiency syndrome (AIDS) or other infectious diseases from intravenous injections. Inhalation is a very potent route of drug administration and is characterized by fast absorption from the nasal mucosa and the extensive lung capillaries. Inhalation results in an immediate elevation of arterial blood drug concentration and a higher bioavailability by avoiding drug metabolism in the liver. The fact that smoking or inhalation provides rapid delivery of drugs to the brain may result in an immediate reinforcing effect of the drug and further contribute to its abuse liability or risk of dependency.

> **Oral Ingestion**, ingesting a drug by mouth, is the most common and simplest way of taking a drug.

Injection

A drug can be administered by injection (given parenterally) into a vein, into the skin, under the skin, or into a muscle using a hypodermic syringe and needle. For drugs given via any of these techniques, the absorption time is very fast. Drugs that are commonly injected include amphetamine-type stimulants, buprenorphine, heroin and sedatives.

Intravenous injection (IV)[2]

This method involves putting a drug through a needle directly into the vein. Drugs given intravenously enter the bloodstream directly and are absorbed faster. Therefore, drugs are given in this way when a rapid effect is needed. IV injection is the preferred route of administration with cocaine and heroin users. Because the drug is not significantly diluted, when a drug is injected into a vein, it reaches the brain by way of the lungs in a matter of seconds. For IV injecting, sometimes called **mainlining**, the arms are the site of first choice. Other sites of the body used for injecting are the hands, fingers, legs, and feet.

Injections into the skin, sometimes called an **intradermal injection**, is given in the dermal layer of the skin just below the top layer, which is called the **epidermis**. The most common site for this type of injection is the lower arm.

> **Intravenous injection**
> Putting a drug through a needle directly into the vein.
>
> **Mainlining**
> Injecting a drug of abuse intravenously.
>
> **Intradermal injection**
> Drug injection given in the dermal layer of the skin. Epidermis - The top layer of skin.
>
> **Epidermis**
> The top layer of skin.

Other methods of injection[4]

- Intramuscular injection (IM) - The drugs are injected by needle into the muscle.[4]
- Subcutaneous injection (SC) - The drugs are injected by needle just below the skin.[4]
- Intrathecal injection (IT) - The drugs are injected by needle into the spinal fluid.[4]

Subcutaneous injec-tions (injections under the skin) go into the fatty tissue just below the skin. Many drugs are injected subcutaneously.

Transdermal Patch[5]
Transdermal patch an adhesive patch applied to the skin in order to administer a time-released dose of medication.

Intranasal[6]
Intranasal use, or snorting is the process of inhaling cocaine powder through the nostrils which leads to quicker absorption through the nasal tissue.

- Rectal suppository is a suppository intended to be inserted into the rectum.
- Vaginal suppository is a suppository intended to be inserted into the vagina.

The method of administering a drug can play a major role in the drug's rate of action, which in turn affects its abuse liability or potential as a treatment medication. Smoked drugs that are delivered through a "crack" pipe or a cigarette and injected drugs that are administered through a hypodermic needle reach the brain very rapidly. This rapid action is an important factor in the strong effect and high abuse liability of such drugs. Drug abuse treatment medications are often administered orally or through a patch affixed to the skin and take longer to reach the brain. This slower rate of action is an important factor in the milder effect and low abuse liability of treatment medications.

Many complex factors affect a compound's rate of action in the human body. These factors include the compound's structure and properties, how quickly the body absorbs it and transports it to the brain, and how rapidly it crosses the blood-brain barrier and binds to a sufficient number of targeted brain sites called receptors to produce its effects. Because the method of administering a drug affects a number of these factors, method can play a major role in the drug's rate of action and abuse liability.

By using different methods of administering a compound, scientists can slow the rate of action and produce milder effects that may substitute for the stronger effects of an abused substance. Such a compound may be an effective medication for patients in treatment.

The effect that method of administration has on rate of action may be best illustrated by the transdermal nicotine patch, which uses the abused substance itself nicotine to ease withdrawal symptoms and aid smoking cessation. However, unlike smoking a cigarette, which delivers an almost instantaneous burst of nicotine-laden blood to the brain, administering nicotine through a patch affixed to the skin slows its rate of action and produces more sustained, lower peak levels of nicotine in the blood. This lower, slower dose of nicotine has proven to be effective in short-term treatment of nicotine dependence and aiding smoking cessation.

The rate at which a psycho-active drug occupies, or binds to, the receptors for that drug determines the intensity of its rewarding effects and its abuse liability, according to a hypothesis discussed at several scientific forums by Drs. George Uhl and David Gorelick of the National Institute of Drug Abuse (NIDA) Division of Intramural Research in Baltimore and Dr. Mary Jeanne Kreek of the Rockefeller University in New York. According to this rate hypothesis, the faster a drug such as heroin or cocaine occupies enough brain receptors to

> Smoked drugs that are delivered through a "crack" pipe or a cigarette and injected drugs that are administered through a hypodermic needle reach the brain very rapidly.

produce a psychoactive effect, the greater the euphoria users experience, the more they "like" the drug, and the more liable they are to abuse it.

What Makes a Faster-acting Psychoactive Drug Produce More Euphoria Than a Slower-acting One?

"The brain has many adaptive mechanisms, and if you disturb the brain slowly, it often can catch up or compensate," explains Dr. George Uhl, who directs the Molecular Neurobiology Branch of the NIDA Division of Intramural Research in Baltimore. "But, if you do things rapidly, often it can't. For example, going suddenly from a dark room into bright sunlight will result in a temporary loss of vision because the light sensitivity mechanisms of the eye and the brain cannot adapt that quickly. But, if you proceed from dark to light slowly by stages, the mechanism works in a fairly automatic way, and you are able to see normally."

"If a drug acts slowly, the brain is able to compensate for the changes that the drug produces," Dr. Uhl says. "However, when a drug's onset of action is very rapid, it may be able to overwhelm the brain's adaptive mechanisms, thus producing a bigger boost in its pleasure circuits. Smoked and intravenous cocaine, for example, have fast rates of action. They reach the brain within seconds and rapidly flood the brain's 'pleasure pathway' with excess dopamine, a brain chemical that helps transmit pleasurable feelings. If the brain's adaptive mechanisms cannot respond quickly enough to the sudden excess of dopamine, the euphoric rush ensues."

Duration of Action

Duration of action, or how long the drug occupies a receptor once it gets there, also plays an important role in drug abuse and treatment. For example, cocaine has both a fast rate of action that produces euphoria and a rapid offset, or short duration of action, that allows frequent abuse.

Compounds with these traits may foster cravings for more drug and stronger conditioning of the drug-taking habit, according to a hypothesis proposed by Dr. Nora Volkow of Brookhaven National Laboratory in Upton, New York. Brain imaging studies conducted by Dr. Volkow indicate that while the fast rate at which cocaine acts on the brain plays a major role in its rewarding effects, it is cocaine's extremely rapid removal from the brain that both promotes and enables its frequent reuse and abuse.

The Central Nervous System

The brain and spinal cord make up the **central nervous system**. The human brain is the most complex organ in the body. This three-pound mass of gray and white matter sits at the center of all human activity—you need it to drive a car, to enjoy a meal, to breathe, to create an artistic masterpiece, and to enjoy everyday activities. In brief, the brain regulates your basic body functions; enables you to interpret and respond to everything you experience; and shapes your thoughts, emotions, and behavior.

The brain consists of several large regions, each responsible for some of the activities vital for living. These include the brainstem, cerebellum, limbic system, diencephalon, and cerebral cortex. The **brainstem** is the part of the brain that connects the brain and the spinal cord. It controls many basic functions, such as heart rate, breathing, eating, and sleeping. The **cerebellum** (Latin for "little brain") is a region of the brain that plays an important role in motor control.

On top of the brainstem and buried under the cortex, there is a set of more evolutionarily primitive brain structures called the **limbic system**

Duration of action, or how long the drug occupies a receptor once it gets there, also plays an important role in drug abuse and treatment.

The human brain is the most complex organ in the body.

The cerebellum (Latin for little brain) is a region of the brain that plays an important role in motor control.

Brainstem
The part of the brain that connects the brain and the spinal cord.

Cerebellum
A region of the brain that plays an important role in motor control.

Limbic system
The limbic system structures are involved in many of our emotions and motivations, particularly those that are related to survival, such as fear, anger, and sexual behavior.

Brain diagram labels: Corpus callosum, Pineal gland, Thalamus, Parietal lobe, Frontal lobe, Superior colliculus, Hypothalamus, Cerebral aqueduct, Mammillary body, Inferior colliculus, Optic chiasma, Occipital lobe, Pituitary, Cerebral peduncle, Cerebellum, Pons, Fourth ventricle, Medulla oblongata

Diencephalon
Located beneath the cerebral hemispheres, contains the thalamus and hypothalamus.

Thalamus
Is involved in sensory perception and regulation of motor functions (i.e., movement).

Hypothalamus
A region of the brain that helps regulate hormone activity, directs autonomic nervous system functions, and influences or manages many critical functions including sleep.

Cerebral cortex
Is divided into right and left hemispheres, encompasses about two-thirds of the human brain mass and lies over and around most of the remaining structures of the brain.

The brain consists of several large regions, each responsible for some of the activities vital for living. These include the brainstem, cerebellum, limbic system, diencephalon, and cerebral cortex.

(e.g., amygdala and hippocampus). The limbic system structures are involved in many of our emotions and motivations, particularly those that are related to survival, such as fear, anger, and sexual behavior. The limbic system also is involved in feelings of pleasure that are related to our survival, such as those experienced from eating and sex.

The **diencephalon**, which is also located beneath the cerebral hemispheres, contains the thalamus and hypothalamus. The **thalamus** is involved in sensory perception and regulation of motor functions (i.e., movement). The **hypothalamus** is a very small but important component of the diencephalon. It is a region of the brain that helps regulate hormone activity, directs autonomic nervous system functions, and influences or manages many critical functions including sleep.

The **cerebral cortex**, which is divided into right and left hemispheres, encompasses about two-thirds of the human brain mass and lies over and around most of the remaining structures of the brain. It is the most highly developed part of the human brain and is responsible for thinking, perceiving, and producing and understanding language.

How Does the Brain Basically Communicate?

The brain is a communications center consisting of billions of neurons, or nerve cells. Networks of neurons pass messages back and forth to different structures within the brain, the spinal column, and the peripheral nervous system. These nerve networks coordinate and regulate everything we feel, think, and do.

- **Neuron to neuron.** Also called a "nerve cell," a neuron is a cell of the nervous system, which conducts nerve impulses. Neurons consist of an **axon** and several **dendrites**. Neurons are connected by **synapses**. Each nerve cell in the brain sends and receives messages in the form of electrical impulses. Once a cell receives and processes a message, it sends it on to other neurons.
- **Neurotransmitters.** Also known as the brain's chemical messengers, neurotransmitters are chemicals that carry, or transmit, messages between neurons.
- **Receptors.** Receptors are the brain's chemical receivers. The neurotransmitter attaches to a specialized site on the receiving cell called a receptor. A neurotransmitter and its receptor operate like a key and lock, an exquisitely specific mechanism that ensures that each receptor will forward the appropriate message only after interacting with the right kind of neurotransmitter.
- **Transporters.** The brain's chemical recyclers, transporters are located on the cell that releases the neurotransmitter. They recycle these neurotransmitters (i.e., bring them back into the cell that released them), thereby shutting off the signal between neurons.

Neuron
Also called a "nerve cell," a neuron is a cell of the nervous system, which conducts nerve impulses and release neurotransmitters.

Axon
The long, thread-like part of a neuron, or nerve cell, along which nerve signals are conducted.

Dendrites
Short fibers of a neuron that receive transmitter signals.

Synapse
The zone of junction between nerve cells through which they "communicate".

Neurotransmitter
A chemical substance that carries impulses from one nerve cell to another.

Receptor
A cell that detects different forms of energy and conveys it into the electrochemical signals used by the nervous system.

Transporters
A protein that recycles neurotransmitters.

Dopamine
A neurotransmitter present in regions of the brain that control movement, emotion, motivation, and feelings of pleasure.

Drugs are chemicals that tap into the brain's communication system and disrupt the way nerve cells normally send, receive, and process information. There are at least two ways that drugs are able to do this: (1) by imitating the brain's natural chemical messengers, and/or (2) by over stimulating the "reward circuit" of the brain. Some drugs, such as marijuana and heroin, have a similar structure to neurotransmitters, which are naturally produced by the brain. Because of this similarity, these drugs are able to "fool" the brain's receptors and activate nerve cells to send abnormal messages. Generally, each neurotransmitter can bind only to a very specific matching receptor.

Other drugs, such as cocaine or methamphetamine, can cause the nerve cells to release abnormally large amounts of natural neurotransmitters, or prevent the normal recycling of these brain chemicals, which is needed to shut off the signal between neurons. This disruption produces a greatly amplified message that ultimately disrupts normal communication patterns. The difference in effect can be described as the difference between someone whispering into your ear and someone shouting into a microphone.

Nearly all drugs, directly or indirectly, target the brain's reward system by flooding the circuit with dopamine. **Dopamine** is a neurotransmitter present in regions of the brain that control movement, emotion, motivation, and feelings of pleasure. The overstimulation of this system, which normally responds to natural behaviors that are linked to survival (e.g., eating, spending time with loved ones), produces euphoric effects. This reaction sets in motion a pattern that "teaches" people to repeat the behavior of abusing drugs.

Our brains are wired to ensure that we will repeat life-sustaining activities by associating those activities with pleasure or reward. Whenever this reward circuit is activated, the brain notes that something important is happening that needs to be remembered and teaches us to do it again and again without thinking about it. Because drugs of abuse stimulate the same circuit, we learn to abuse drugs in the same way.

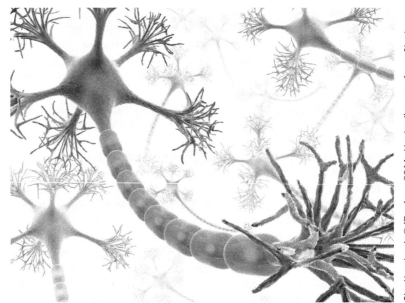

Also called a "nerve cell," a neuron is a cell of the nervous system, which conducts nerve impulses; consisting of an axon and several dendrites.

When some drugs of abuse are taken, they can release two to ten times the amount of dopamine that natural rewards do. In some cases, this occurs almost immediately (as when drugs are smoked or injected), and the effects can last much longer than those produced by natural rewards. The resulting effects on the brain's pleasure circuit dwarfs those produced by naturally rewarding behaviors such as eating and sex. The effect of such a powerful reward strongly motivates people to take drugs again and again. This is why scientists sometimes say that drug abuse is something we learn to do very, very well.

> When some drugs of abuse are taken, they can release two to ten times the amount of dopamine that natural rewards do.

As a person continues to abuse drugs, the brain adapts to the overwhelming surges in dopamine by producing less dopamine or by reducing the number of dopamine receptors in the reward circuit. As a result, dopamine's effect on the reward circuit is lessened, reducing the abuser's ability to enjoy the drugs and the things that previously brought pleasure. This decrease compels those addicted to drugs to keep abusing drugs in order to attempt to bring their dopamine function back to normal. Additionally, they may then require larger amounts of the drug than they first did to achieve the dopamine high, an effect known as tolerance.

Glutamate
A neurotransmitter that influences the reward circuit and the ability to learn.

Long-term abuse causes changes in other brain chemical systems and circuits as well. **Glutamate** is a neurotransmitter that influences the reward circuit and the ability to learn. When the optimal concentration of glutamate is altered by drug abuse, the brain attempts to compensate, which can impair cognitive function. Drugs of abuse facilitate nonconscious (i.e., conditioned) learning, which leads the user to experience uncontrollable cravings when they see a place or person they associate with the drug experience, even when the drug itself is not available. Brain imaging studies of drug-addicted individuals show changes in areas of the brain that are critical to judgment, decision making, learning and memory, and behavior control. Together, these changes can drive an abuser to seek out and take drugs compulsively despite adverse consequences. In other words, to become addicted to drugs.

What Other Brain Changes Occur with Abuse?

Chronic exposure to drugs of abuse disrupts the way critical brain structures interact to control and inhibit behaviors related to drug abuse. Just as continued abuse may lead to tolerance or the need for higher drug dosages to produce an effect, it also may lead to addiction, which can drive an abuser to seek out and take drugs compulsively. Drug addiction erodes a person's self-control and ability to make sound decisions, while sending intense impulses to take drugs.

All drugs of abuse act in the brain to produce their euphoric effects; however, some of them also have severe negative consequences in the brain such as seizures, stroke, and widespread brain damage that can affect all aspects of daily life. Drug use also can cause brain changes that lead to problems with memory, attention, and decision-making.

Drugs that can cause neurological problems include the following:

- Cocaine
- Gamma-hydroxybutyrate
- Inhalants
- Marijuana
- Methylenedioxymethamphetamine
- Methamphetamine
- Nicotine
- Prescription stimulants
- Rohypnol[11]

The Blood Brain Barrier

The **blood-brain barrier** (BBB) is a thin layer of tightly packed cells that line the brain's blood vessels, shielding the organ from harmful chemical intruders while ushering in needed substances such as glucose, the molecular fuel used by all cells. The function of the BBB is to maintain the fluid level of the brain and spinal cord. Should there be a breakdown of the thin layer of the BBB, the fluid level is changed in the brain resulting in abnormalities. Brain cells and tissues are also at risk from the breakdown of the BBB leading to brain damage.

Drug Dependence

Psychological dependence centers on the user's need of a drug to reach a level of functioning or feeling of well-being. Because this term is particularly subjective and almost impossible to quantify, it is of limited usefulness in making a diagnosis.

Physical dependence is not equivalent to dependence or addiction and may occur with the chronic use of any substance, legal or illegal, even when taken as prescribed. Physical dependence occurs because the body naturally adapts to chronic exposure to a substance (e.g., caffeine, a prescription drug), and when that substance is taken away, symptoms can emerge while the body readjusts to the loss of the substance. Physical dependence can lead to cravings for the drug to relieve the withdrawal symptoms. **Drug dependence** and addiction refer to drug or substance use disorders, which may include physical

---

Drug use also can cause brain changes that lead to problems with memory, attention, and decision-making.

Blood-brain barrier
A thin layer of tightly packed cells that line the brain's blood vessels.

Psychological dependence
The user's need of a drug to reach a level of functioning or feeling of well-being.

Physical dependence
Can lead to craving for the drug to relieve the withdrawal symptoms.

Drug dependence
A state in which the use of a drug is necessary for either physical or psychological well being.

> For alcoholics, the onset of withdrawal symptoms is 24–48 hours after blood alcohol level drops.

Cross dependence
Dependence on a drug that can be relieved only by other similar drugs.

Tolerance
A reduction in the response to a drug after repeated exposure to the drug.

Cross-tolerance
The development of tolerance to one drug that causes tolerance to related drugs.

Metabolites
Substances that result from metabolism.

dependence but also must meet additional criteria. **Cross dependence** is dependence on a drug that can be relieved only by other similar drugs.[13]

Tolerance occurs when the body changes in response to taking a drug causing the drug to no longer be affective at the same repeated dosage. A higher dosage of the drug is needed to maintain the level of response previously achieved. Tolerance may be a chief factor in drug addiction.

Cross-tolerance is defined as the development of tolerance to one substance that causes tolerance to related substances. A tolerance to morphine may cause a cross-tolerance to related drugs like heroin. For example, cross-tolerance between alcohol and nicotine has been well documented in animal models, though difficult to evaluate in humans.

Withdrawal describes the various symptoms that occur after long-term use of a drug is reduced or stopped abruptly. Length of withdrawal and symptoms vary with the type of drug. For example, physical symptoms of heroin withdrawal may include restlessness, muscle and bone pain, insomnia, diarrhea, vomiting, and cold flashes. These physical symptoms may last for several days, but the general depression, or dysphoria (opposite of euphoria), that often accompanies heroin withdrawal may last for weeks. In many cases, withdrawal can be treated easily with medications to ease the symptoms, but treating withdrawal is not the same as treating addiction. For alcoholics, the onset of withdrawal symptoms is twenty-four to forty-eight hours after their blood alcohol level drops. Duration is five to seven days with characteristics of nausea, vomiting, and diarrhea; seizures; delirium; a rise in blood pressure and heart rate; and body temperature. The onset of withdrawal symptoms for the Cannabis (marijuana) user may be a few days. Duration may last up to several weeks with characteristics of irritability, appetite disturbance, sleep disturbance, nausea, concentration problems, and diarrhea.

Drugs can leave the body through different routes of elimination which can include:

- Skin (perspiration)
- Kidney's (urine)
- Lungs (exhalation)
- Bile (eliminated through feces)

Other forms of drug elimination are:

- Saliva
- Breast milk (which may affect the breastfeeding infant)

As drugs are taken into the body, they are metabolized or chemically altered creating new substances called metabolites. The **metabolites** may be eliminated from the body by excretion or changed further through metabolism into another form and then used by the body. Many drugs are metabolized in the liver where they may be changed into active or inactive substances by particular enzymes. The group of **cytochrome P-450** enzymes is responsible for the majority of drug metabolism. These enzymes control how fast a drug may be metabolized by the body.

The metabolic enzyme systems are not fully developed in newborns, causing them to be unable to metabolize some drugs. In older adults, the enzyme systems no longer function as well causing them also to be unable to metabolize drugs as well as younger adults and children. Typically older adults and newborns must receive smaller doses of drugs.

> **Cytochrome P-450**
> A system of enzymes, located primarily in the liver, that participate in the break-down of drugs.

References

National Institute on Drug Abuse (NIDA). NIDA for Teens, Glossary. 2011. http://teens.drugabuse.gov/utilities/glossary.php

National Institute on Drug Abuse (NIDA). "Testimony at the United States Sentencing Commission Public Hearing." 2006. http://www.nida.nih.gov/testimony/11-14-06testimony.html

National Institute on Drug Abuse. NIDA Notes: Inhalation Studies With Drugs of Abuse [Monograph].1990; 173, pp. 201-224. http://archives.drugabuse.gov/pdf/Monographs/Monograph173/201-224_Meng.pdf

National Cancer Institute (NCI). Young People with Cancer: A Handbook for Parents. 2003. http://www.cancer.gov/cancertopics/coping/youngpeople/page5.

National Institute on Health (NIH), Appendix 2, Rx-Norm Dose Definitions. 2011.http://www.nlm.nih.gov/research/umls/rxnorm/docs/2010/appendix2.html

National Institute on Health (NIH), Appendix 2, Rx-Norm Dose Definitions. 2011.http://www.nlm.nih.gov/research/umls/rxnorm/docs/2010/appendix2.html

National Institute on Health (NIH), Appendix 2, Rx-Norm Dose Definitions. 2011.http://www.nlm.nih.gov/research/umls/rxnorm/docs/2010/appendix2.html

Mathias R. Rate and duration of drug activity play major roles in drug abuse, addiction, and treatment. NIDA Notes. 1997; 12 (2). http://archives.drugabuse.gov/nida_notes/nnvol12n2/NIDASupport.html. Accessed April 22, 2011.

Mind over matter: teacher's guide. NIDA for Teens. http://teens.drugabuse.gov/mom/tg_brain.php. Accessed April 22, 2011.

Drugs and the brain. National Institute on Drug Abuse (NIDA). 2010 http://www.nida.nih.gov/scienceofaddiction/brain.html. Accessed April 22, 2011.

National Institute on Drug Abuse (NIDA).Neurological Effects. 2010. http://www.nida.nih.gov/consequences/neurological/

National Institute on Drug Abuse. The blood brain barrier. NIDA Notes. 2006;20(6).

Frequently asked questions. National Institute on Drug Abuse (NIDA). http://www.nida.nih.gov/tools/faq.html#Anchor-What-45736. Accessed April 22, 2011.

Funk D, Marinelli PW, Le AD. Biological processes underlying co-use of alcohol and nicotine: neuronal mechanisms, cross-tolerance, and genetic factors. National Institute on Alcohol Abuse and Alcoholism. http://pubs.niaaa.nih.gov/publications/arh293/186-192.htm. Accessed April 22, 2011.

Kopacek KB. Elimination. The Merck Manuals Online Medical Library. Revised November 2007. http://www.merckmanuals.com/home/sec02/ch011/ch011f.html. Accessed April 22, 2011.

Kopacek KB. Metabolism. The Merck Manuals Online Medical Library. Revised November 2007. http://www.merckmanuals.com/home/sec02/ch011/ch011e.html. Accessed April 22, 2011.

Name _____ Section _____ Date _____

IN-CLASS QUIZ DRUGS: AND YOUR BODY CHAPTER 8

1. List the four ways drugs are introduced into the body.
   A.
   B.
   C.
   D.
2. The user's need of a drug to reach a level of functioning or feeling of wellbeing is called:
   A. Physical dependence
   B. Repeated dependence
   C. Psychological dependence
   D. Cross dependence
3. When some drugs of abuse are taken, they can release two to _____ times the amount of dopamine that natural rewards do.
   A. ten
   B. nine
   C. eight
   D. seven
4. A reduction in the response to a drug after repeated exposure to the drug is called:
   A. Cross dependence
   B. Tolerance
   C. Drug dependence
   D. Repeated response
5. _____ is the most common and simplest way of taking a drug.
   A. Absorption
   B. Inhalation
   C. Injection
   D. Oral Ingestion
6. The _____ is involved in sensory perception and regulation of motor functions.
   A. cerebellum
   B. diencephalon
   C. hypothalamus
   D. thalamus
7. For alcoholics, the onset of withdrawal symptoms is _____ hours after blood alcohol level drops.
   A. twelve to thirty-six
   B. twelve to forty-eight
   C. twenty-four to thirty-six
   D. twenty-four to forty-eight
8. A chemical substance that carries impulses from one nerve cell to another is called a:
   A. Axon
   B. Dendrite
   C. Neurotransmitter
   D. Synapse
9. A cell that detects different forms of energy and conveys it into the electrochemical signals used by the nervous system is called a:
   A. Axon
   B. Glutamate
   C. Receptor
   D. Transporter
10. A neurotransmitter that influences the reward circuit and the ability to learn is called:
    A. Dopamine
    B. Glutamate
    C. Neuron
    D. Synapse

Chapter 9
Psychoactive Drug Types

OBJECTIVES

Students will be able to:
- Identify types of psychoactive drugs.
- Explain the basic neuroscience of the following psychoactive drug types: alcohol, tobacco, marijuana, and prescription stimulants.
- Explain the dangers of misusing or abusing each psychoactive drug type.
- Identify the drug enforcement agency schedule of a psychoactive substance.
- Distinguish between the alcohol content of various beverages.
- Chart the progression of alcohol's sedative effect on the brain and body.

Adaptation of Health and Fitness: A Guide to a Healthly Lifestyle, 5/e by Laura Bounds, Gayden Darnell, Kirstin Brekken Shea and Dottiede Agnor. Copyright © 2012 by Kendall Hunt Publishing Company.

> *"First we form habits, then they form us. Conquer your bad habits, or they'll eventually conquer you."*
>
> —Dr. Rob Gilbert

Introduction

America has always had some opposition to the non-medicinal use of drugs. Alcohol and tobacco created outcries throughout the Country during colonial times and through the Civil War, which provoked prohibition legislation. Warnings of alcohol and tobacco use did not seem to deter the prevalence in American society.

Early prohibitionists were the precursors to the twentieth century "war on drugs" but it was hard to categorize the variety of substances until Congress passed The Controlled Substance Act in 1970.

Richard Evans, a Professor from the University of Houston, created a model that included teaching students to resist social influences and peer pressure. The slogan "Just Say No" was adopted and The National Institutes of Health supported this model. This program that emerged from the substance abuse model created by Evans, became a campaign throughout college campuses. First Lady, Nancy Reagan, became involved in the program in 1980 during her husband's presidency.

This campaign had a positive outcome with a significant decline in drug use during the late 70's and 80's. However, illicit drug use continues to rise in our country.

The World Health Organization's survey of legal and illegal drug use in 17 countries, including the Netherlands and other countries with less stringent drug laws, shows Americans report the highest level of cocaine and marijuana use.

Despite tough anti-drug laws, the United States has the highest level of illegal drug use in the world.

> "Understanding what drugs can do to your children, understanding peer pressure and understanding why they turn to drugs is ... the first step in solving the problem." Nancy Reagan

Alcohol

Alcohol has been prevalent in our society for centuries. Except for the Prohibition Era in the United States from 1917 to 1932, when alcohol was considered an illegal substance, it has become the legal and accepted drug of choice. 131.3 million people, approximately 51.8 percent of Americans twelve and over, reported being current drinkers in the 2010 National Survey on Drug Use and Health (SAMHSA, 2010).

What Is Alcohol?

Ethyl alcohol, or ethanol, is a central nervous system depressant and intoxicating component found in beer, wine, and liquor. The chemical formula is C_2H_5OH.

Distributors create ethyl alcohol through a process called fermentation. **Fermentation** is usually the oxidative decomposition of a simple carbohydrate by enzymes in the yeast organism. Depending on the type of beverage, specific yeasts and carbohydrates are used to produce alcohol.

Types of Drinks

Distilled spirits or hard liquor, include scotch, gin, rum, vodka, tequila, and whiskey. The alcohol content varies according to the proof of the beverage, which is twice the percent of alcohol. For example, if whiskey is 80 proof, then the beverage is 40 percent alcohol by volume. The average mixed drink contains a 1.0 to 1.5 ounce shot of hard liquor.

Wine usually averages 12 percent alcohol by volume and wine coolers average approximately 5 percent alcohol by volume. The average glass of wine is 4.0 to 5.0 ounces. Wine coolers are usually served in 12 ounce bottles.

Beer is typically served in 12 ounce cans or bottles. The average alcohol content of beer is 4.5 to 5.0 percent by volume. To be considered a beer, the alcohol content must not exceed 5 percent by weight by volume. If the amount of alcohol is greater, it is considered an ale.

```
12 oz. Beer            5 oz. Wine
× .05                  × .12
─────────              ─────────
.60 oz. Alcohol        .60 oz. Alcohol

12 oz. Wine Cooler     1.5 oz Whiskey
× .05                  × .40
─────────              ─────────
.60 oz. Alcohol        .60 oz. Alcohol
```

Alcoholic content varies according to the type of drink and the proof of the beverage.

The alcoholic content of some other typical drinks:

86 Proof Liquor	1 oz.	.43 oz.
Light Beer	12 oz	.46 oz.
Champagne	4 oz.	.58 oz.
Malt liquor	12 oz.	.75 oz.
Margarita	12 oz.	.75 oz.

Basic Neuroscience

Ethanol binds directly to neuron receptors and changes neurotransmission. Information is relayed slowly, causing a sedative affects on the brain and body. Long-term abuse irritates and sedates dendrites (nerve endings), causing mood and behavioral changes such as depression, agitation, memory loss, and possibly seizures. With abstinence from alcohol, the dendrites can be repaired and neurotransmitter balance can be restored.

Reading Quiz Preparation

Investigate the immediate and chronic influence of alcohol on the brain and body by visiting *CollegeDrinkingPrevention.gov* Click on the Interactive Body and research alcohol's impact on the following body systems: nervous, circulatory, respiratory, digestive, urinary, and lymphatic.

Blood Alcohol Concentration

Blood Alcohol Concentration (BAC) is a measure of the concentration of alcohol in blood, expressed in grams per 100 ml. An example would be 100 mg of alcohol in 10 ml of blood would be reported as .10 percent. The higher the alcohol content of the drink, the higher

Alcohol slows down the nervous system, impairs vision, and increases the risk of certain cancers, heart, and blood pressure problems.

How to Calculate Your Estimated Blood Alcohol Content/BAC

Showing estimated percent of alcohol in the blood by number of drinks in relation to body weight. This percent can be estimated by:

1. Count your drinks (1 drink *equals* 1 ounce of 100-proof liquor, one five ounce glass of table wine or one 12-ounce bottle of regular beer).
2. Use the chart below and under number of "drinks" and opposite "body weight" find the percent of blood alcohol listed.
3. Subtract from this number the percent of alcohol "burned up" during the time elapsed since your first drink. This figure is .015% per hour. (Example: 180 lb. man—8 drinks in 4 hours / .167% minus (.015 × 4) = .107%

Drinks

Body weight	1	2	3	4	5	6	7	8	9	10	11	12
100 lb.	.038	.075	.113	.150	.188	.225	.263	.300	.338	.375	.413	.450
110 lb.	.034	.066	.103	.137	.172	.207	.241	.275	.309	.344	.379	.412
120 lb.	.031	.063	.094	.125	.156	.188	.219	.250	.281	.313	.344	.375
130 lb.	.029	.058	.087	.116	.145	.174	.203	.232	.261	.290	.320	.348
140 lb.	.027	.054	.080	.107	.134	.161	.188	.214	.241	.268	.295	.321
150 lb.	.025	.050	.075	.100	.125	.151	.176	.201	.226	.251	.276	.301
160 lb.	.023	.047	.070	.094	.117	.141	.164	.188	.211	.234	.258	.281
170 lb.	.022	.045	.066	.088	.110	.132	.155	.178	.200	.221	.244	.265
180 lb.	.021	.042	.063	.083	.104	.125	.146	.167	.188	.208	.229	.250
190 lb.	.020	.040	.059	.079	.099	.119	.138	.158	.179	.198	.217	.237
200 lb.	.019	.038	.056	.075	.094	.113	.131	.150	.169	.188	.206	.225
210 lb.	.018	.036	.053	.071	.090	.107	.125	.143	.161	.179	.197	.215
220 lb.	.017	.034	.051	.068	.085	.102	.119	.136	.153	.170	.188	.205
230 lb.	.016	.032	.049	.065	.081	.098	.115	.130	.147	.163	.180	.196
240 lb.	.016	.031	.047	.063	.078	.094	.109	.125	.141	.156	.172	.188

Source: National Highway Traffic Safety Administration.

BAC it will produce. Additional factors influencing a person's BAC are body weight, size of the drink, time spent drinking, and food (see table below). Gender is also a factor in determining one's BAC.

The human body absorbs alcohol quickly. From the stomach and small intestine, alcohol enters the blood stream. With a full stomach, alcohol enters the blood stream at a slower rate when compared to an empty stomach.

The liver is the only organ capable of metabolizing alcohol. Ninety percent of alcohol is metabolized through the oxidation process of the liver, at a rate of .015 percent per hour. The lungs and kidneys eliminate the remaining ten percent. For the average individual, the rate of elimination will reduce a given BAC by .015 per hour.

Alcohol dehydrogenase converts alcohol to acetaldehyde. Alcohol is then metabolized at approximately 0.25 to 0.30 ounces per hour, regardless of the blood alcohol concentration. The rate of metabolism is based on the activity of alcohol dehydrogenase, working at its own pace (Ray and Kisr, 1999).

Women do not process alcohol as well as men. There are two biological factors contributing to this principle. Women have a much higher percentage of fat mass compared to men. Alcohol is soluble in oils, so molecules permeate the membranes of cells quickly and easily. This results in a faster rate of absorption and a higher BAC for women. Additionally, men produce more alcohol

FIGURE 9.1
Alcohol Absorption and Elimination

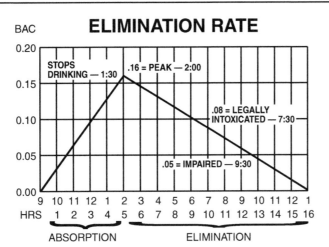

Source: Texas Commission on Alcohol and Drug Abuse

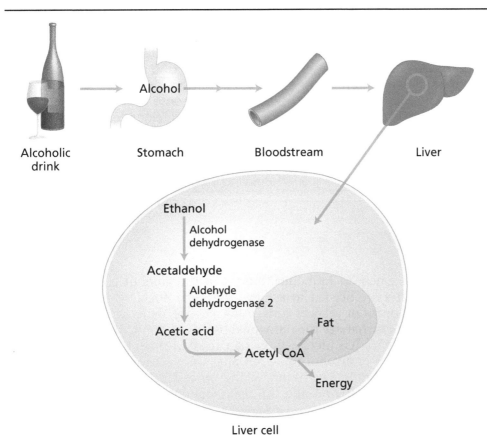

Alcohol is absorbed in the stomach. From the stomach the alcohol enters the bloodstream and then travels through the bloodstream into the liver, where liver cells break it down. When alcohol is broken down by the liver, it is either used as a source of energy or, more likely, converted to fat.

From *Nutrition: Real People, Real Choices, 2nd Edition* by Susan J. Hewlings and Dennis M. Medeiros. Copyright © 2011 by Kendall Hunt Publishing Company. Reprinted by permission.

dehydrogenase than women; therefore, men can eliminate alcohol at a slightly faster rate (Dennis and the Texas Commission on Alcohol and Drug Abuse, 2005).

Intoxication

Intoxication is defined as a transient state of physical and mental disruption due to the presence of a toxic substance, such as alcohol (Maisto, 2005). As BAC increases, the central nervous system alters behavior and physical function. Change can occur as low as 0.02 BAC in some people, while everyone is impaired to some degree at 0.05 BAC.

Blood Alcohol Level and Behavioral Effects

BAC%	Behavioral Effect(s)
0.02–0.03	Alcohol starts to relax the drinker.
0.04	Often drinkers at this stage are relaxed, happier, chatty.
0.05	Skills begin to diminish, as far as judgment, attention, and control. Decision-making ability (such as whether to drive) is impaired, as are sensory-motor skills.
0.08	All states recognize this as the legal point of intoxication. Driving skills and coordination are impaired.
0.10–0.125	Several effects are likely at this point, such as balance, vision, speech, and control problems. Reaction time increases.
0.12–0.15	Peripheral vision is diminished, so less detail is visible. Balance and coordination are problematic. A sense of tiredness, displays of unstable emotions, diminished perception, memory, and comprehension are seen at this level. A person may vomit if not accustomed to drinking or if this BAC level has been reached too quickly.
0.18–0.25	Often the drinker is in a state of apathy and lethargy, and is less likely to feel pain. Vision is certainly diminished in the areas of color, form, motion and dimensions. Drinkers are intense with emotions, confused, dizzy, and disoriented. Walking is often difficult or impossible as muscle coordination is diminished and speech is slurred.
0.25–0.30	Drinkers may lose consciousness during this stage. They have almost completely lost motor functions, have little response to stimuli, can't stand or walk, and experience vomiting and incontinence.
0.30–0.50	Once the BAC reaches 0.45 percent, alcohol poisoning is almost always fatal, and death may occur at a level of 0.37 percent. Unconsciousness, diminished or absent reflexes, lower than normal body temperature, circulatory and respiratory problems and incontinence commonly occur.

From *Nutrition: Real People, Real Choices*, 2nd Edition by Susan J. Hewlings and Dennis M. Medeiros. Copyright © 2011 by Kendall Hunt Publishing Company. Reprinted by permission.

Tolerance is when an individual adapts to the amount consumed so that larger quantities are needed to achieve the same effect. This can take place over several months or years of consuming alcohol, depending on the amount consumed and at what age the individual begins to drink. At some point, after a person's tolerance has increased over a period of time, it begins to drop, allowing the effects of alcohol to be felt after only a few drinks. This **reverse tolerance** is caused by the natural aging process or liver disease after years of abusive drinking (Dennis and the Texas Commission of Alcohol and Drug Abuse, 2005).

Drinking Problems

Nearly 14 million Americans, one in every thirteen adults, abuse alcohol or are alcoholics. Rates of alcohol problems are highest among young adults ages 18 to 29 and lowest in adults ages 65 and older (NIAAA, 2006).

The National Institute on Alcohol Abuse and Alcoholism (NIAAA) found that the earlier young people begin to drink alcohol, the more likely they are to become an alcohol abuser or alcoholic. According to the report:

- Young people who start drinking before age 15 are four times more likely to become an alcoholic than if they start after age 21.
- Forty percent who drink before age 15 become alcohol dependent; 10 percent if they wait until 21.
- Fourteen percent decreased risk of alcoholism for each year drinking is delayed until age 21.

Michael Wagener Story

On August 3, 1999, Michael Wagener, a student at Texas A&M University celebrating his 21st birthday, died as a result of alcohol poisoning. He was an intelligent and insightful young man with many friends and his whole life ahead of him. He was not an alcoholic, nor did he abuse alcohol. Michael was typically a responsible drinker.

On August 2nd, the eve of his 21st birthday, friends joined Michael at a local establishment. While having a few beers, some friends bought him a couple of shots for his birthday. His friends had bought him eight or nine (four-ounce) shots in a matter of thirty to forty-five minutes. Michael had many friends who wanted to share in his celebration; no one wanted him to die.

By the time he was taken home, Michael's body had begun to shut down. He could no longer move and had to be carried into the house. His friends thought they had taken all the precautions: designated driver, turn him on his side in case he vomits. They even stayed the night to ensure his safety.

At 7:00 a.m. his mother called to wish Michael a happy birthday. The call stirred his friends. At 7:10 a.m. the call was made to 911—Michael never woke up. One fun-filled night of celebration turned deadly.

This can happen to anyone. Consuming excessive amounts of alcohol, even one night, can kill you. We often think the only way alcohol can kill is if someone drinks and drives or abuses alcohol for many years. Educating yourself about alcohol will help you make informed decisions and hopefully prevent this tragedy from occurring again.

Alcoholism

Alcoholism, also known as alcohol dependence, is a chronic, progressive disease with symptoms that include a strong need to drink and continued drinking despite repeated negative alcohol-related consequences. There are four symptoms generally associated with alcoholism:

1. a craving or a strong need to drink,
2. impaired control or the inability to limit one's drinking,
3. a physical dependence accompanied by withdrawal symptoms such as nausea, sweating, shakiness, and anxiety when alcohol use is stopped, and
4. an increased tolerance.

Can alcoholism be hereditary? Alcoholism has a biological base. The tendency to become an alcoholic is inherited. Men and women are four times more likely to become alcoholics if their parents were (NIAAA, 2008). Currently, researchers are finding the genes that influence vulnerability to alcohol. A person's environment may also play a role in drinking and the development of alcoholism. This is not destiny. A child of an alcoholic parent will not automatically develop alcoholism, and a person with no family history of alcohol can become alcohol dependent.

There are ways to avoid becoming alcohol dependent. It is important to know your limit and stick to it. If choosing to drink, drink slowly and alternate an alcoholic beverage with a non-alcoholic beverage, eat while drinking, and most importantly find more effective ways of dealing with problems instead of turning to alcohol.

If you feel this is a problem, the sooner you stop the better the chances of avoiding serious psychological effects.

- Admit to your drinking—first step in avoiding serious problems.
- Change your lifestyle—try to stay out of situations where alcohol is prominent until you can control your drinking.
- Get involved in self-help groups.

How can you tell if someone has a drinking problem? An individual does not have to be an alcoholic to have problems with alcohol. Problems linked to abuse are neglecting work, school, or family responsibilities. Legal issues such as alcohol violations and drinking-and-driving-related problems can also be a result of alcohol abuse. There are many "red flags" that can point to a problem with alcohol. One way is to answer these questions developed by Dr. John Ewing:

- Have you ever felt you should CUT down on your drinking?
- Have people ANNOYED you by criticizing your drinking?
- Have you ever felt bad or GUILTY about your drinking?
- Have you ever had a drink first thing in the morning to steady your nerves or to get rid of the hangover ("EYE OPENER")?

To help remember these questions, notice that the first letter of each key word spells CAGE. One "yes" answer suggests a possible alcohol problem. More than one "yes" means it is highly likely that a problem exists (Ewing, 1995).

Other signs and symptoms also could indicate that a person could be misusing or abusing alcohol or other drugs. One or two of them does not necessarily point to a problem, but several, combined with the right circumstances, need to be addressed. Some of these signs may include a grade decline or a sudden drop in grades, frequently missing class because of hangovers, binge drinking, legal problems associated with alcohol, or a significant increase in tolerance to alcohol. Other major signs of a drinking problem could be frequently drinking alone, drinking to forget about personal problems, or avoiding activities where alcohol is not available. Another more serious physical sign of alcohol abuse is a **blackout**. This occurs when an individual has amnesia about events after drinking, even though there was no loss of consciousness.

Chronic Effects

Drinking too much alcohol can cause a wide range of chronic health problems including liver disease, cancer, heart disease, nervous system problems, as well as alcoholism (see Figure 9.2).

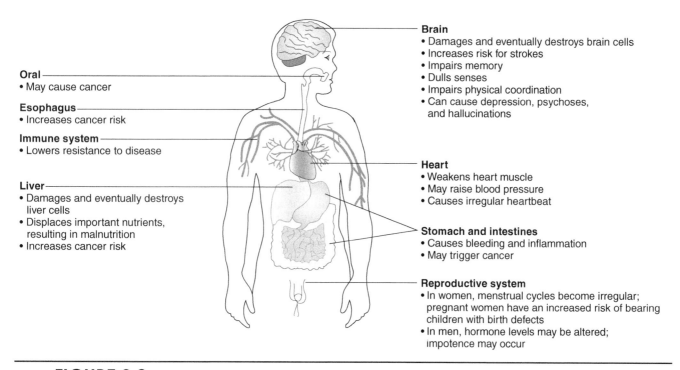

FIGURE 9.2

Long-term Risks Associated with Alcohol Abuse

Source: Adapted from "Long-Term Risk Associated with Alcohol Abuse" *Lifetime Physical Fitness & Wellness* by Hoeger and Hoeger, Wadsworth Publishers.

Although moderate amounts of alcohol may not be harmful, there are some major health issues associated with chronic alcohol use and abuse.

- **Liver disease** is commonly associated with alcohol abuse. The liver has many vital functions in the body. It is a common mistake for people to think that only those individuals who abuse alcohol can harm the liver. Individuals who are heavy social drinkers may run the risk of liver damage as well.
- **Hepatotoxic trauma** or "fatty liver" is the most common alcohol-related disorder causing enlargement of the liver. Some damage can be reversed if alcohol is completely avoided.
- **Alcoholic hepatitis** is an enlarged and tender liver with an elevation of white blood cells. Symptoms can include nausea, vomiting, abdominal pain, fever, and jaundice. If alcohol use continues, this could progress to cirrhosis.
- **Alcohol cirrhosis** results from continued alcohol use and may cause permanent scar tissue to form when the liver cells are damaged. This problem usually occurs in 10 to 15 percent of people who consume large quantities and can develop in as little as five years of heavy drinking.
- **Alcohol pellagra** is a deficiency of protein and niacin. Symptoms may include skin inflammation, gastrointestinal disorders, diarrhea, and mental and nervous disorders.
- **Malnutrition** occurs from a lack of needed nutrients through prolonged alcohol consumption, by depressing the appetite and attacking the lining of the stomach. Heavy drinkers do not get the calories they need, which triggers increased mineral loss and increases fatty acids because of the interference of the transfer of glucose into energy.
- **Polyneuritis** is a condition caused by thiamin deficiency, which causes inflammation of several nerves and causes the drinker to become weak and have a tingling sensation.
- **Cancers**—It is established that 2 to 4 percent of all cancer cases could be caused by alcohol use. Cancer of the upper digestive tract such as mouth, esophageal, pharynx, and larynx can be attributed to alcohol use. Liver cancer as well as breast cancer may be caused by excessive alcohol consumption. Studies indicated that a woman's risk of developing breast cancer increases with age and alcohol consumption (JAMA, 1995).
- **Fetal alcohol syndrome.** The alcohol crosses the placenta but experts don't know exactly how drinking causes problems for the fetus. It may directly affect the fetus or it may be acetaldehyde, the metabolic by-product of alcohol that is harmful to the fetus. Some researchers believe that alcohol effects on the placenta cause blood flow and nutrient deficiencies. Whatever the reason, drinking during pregnancy clearly puts infants at risk for birth defects (Herman, 2003).

Neurological disorders associated with alcohol use are:

- **Wernickes disease** is caused by a thiamine deficiency. Some symptoms include decreased mental functions, double vision, and involuntary oscillation of the eyeballs.
- **Korsakoff's syndrome** is caused by a B complex vitamin deficiency. Symptoms are amnesia, personality alterations, and a loss of reality. This person may become apathetic and have difficulty walking.

Organizations For information regarding alcohol use and abuse contact:

- National Institute on Alcohol Abuse and Alcoholism (NIAAA) www.niaaa.org
- *Alcoholics Anonymous (AA)* is an organization designed to support and help individuals become sober and stay sober. AA has over 19,000 affiliated groups and more than 350,000 members across the United States AA (212) 870-3400 www.alcoholic-anonymous.org

- *Al-Anon* and *Alateen* are organizations designed to help family members of alcoholics to cope with problems. 800-344-2666 www.al-anon-alateen.org

The dangers of alcohol consumption are a major problem in our society. Drinking too much alcohol can cause a range of very serious problems, in addition to the obvious health issues. Alcohol is a contributing factor in motor vehicle accidents, violence, and school/work problems, as well as family problems.

Drinking and Driving

Driving under the influence of alcohol is the most frequently committed and deadliest crime in America. In the Federal Bureau of Investigation's (FBI) Uniform Crime Report, more than 1.4 million people were arrested in 2009 for alcohol-impaired driving. The National Highway Traffic Safety Administration (NHTSA) reported in 2009 that 10,839 people were killed in alcohol impaired driving crashes. This is 32 percent of the nation's total traffic fatalities for the year. The 10,839 deaths in 2009 represent an average of one alcohol-related fatality every 22 minutes (NHTSA, 2011). Even small amounts of alcohol impair driving (see Table 9.1)

Drunk driving is no accident; it is a crime. The greatest tragedy is that these crashes are preventable, predictable, and 100 percent avoidable.

Although most drivers involved in fatal crashes have no prior convictions for DUI, about one-third of all drivers arrested for DUI are repeat offenders, which greatly increases their risk of causing a drunk driving accident. As a nation, we have seen a downward trend in alcohol-related fatalities. Today, all states have lowered their legal level of intoxication to .08 BAC. All states

---

A Snapshot of Annual High-Risk College Drinking Consequences (NIAAA, 2010)

Academic Problems: About 25 percent of college students report academic consequences of their drinking including missing classes, falling behind, doing poorly on exams or papers, and receiving lower grades overall.

Police Involvement: About 5 percent of 4-year college students are involved with the police or campus security as a result of their drinking, and 110,000 students between the ages of 18 and 24 are arrested for an alcohol-related violation such as public drunkenness or driving under the influence.

Alcohol Abuse and Dependence: 31 percent of college students met criteria for a diagnosis of alcohol abuse and 6 percent for a diagnosis of alcohol dependence in the past 12 months, according to questionnaire-based self-reports about their drinking.

TABLE 9.1 ♦ Some Likely Effects on Driving

Blood Alcohol Concentration (BAC) Levels		Effect
.15%	About 7 beers	Serious difficulty controlling the car and focusing on driving
.10%	About 5 beers	Marked slowed reaction time Difficulty staying in lane and braking when needed
.08%	About 4 beers	Trouble controlling speed Difficulty processing information and reasoning.
.05%	About 3 beers	Reduced coordination and ability to track moving objects Difficulty steering
.02%	About 2 beers	Loss of judgment Trouble doing two tasks at the same time

Source: Adapted from The ABCs of BAC, National Highway Traffic Safety Administration, 2005, and How to Control Your Drinking, WR Miller and RF Munoz, University of New Mexico, 1982.

have some form of the zero tolerance law, as well as an open container law. These laws, in addition to stricter enforcement of existing laws, have helped in changing behavior. High school and university education programs, such as non-alcoholic activities for prom nights and designated driver organizations, have also contributed in raising awareness to combat such a serious problem.

The NHTSA and the Advertising Council's Innocent Victims public service campaign stresses the need to get the keys from someone who is about to drive. Here are some tips:

- If it is a close friend, try to use a soft, calm approach. Suggest to them that they have had too much to drink and it would be better if someone else drove, or call a cab.
- Be calm. Joke about it. Make light of it.
- Try to make it sound like you are doing them a favor.
- If it is somebody you do not know well, speak to their friends; usually they will listen.
- If it is a good friend, tell them if they insist on driving, you are not going with them.
- Locate their keys while they are preoccupied and take them away. Mostly they will think they lost them and will be forced to find another mode of transportation.
- Avoid embarrassing the person or being confrontational.

Alcohol Use in College

The legal drinking age in all states is 21 years old, but that does not mean individuals under 21 do not consume alcohol. Studies suggest that substance use, including alcohol, tobacco, and other drug use, is common among college-aged youth. Students who use any of these substances are at significantly greater risk than non-substance using peers to: drive after drinking and with a driver who has been drinking, and are less likely to use a seatbelt. These consistently poor and risky choices increase their risk of being in a motor vehicle crash and having crash-related injuries (Everett, 1999). College students and administrators struggle with the problems associated with alcohol abuse, binge drinking, and drunk driving (see Figure 9.3). These actions put students at risk for many serious problems, such as date rape violence and possibly death.

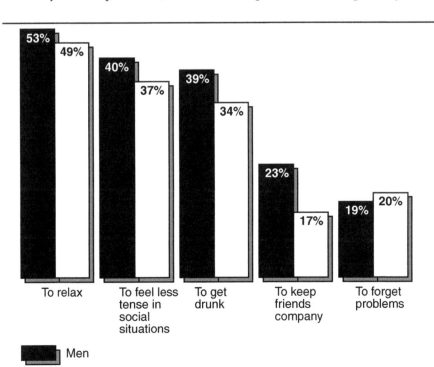

FIGURE 9.3

Why College Freshmen Drink

Source: Adapted from Weschler and McFadden for AAA Foundation for Traffic Safety. Survey of 1,669 college freshmen at 14 Massachusetts institutions.

The National Institute on Alcohol Abuse and Alcoholism (NIAAA) reports that 1,825 college students die annually from alcohol-related unintentional injuries, including motor vehicle crashes, with 3.36 million students driving under the influence of alcohol. Another 599,000 between the ages of 18 and 24 are injured, and approximately 696,000 students per year are assaulted by a drinking student. Also, approximately 400,000 students between 18 and 24 years old reported having unprotected sex as a result of drinking. More than 97,000 students are victims of alcohol-related sexual assault or date rape. More than 150,000 college students develop an alcohol-related health problem, with 1.5 percent attempting suicide because of alcohol (NIAAA, 2010).

Binge Drinking

College presidents agree that **binge drinking** is the most serious problem on campus. The National Institute on Alcohol Abuse and Alcoholism defines **binge drinking** as a pattern of drinking that brings a person's blood alcohol concentration to .08 or above. This typically happens when men consume five or more drinks and women consume four or more drinks in about two hours (NIAAA, 2010). Binge drinkers usually experience more alcohol-related problems than their non-drinking counterparts. These problems affect their health, education, safety, and interpersonal relationships. According to the Harvard School of Public Health College Alcohol Study, these problems include driving after drinking, damaging property, getting injured, missing classes, and getting behind in school work. According to the same Harvard study, one in five students surveyed experienced five or more different alcohol-related problems and more than one-third of the students reported driving after drinking.

The study also found that the vast majority of non-binge drinking students are negatively affected by the behavior of binge drinkers. It was reported that four out of five students who were non-binge drinkers and who lived on campus experienced secondary effects of binge drinking such as being the victim of a sexual assault or an unwanted sexual advance, having property vandalized, and having sleep or study interrupted.

Alcohol Poisoning

> For more information on 911 Lifeline Legislation visit awareawakealive.org

The most serious consequence of binge drinking is **alcohol poisoning.** This results when an overdose of alcohol is consumed. When excessive amounts of alcohol are consumed, the brain is deprived of oxygen, which causes it to shut down the breathing and heart rate functions. Many think that the only deadly mix is alcohol and driving, but an alcohol overdose can be lethal. It can happen to anyone.

Some symptoms of alcohol poisoning are:

- Person does not respond to talking, shouting, or being shaken.
- Person cannot stand up.
- Person has slow, labored, or abnormal breathing—less than eight breaths/minute or ten or more seconds between each breath.
- Person's skin feels clammy.
- Person has a rapid pulse rate and irregular heart rhythm.
- Person has lowered blood pressure.
- Vomiting.

If you think a friend is experiencing alcohol poisoning, seek medical attention immediately. Stay with the person until help arrives. Turn the victim onto one side in case of vomiting. Choking to death on one's own vomit after an alcohol overdose is quite common. Death by asphyxiation occurs when alcohol depresses and inhibits the gag reflex to the point that the person cannot vomit properly. **Do not leave the victim alone.** Be honest in telling medical staff exactly how much alcohol the victim consumed. This is an extreme medical emergency and one that is a matter of life and death. Some states have passed legislation providing limited immunity for a minor who calls 911 for someone who is a possible victim of alcohol poisoning.

Colleges are attempting to make progress in preventing some of these problems. Many sororities and fraternities as well as other student organizations have taken action by banning alcohol at many functions. By implementing alcohol awareness programs, stronger hazing policies, and tougher enforcements on drinking violations and alcohol restrictions on campus and with the student body, some of these tragedies may be prevented.

Laws Relating to Alcohol

In every state in the United States, it is illegal for a person under the age of 21 to attempt to purchase, possess, or consume alcohol. In Texas, this violation is a **minor in possession** (MIP). This offense is punishable by fines, community service, loss of driver's license, alcohol awareness class, and possibly jail.

The most serious consequence of binge drinking is alcohol poisoning. Alcohol overdose can be lethal.

The **Zero Tolerance Law** prohibits the use of alcohol by a minor operating a motor vehicle. In Texas, it is illegal for a minor to operate a motor vehicle in a public place with *any* **detectable** amount of alcohol. This violation is referred to as **driving under the influence** (DUI). This may be determined by a blood/breath test or simply smelling alcohol on the minor's breath. The penalties are very similar to MIP with fine, loss of license, education courses, and community service.

By operating a motor vehicle in a public place, the driver has given consent to take a breath/blood test to determine alcohol in his/her system. Refusing or failing the test is considered a violation, and penalties will result in loss of license, regardless of the outcome of the violation.

In many states, the legal definition of **driving while intoxicated (DWI)** is not having normal use of your mental or physical faculties because of alcohol or other drugs; or a blood alcohol concentration of .08 or more. It is, however, illegal in all states to drink and drive. In addition, in most states, it is also illegal for anyone in the vehicle to possess an **open container** of alcohol regardless of age.

When someone is injured in an alcohol-related motor vehicle accident, the intoxicated driver can be charged with **intoxication assault.** If there were a fatality in a drinking-and-driving accident, the offense would be elevated to **intoxication manslaughter**. Each state may have different terminology, but the offense is the same throughout the country. In Texas, both offenses are considered felonies.

Tobacco

The U.S. Surgeon General reported in 1970 that cigarette smoking is dangerous to your health. Over the years we have come to realize just how dangerous. Cigarette smoking is the leading preventable cause of death in the United States, responsible for one in five deaths annually. Through study after study, reports have proven that tobacco use is one of the biggest public health issues that faces the world today (CDC, 2006) (see Figure 9.4). Tobacco is a risk factor for six of the eight leading causes of death, with smokers dying thirteen to fourteen years earlier than non-smokers.

Tobacco Components

The toxic components of tobacco include tar, nicotine, and carbon monoxide. **Tar** is a by-product of burning tobacco. Its composition is a dark, sticky substance that can be condensed from cigarette smoke. Tar contains many potent carcinogens and chemicals that irritate tissue in the lungs and promote

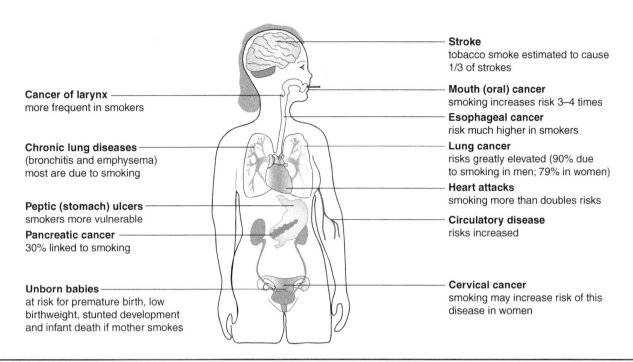

FIGURE 9.4

The Health Effects of Smoking

Source: Adapted from "The Health Effects of Smoking" Lifetime Physical Fitness & Wellness by Hoeger and Hoeger, Wadsworth Publishing.

chronic bronchitis and emphysema. These substances paralyze and destroy the cilia that line the bronchi, causing "smoker's cough." Long-term exposure of extremely toxic tar to lung tissue can lead to the development of cancer.

Nicotine is a colorless, oily compound that is extremely poisonous in concentrated amounts. This highly addictive drug is a major contributor to heart and respiratory diseases causing short-term increases in blood pressure, heart rate, and blood flow from the heart, resulting in narrowing of the arteries. A strong dependence on nicotine can occur after as little as three packs of cigarettes, and it is more addictive than cocaine or heroin. Because of its addictive effects, the Food and Drug Administration (FDA) has determined nicotine should be regulated.

At first, nicotine acts as a stimulant and then it tends to tranquilize the nervous system. The effects depend largely on how one chooses to smoke. Shallow puffs seem to increase alertness because low doses of nicotine facilitate the release of acetylcholine, which creates feelings of alertness. Long, deep drags tend to relax the smoker because high doses of nicotine block the flow of acetylcholine. Ninety percent of the nicotine inhaled while smoking is absorbed into the body, while 20 to 30 percent of nicotine is absorbed if the smoke is drawn only into the mouth, not the lungs.

Other side effects include inhibiting formation of urine, discoloration of the fingers, dulling the taste buds, and irritating the membranes in the mouth and throat. Because nicotine constricts blood vessels, it causes the skin to be clammy and have a pallid appearance, as well as reducing body temperature. The highly addictive nature of nicotine can cause withdrawal symptoms to occur quite suddenly. These symptoms include irritability, anxiousness, hostility, food cravings, headaches, and the inability to concentrate.

Carbon monoxide is an odorless, tasteless gas that is highly toxic. It reduces the amount of oxygen the blood can carry, causing shortness of breath. Carbon monoxide ultimately damages the inner walls of the arteries, thus encouraging a buildup of fat on the walls of the arteries; this is called atherosclerosis. Over time, this causes the arteries to narrow and harden, which may lead to a heart attack.

Approximately 1 percent of cigarette smoke and 6 percent of cigar smoke is carbon monoxide. It impairs normal function of the nervous system and is partially responsible for the increased risk of heart attacks and strokes in smokers.

Types of Tobacco Use

Cigarette smoking greatly impairs the respiratory system and is a major cause of chronic obstructive pulmonary diseases (COPD), including emphysema and chronic bronchitis.

Problems associated with cigarette smoking (see Figure 9.2) include mouth, throat, and other types of cancer, cirrhosis of the liver, stomach, and duodenal ulcers, gum and dental disease, decreased HDL cholesterol and decreased platelet survival and clotting time, as well as increased blood thickness.

Cigarette smoking increases problems such as heart disease, atherosclerosis, and blood clots. It increases the amount of fatty acids, glucose, and various hormones in the blood, cardiac arrhythmia, allergies, diabetes, hypertension, peptic ulcers, and sexual impotence. Smoking doubles the risk of heart disease, and those who smoke have only a 50 percent chance of recovery. Smokers also have a 70 percent higher death rate from heart disease than non-smokers (CDC, 2006). Smoking also causes cardiomyopathy, a condition that weakens the heart's ability to pump blood.

Life expectancy of smokers parallels smoking habits in that the younger one starts smoking and the longer one smokes, the higher the mortality rate. Also, the deeper smoke is inhaled and the higher the tar and nicotine content, the higher the mortality rate. On average, smokers die 13–14 years earlier than non-smokers (CDC, 2012).

The risk and mortality rates for lip, mouth, and larynx cancers for **pipe and cigar smoking** are higher than for cigarette smoking. Pipe smoke, which is 2 percent carbon monoxide, is more irritating to the respiratory system than cigarette smoking, but for those who do not inhale, the risk for developing cancer is just as likely.

Cigars have recently gained popularity in the United States among younger men and women with approximately 4.5 billion cigars consumed yearly.

Clove cigarettes are erroneously believed to be safer because they do not contain as much tobacco. In actuality, clove cigarettes are most harmful because they contain **eugenol**, which is an active ingredient of clove. Eugenol deadens sensations in the throat, which allows smokers to inhale more deeply and hold smoke in the lungs longer. Clove cigarettes also contain twice as much tar, nicotine, and carbon monoxide as most moderate brands of American cigarettes.

© Gemenacom, 2014. Used under license from Shutterstock, Inc.

Once thought of as a less harmful way to smoke tobacco, **water pipes** have regained popularity in many cities across the United States. In 2005 the World Health Organization (WHO) published its findings emphasizing the harmful effects of waterpipe smoking, also known around the world as narghile, shisha, goza, and hookah. For centuries, smokers have been lead to believe that water pipe smoking is a safer alternative to cigarette smoking, but research shows that this method exposes the smoker to high rates of lung cancer and heart disease, as well as other tobacco-related diseases. There is also a high risk of communicable diseases like tuberculosis and hepatitis because of shared mouthpieces (WHO, 2005).

The younger someone is when they start smoking, and the longer they continue to smoke, the greater their chance of dying from a smoking-related illness.

Research Report Series

Tobacco/Nicotine

from the director:

Tobacco use kills approximately 440,000 Americans each year, with one in every five U.S. deaths the result of smoking. Smoking harms nearly every organ in the body, causes many diseases, and compromises smokers' health in general. Nicotine, a component of tobacco, is the primary reason that tobacco is addictive, although cigarette smoke contains many other dangerous chemicals, including tar, carbon monoxide, acetaldehyde, nitrosamines, and more.

An improved overall understanding of addiction and of nicotine as an addictive drug has been instrumental in developing medications and behavioral treatments for tobacco addiction. For example, the nicotine patch and gum, now readily available at drugstores and supermarkets nationwide, have proven effective for smoking cessation when combined with behavioral therapy.

Advanced neuroimaging technologies make it possible for researchers to observe changes in brain function that result from smoking tobacco. Researchers are now also identifying genes that predispose people to tobacco addiction and predict their response to smoking cessation treatments. These findings—and many other recent research accomplishments—present unique opportunities to discover, develop, and disseminate new treatments for tobacco addiction, as well as scientifically based prevention programs to help curtail the public health burden that tobacco use represents.

We hope this Research Report will help readers understand the harmful effects of tobacco use and identify best practices for the prevention and treatment of tobacco addiction.

Nora D. Volkow, M.D.
Director
National Institute on Drug Abuse

What Are the Extent and Impact of Tobacco Use?

According to the 2010 National Survey on Drug Use and Health, an estimated 69.6 million Americans aged 12 or older reported current use of tobacco—58.3 million (23.0 percent of the population) were current cigarette smokers, 13.2 million (5.2 percent) smoked cigars, 8.9 million (3.5 percent) used smokeless tobacco, and 2.2 million (0.8 percent) smoked pipes, confirming that tobacco is one of the most widely abused substances in the United States. Although the numbers of people who smoke are still unacceptably high, according to the Centers for Disease Control and Prevention there has been a decline of almost 50 percent since 1965.

Source: U.S. Department of Health and Human Services / National Institutes of Health.

Research Report Series

Tobacco/Nicotine

NIDA's 2011 Monitoring the Future survey of 8th-, 10th-, and 12th-graders, which is used to track drug use patterns and attitudes, has also shown a striking decrease in smoking trends among the Nation's youth. The latest results indicate that about 6 percent of 8th-graders, 12 percent of 10th-graders, and 19 percent of 12th-graders had used cigarettes in the 30 days prior to the survey—the lowest levels in the history of the survey.

The declining prevalence of cigarette smoking among the general U.S. population, however, is not reflected in patients with mental illnesses. The rate of smoking in patients suffering from post-traumatic stress disorder, bipolar disorder, major depression, and other mental illnesses is twofold to fourfold higher than in the general population; and among people with schizophrenia, smoking rates as high as 90 percent have been reported.

Tobacco use is the leading preventable cause of death in the United States. The impact of tobacco use in terms of morbidity and mortality to society is staggering.

Economically, more than $96 billion of total U.S. healthcare costs each year are attributable directly to smoking. However, this is well below the total cost to society because it does not include burn care from smoking-related fires, perinatal care for low-birthweight infants of mothers who smoke, and medical care costs associated with disease caused by secondhand smoke. In addition to healthcare costs, the costs of lost productivity due to smoking effects are estimated at $97 billion per year, bringing a conservative estimate of the economic burden of smoking to more than $193 billion per year.

How Does Tobacco Deliver Its Effects?

There are more than 7,000 chemicals found in the smoke of tobacco products. Of these, nicotine, first identified in the early 1800s, is the primary reinforcing component of tobacco.

Cigarette smoking is the most popular method of using tobacco; however, many people also use smokeless tobacco products, such as snuff and chewing tobacco. These smokeless products also contain nicotine, as well as many toxic chemicals.

The cigarette is a very efficient and highly engineered drug delivery system. By inhaling tobacco smoke, the average smoker takes in 1–2 milligrams of nicotine per cigarette. When tobacco is smoked, nicotine rapidly reaches peak levels in the bloodstream and enters the brain. A typical smoker will take 10 puffs on a cigarette over a period of 5 minutes that the cigarette is lit. Thus, a person who smokes about 1½ packs (30 cigarettes) daily gets 300 "hits" of nicotine to the brain each day. In those who typically

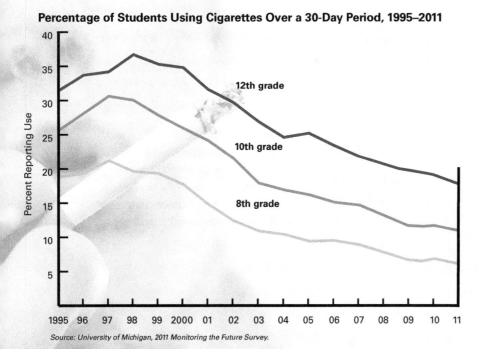

Trends in Prevalence of Cigarette Use for 8th-, 10th-, and 12th-Graders

Percentage of Students Using Cigarettes Over a 30-Day Period, 1995–2011

Source: University of Michigan, 2011 Monitoring the Future Survey.

do not inhale the smoke—such as cigar and pipe smokers and smokeless tobacco users—nicotine is absorbed through the mucosal membranes and reaches peak blood levels and the brain more slowly.

Immediately after exposure to nicotine, there is a "kick" caused in part by the drug's stimulation of the adrenal glands and resulting discharge of epinephrine (adrenaline). The rush of adrenaline stimulates the body and causes an increase in blood pressure, respiration, and heart rate.

Is Nicotine Addictive?

Yes. Most smokers use tobacco regularly because they are addicted to nicotine. Addiction is characterized by compulsive drug-seeking and abuse, even in the face of negative health consequences. It is well-documented that most smokers identify tobacco use as harmful and express a desire to reduce or stop using it, and nearly 35 million of them want to quit each year. Unfortunately, more than 85 percent of those who try to quit on their own relapse, most within a week.

Research has shown how nicotine acts on the brain to produce a number of effects. Of primary importance to its addictive nature are findings that nicotine activates reward pathways—the brain circuitry that regulates feelings of pleasure. A key brain chemical involved in mediating the desire to consume drugs is the neurotransmitter dopamine, and research has shown that nicotine increases levels of dopamine in the reward circuits. This reaction is similar to that seen with other drugs of abuse and is thought to underlie the pleasurable sensations experienced by many smokers. For many tobacco users, long-term brain changes induced by continued nicotine exposure result in addiction.

Nicotine's pharmacokinetic properties also enhance its abuse potential. Cigarette smoking produces a rapid distribution of nicotine to the brain, with drug levels peaking within 10 seconds of inhalation. However, the acute effects of nicotine dissipate quickly, as do the associated feelings of reward, which causes the smoker to continue dosing to maintain the drug's pleasurable effects and prevent withdrawal.

Nicotine withdrawal symptoms include irritability, craving, depression, anxiety, cognitive and attention deficits, sleep disturbances, and increased appetite. These symptoms may begin within a few hours after the last cigarette, quickly driving people back to tobacco use. Symptoms peak within the first few days of smoking cessation and usually subside within a few weeks. For some people, however, symptoms may persist for months.

Although withdrawal is related to the pharmacological effects of nicotine, many behavioral factors can also affect the severity of withdrawal symptoms. For some people, the feel, smell, and sight of a cigarette and the ritual of obtaining, handling, lighting, and smoking the cigarette are all associated with the pleasurable effects of smoking and can make withdrawal or craving worse. Nicotine replacement therapies such as gum, patches, and inhalers may help alleviate the pharmacological aspects of withdrawal; however,

Tobacco plants

> **Most smokers identify tobacco use as harmful and express a desire to reduce or stop using it, and nearly 35 million want to quit each year.**

cravings often persist. Behavioral therapies can help smokers identify environmental triggers of craving so they can employ strategies to prevent or circumvent these symptoms and urges.

Nicotine replacement therapies such as gum, patches, and inhalers may help alleviate the pharmacological aspects of withdrawal.

Are There Other Chemicals That May Contribute to Tobacco Addiction?

Yes, research is showing that nicotine may not be the only ingredient in tobacco that affects its addictive potential. Using advanced neuroimaging technology, scientists can see the dramatic effect of cigarette smoking on the brain and are finding a marked decrease in the levels of monoamine oxidase (MAO), an important enzyme that is responsible for the breakdown of dopamine. This change is likely caused by some ingredient in tobacco smoke other than nicotine, because we know that nicotine itself does not dramatically alter MAO levels. The decrease in two forms of MAO (A and B) results in higher dopamine levels and may be another reason that smokers continue to smoke—to sustain the high dopamine levels that lead to the desire for repeated drug use.

Animal studies by NIDA-funded researchers have shown that acetaldehyde, another chemical found in tobacco smoke, dramatically increases the reinforcing properties of nicotine and may also contribute to tobacco addiction. The investigators further report that this effect is age-related: adolescent animals display far more sensitivity to this reinforcing effect, which suggests that the brains of adolescents may be more vulnerable to tobacco addiction.

What Are the Medical Consequences of Tobacco Use?

Cigarette smoking kills an estimated 440,000 U.S. citizens each year—more than alcohol, illegal drug use, homicide, suicide, car accidents, and AIDS combined. Between 1964 and 2004, more than 12 million Americans died prematurely from smoking, and another 25 million U.S. smokers alive today will most likely die of a smoking-related illness.

Cigarette smoking harms nearly every organ in the body. It has been conclusively linked to cataracts and pneumonia, and accounts for about one-third of all cancer deaths. The overall rates of death from cancer are twice as high among smokers as nonsmokers, with heavy smokers having rates that are four times greater than those of nonsmokers. Foremost among the cancers caused by tobacco use is lung cancer—cigarette smoking has been linked to about 90 percent of all cases of lung cancer, the number one cancer killer of both men and women. Smoking is also associated with cancers of the mouth, pharynx, larynx, esophagus, stomach, pancreas, cervix, kidney, bladder, and acute myeloid leukemia.

In addition to cancer, smoking causes lung diseases such as chronic bronchitis and emphysema, and it has been found to exacerbate asthma symptoms in adults and children. About 90 percent of all deaths from chronic obstructive pulmonary diseases are attributable to cigarette smoking. It has also been well-documented that smoking substantially increases the risk of heart disease, including stroke,

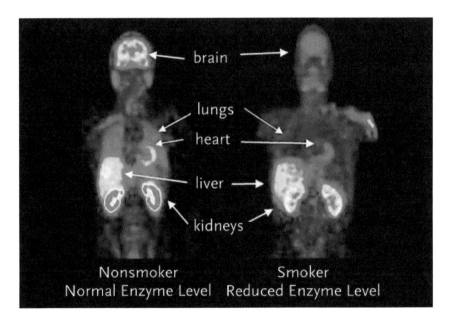

Using advanced neuroimaging technology, scientists can see the dramatic effect of cigarette smoking on the brain and body and are finding a marked decrease in the levels of monoamine oxidase (MAO), an important enzyme that is responsible for the breakdown of dopamine. (Source: Fowler et al., 2003)

Sticky, brown tar coats the lungs of tobacco smokers. Along with thousands of other damaging chemicals, tar can lead to lung cancer and acute respiratory diseases.

Are There Safe Tobacco Products?

The adverse health effects of tobacco use are well known, yet many people do not want to quit or have difficulty quitting. As a result, there has been a recent surge in the development of tobacco products that claim to reduce exposure to harmful tobacco constituents or to have fewer health risks than conventional products. These "potentially reduced exposure products" (PREPs), which include cigarettes and smokeless tobacco (e.g., snuff, tobacco lozenges), have not yet been evaluated sufficiently to determine whether they are indeed associated with reduced risk of disease. Recent studies indicate that the levels of carcinogens in these PREPs range from relatively low to comparable to conventional tobacco products. These studies conclude that medicinal nicotine (found in the nicotine patch and gum) is a safer alternative than these modified tobacco products.

heart attack, vascular disease, and aneurysm. Smoking causes coronary heart disease, the leading cause of death in the United States: cigarette smokers are 2–4 times more likely to develop coronary heart disease than nonsmokers.

Exposure to high doses of nicotine, such as those found in some insecticide sprays, can be extremely toxic as well, causing vomiting, tremors, convulsions, and death. In fact, one drop of pure nicotine can kill a person. Nicotine poisoning has been reported from accidental ingestion of insecticides by adults and ingestion of tobacco products by children and pets. Death usually results in a few minutes from respiratory failure caused by paralysis.

Although we often think of medical consequences that result from direct use of tobacco products, passive or secondary smoke also increases the risk for many diseases. Environmental tobacco smoke is a major source of indoor air contaminants; secondhand smoke is estimated to cause approximately 3,000 lung cancer deaths per year among nonsmokers and contributes to more than 35,000 deaths related to cardiovascular disease. Exposure to tobacco smoke in the home is also a risk factor for new cases

Tobacco Use and Comorbidity

There is clear evidence of high rates of psychiatric comorbidity, including other substance abuse, among adolescents and adults who smoke. For example, it has been estimated that individuals with psychiatric disorders purchase approximately 44 percent of all cigarettes sold in the United States, which undoubtedly contributes to the disproportionate rates of morbidity and mortality in these populations. In addition, studies have shown that as many as 80 percent of alcoholics smoke regularly, and that a majority of them will die of smoking-related, rather than alcohol-related, disease.

In young smokers, the behavior appears to be strongly associated with increased risk for a variety of mental disorders. In some cases—such as with conduct disorders and attention-deficit hyperactivity disorder—these disorders may precede the onset of smoking, while in others—such as with substance abuse—the disorders may emerge later in life. Whether daily smoking among boys and girls is the result or the cause of a manifest psychiatric condition, it is troubling that so very few adolescents have their nicotine dependence diagnosed or properly treated. Preventing the early onset of smoking and treating its young victims are critical primary-care priorities, the fulfillment of which could have a dramatic impact on our ability to prevent or better address a wide range of mental disorders throughout life.

Among adults, the rate of major depressive episodes is highest in nicotine-dependent individuals, lower in nondependent current smokers, and lowest in those who quit or never started smoking. Furthermore, there is evidence showing that, for those who have had more than one episode, smoking cessation may increase the likelihood of a new major depressive episode. Adult tobacco use also increases risk for the later development of anxiety disorders, which may be associated with an increased severity of withdrawal symptoms during smoking cessation therapy. But the most extensive comorbidity overlap is likely the one that exists between smoking and schizophrenia, since, in clinical samples, the rate of smoking in patients with schizophrenia has ranged as high as 90 percent.

and increased severity of childhood asthma. Additionally, dropped cigarettes are the leading cause of residential fire fatalities, leading to more than 1,000 deaths each year.

Smoking and Pregnancy—What Are the Risks?

In the United States, it is estimated that about 16 percent of pregnant women smoke during their pregnancies. Carbon monoxide and nicotine from tobacco smoke may interfere with the oxygen supply to the fetus. Nicotine also readily crosses the placenta, and concentrations in the fetus can be as much as 15 percent higher than maternal levels. Nicotine concentrates in fetal blood, amniotic fluid, and breast milk. Combined, these factors can have severe consequences for the fetuses and infants of smoking mothers. Smoking during pregnancy caused an estimated 910 infant deaths annually from 1997 through 2001, and neonatal care costs related to smoking are estimated to be more than $350 million per year.

The adverse effects of smoking during pregnancy can include fetal growth retardation and decreased birthweight. The decreased birthweights seen in infants of mothers who smoke reflect a dose-dependent relationship—the more the woman smokes during pregnancy, the greater the reduction of infant birthweight. These newborns also display signs of stress and drug withdrawal

consistent with what has been reported in infants exposed to other drugs. In some cases, smoking during pregnancy may be associated with spontaneous abortions and sudden infant death syndrome (SIDS), as well as learning and behavioral problems and an increased risk of obesity in children.

Smoking and Adolescence

In 2010, about 2.6 million American adolescents (aged 12–17) reported using a tobacco product in the month prior to the survey. In that same year, it was found that nearly 60 percent of new smokers were under the age of 18 when they first smoked a cigarette. Of smokers under age 18, more than 6 million will likely die prematurely from a smoking-related disease.

Tobacco use in teens is not only the result of psychosocial influences, such as peer pressure; recent research suggests that there may be biological reasons for this period of increased vulnerability. There is some evidence that intermittent smoking can result in the development of tobacco addiction in some teens. Animal models of teen smoking provide additional evidence of an increased vulnerability. Adolescent rats are more susceptible to the reinforcing effects of nicotine than adult rats, and take more nicotine when it is available than do adult animals.

Adolescents may also be more sensitive to the reinforcing effects of nicotine in combination with other chemicals found in cigarettes, thus increasing susceptibility to tobacco addiction. As mentioned earlier, acetaldehyde increases nicotine's addictive properties in adolescent, but not adult, animals. A recent study also suggests that specific genes may increase risk for addiction among people who begin smoking during adolescence. NIDA continues to actively support research aimed at increasing our understanding of why and how adolescents become addicted, and to develop prevention and treatment strategies to meet their specific needs.

In addition, smoking more than one pack a day during pregnancy nearly doubles the risk that the affected child will become addicted to tobacco if that child starts smoking.

Are There Gender Differences in Tobacco Smoking?

Several avenues of research now indicate that men and women differ in their smoking behaviors. For instance, women smoke fewer cigarettes per day, tend to use cigarettes with lower nicotine content, and do not inhale as deeply as men. However, it is unclear whether this is due to differences in sensitivity to nicotine or other factors that affect women

Large-scale smoking cessation trials show that women are less likely to initiate quitting and may be more likely to relapse if they do quit.

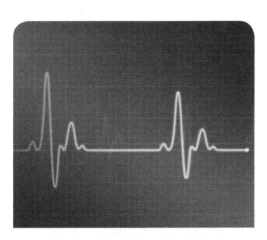

Smoking cessation can have immediate health benefits. For example, within 24 hours of quitting, blood pressure and chances of heart attack decrease.

differently, such as social factors or the sensory aspects of smoking.

The number of smokers in the United States declined in the 1970s and 1980s, remained relatively stable throughout the 1990s, and declined further through the early 2000s. Because this decline in smoking was greater among men than women, the prevalence of smoking is only slightly higher for men today than it is for women. Several factors appear to be contributing to this narrowing gender gap, including women being less likely than men to quit. Large-scale smoking cessation trials show that women are less likely to initiate quitting and may be more likely to relapse if they do quit. In cessation programs using nicotine replacement methods, such as the patch or gum, the nicotine does not seem to reduce craving as effectively for women as for men. Other factors that may contribute to women's difficulty with quitting are that withdrawal may be more intense for women or that women are more concerned about weight gain.

Although postcessation weight gain is typically modest (about 5–10 pounds), concerns about this may be an obstacle to treatment success. In fact, NIDA research has found that when women's weight concerns were addressed during cognitive-behavioral therapy, they were more successful at quitting than women who were in a program designed only to attenuate postcessation weight gain. Other NIDA researchers have found that medications used for smoking cessation, such as bupropion and naltrexone, can also attenuate postcessation weight gain and could become an additional strategy for enhancing treatment success.

It is important for treatment professionals to be aware that standard regimens may have to be adjusted to compensate for gender differences in nicotine sensitivity and in other related factors that contribute to continued smoking.

Are There Effective Treatments for Tobacco Addiction?

Yes, extensive research has shown that treatments for tobacco addiction do work. Although some smokers can quit without help, many individuals need assistance with quitting. This is particularly important because smoking cessation can have immediate health benefits. For example, within 24 hours of quitting, blood pressure and chances of heart attack decrease. Long-term benefits of smoking cessation include decreased risk of stroke, lung and

other cancers, and coronary heart disease. A 35-year-old man who quits smoking will, on average, increase his life expectancy by 5 years.

Nicotine Replacement Treatments

Nicotine replacement therapies (NRTs), such as nicotine gum and the transdermal nicotine patch, were the first pharmacological treatments approved by the Food and Drug Administration (FDA) for use in smoking cessation therapy. NRTs are used (in conjunction with behavioral support) to relieve withdrawal symptoms—they produce less severe physiological alterations than tobacco-based systems and generally provide users with lower overall nicotine levels than they receive with tobacco. An added benefit is that these forms of nicotine have little abuse potential since they do not produce the pleasurable effects of tobacco products, nor do they contain the carcinogens and gases associated with tobacco smoke. Behavioral treatments, even beyond what is recommended on packaging labels, have been shown to enhance the effectiveness of NRTs and improve long-term outcomes.

The FDA's approval of nicotine gum in 1984 marked the availability (by prescription) of the first NRT on the U.S. market. In 1996, the FDA approved Nicorette gum for over-the-counter (OTC) sales. Whereas nicotine gum provides some smokers with the desired control over dose and the ability to relieve cravings, others are unable to tolerate the taste and chewing demands. In 1991 and 1992, the FDA approved four transdermal nicotine patches, two of which became OTC products in 1996. In 1996 a nicotine nasal spray, and in 1998 a nicotine inhaler, also became available by prescription, thus meeting the needs of many additional tobacco users. All the NRT products—gum, patch, spray, and inhaler—appear to be equally effective.

Twin studies indicate that approximately 40–70 percent of a person's risk of becoming addicted to nicotine depends on his or her genes.

Additional Medications

Although the primary focus of pharmacological treatments for tobacco addiction has been nicotine replacement, other treatments are also available. For example, the antidepressant bupropion was approved by the FDA in 1997 to help people quit smoking and is marketed as Zyban. Varenicline tartrate (Chantix) is a medication that recently received FDA approval for smoking cessation. This medication, which acts at the sites in the brain affected by nicotine, may help people quit by easing withdrawal symptoms and blocking the effects of nicotine if people resume smoking.

Several other non-nicotine medications are being investigated for the treatment of tobacco addiction, including other antidepressants and an antihypertensive medication. Scientists are also investigating the potential of a vaccine that targets nicotine for use in relapse prevention. The nicotine vaccine is designed to stimulate the production of antibodies that would block access of nicotine to the brain and prevent nicotine's reinforcing effects.

Behavioral Treatments

Behavioral interventions play an integral role in smoking cessation treatment, either in conjunction with medication or alone. A variety of methods can assist smokers with quitting, ranging from self-help materials to individual cognitive-behavioral therapy. These interventions teach individuals to recognize high-risk smoking situations, develop alternative coping strategies, manage stress, improve problemsolving skills, and increase social support. Research has also shown that the more therapy is tailored to a person's situation, the greater the chances are for success.

Traditionally, behavioral approaches were developed and delivered through formal settings, such as smoking cessation clinics and community and public health settings. Over the past decade, however, researchers have been adapting these approaches for mail, telephone, and Internet formats, which can be more acceptable and accessible to smokers who are trying to quit. In 2004, the U.S. Department of Health and Human Services (HHS) established a national toll-free number, 800-QUIT-NOW (800-784-8669),

to serve as a single access point for smokers seeking information and assistance in quitting. Callers to the number are routed to their State's smoking cessation quitline or, in States that have not established quitlines, to one maintained by the National Cancer Institute. In addition, a new HHS Web site (www.smokefree.gov) offers online advice and downloadable information to make cessation easier.

Quitting smoking can be difficult. People can be helped during the time an intervention is delivered; however, most intervention programs are short-term (1–3 months). Within 6 months, 75–80 percent of people who try to quit smoking relapse. Research has now shown that extending treatment beyond the typical duration of a smoking cessation program can produce quit rates as high as 50 percent at 1 year.

New Frontiers in Tobacco Research

If so many smokers want to quit, why are few able to do so successfully? To address this question, scientists are increasingly focusing on the powerful role of genetics in addiction. Twin studies indicate that approximately 40–70 percent of a person's risk of becoming addicted to nicotine depends on his or her genes. Although complex diseases like addiction involve large numbers of genes interacting with a wide variety of environmental factors, the contribution of a particular gene can be substantial.

Genetic variants associated with nicotine metabolism, for example, have been shown to influence how people smoke. Slow metabolizers smoke fewer cigarettes per day and have a higher likelihood of quitting, and there is greater abstinence among individuals receiving nicotine patch therapy. A recent NIDA-funded study identified a variant in the gene for a nicotinic receptor subunit that doubled the risk for nicotine addiction among smokers. A subsequent study found that this gene variant also increased susceptibility to the severe health consequences of smoking, including lung cancer and peripheral arterial disease. NIDA is currently supporting large-scale genome-wide association studies to uncover additional genetic risk factors in order to better understand tobacco addiction and its adverse effects on health.

In addition to predicting an individual's risk for nicotine addiction, genetic markers can also help predict whether medications (like bupropion) will effectively help a smoker quit. This takes root in the emerging field of pharmacogenomics, which investigates how genes influence a patient's response to drugs and medications. In the future, genetic screening could help clinicians select treatments, adjust dosages, and avoid or minimize adverse reactions, tailoring smoking cessation therapies to an individual's unique genetic inheritance.

Glossary

Addiction: A chronic, relapsing disease characterized by compulsive drug-seeking and abuse despite adverse consequences. It is associated with long-lasting changes in the brain.

Adrenal glands: Glands located above each kidney that secrete hormones, e.g., adrenaline.

Carcinogen: Any substance that causes cancer.

Craving: A powerful, often uncontrollable desire for drugs.

Dopamine: A neurotransmitter present in regions of the brain that regulate movement, emotion, motivation, and feelings of pleasure.

Emphysema: A lung disease in which tissue deterioration results in increased air retention and reduced exchange of gases. The result is difficulty breathing and shortness of breath.

Neurotransmitter: A chemical that acts as a messenger to carry signals or information from one nerve cell to another.

Nicotine: An alkaloid derived from the tobacco plant that is primarily responsible for smoking's psychoactive and addictive effects.

Pharmacokinetics: The pattern of absorption, distribution, and excretion of a drug over time.

Tobacco: A plant widely cultivated for its leaves, which are used primarily for smoking; the *N. tabacum* species is the major source of tobacco products.

Withdrawal: A variety of symptoms that occur after chronic use of an addictive drug is reduced or stopped.

References

Adams, E.K.; Miller, V.P.; Ernst, C.; Nishimura, B.K.; Melvin, C.; and Merritt, R. *Neonatal health care costs related to smoking during pregnancy.* Health Economics 11:193–206, 2002.

Belluzzi, J.D.; Wang, R.; and Leslie, F.M. *Acetaldehyde enhances acquisition of nicotine self-administration in adolescent rats.* Neuropsychopharmacol 30:705–712, 2005.

Benowitz, N.L., *Pharmacology of nicotine: Addiction and therapeutics.* Ann Rev Pharmacol Toxicol 36:597–613, 1996.

Buka, S.L.; Shenassa, E.D.; and Niaura, R., *Elevated risk of tobacco dependence among offspring of mothers who smoked during pregnancy: A 30-year prospective study.* Am J Psychiatry 160:1978–1984, 2003.

Ernst, M.; Moolchan, E.T.; and Robinson, M.L., *Behavioral and neural consequences of prenatal exposure to nicotine.* J Am Acad Child Adolesc Psychiatry 40:630–641, 2001.

Fowler, J.S.; Volkow, N.D.; Wang, G.J.; Pappas, N.; Logan, J.; MacGregor, R.; Alexoff, D.; Shea, C.; Schlyer, D.; Wolf, A.P.; Warner, D.; Zezulkova, I.; and Cilento, R., *Inhibition of monoamine oxidase B in the brains of smokers.* Nature 22:733–736, 1996.

Hatsukami, D.K.; Lemmonds, C.; Zhang, Y.; Murphy, S.E.; Le, C.; Carmella, S.G.; and Hecht, S.S., *Evaluation of carcinogen exposure in people who used "reduced exposure" tobacco products.* J Natl Cancer Inst 96:844–852, 2004.

Henningfield, J.E., *Nicotine medications for smoking cessation.* New Engl J Med 333:1196–1203, 1995.

Johnston, L.D.; O'Malley, P.M.; Bachman, J.G.; and Schulenberg, J.E., *Monitoring the Future National Results on Adolescent Drug Use: Overview of Key Findings, 2011.* NIH Pub. No. 11-6418. Bethesda, MD: National Institute on Drug Abuse, 2011. Available at: www.monitoringthefuture.org.

Kalman, D.; Morissette, S.B.; and George, T.P., *Comorbidity of smoking in patients with psychiatric and substance use disorders.* Am J Addict 14(2):106–123, 2005.

Levin, E.D.; Rezvani, A.H.; Montoya, D.; Rose, J.E.; and Swartzwelder, H.S., *Adolescent-onset nicotine self-administration modeled in female rats.* Psychopharmacol 169:141–149, 2003.

Perkins, K.A.; Donny, E.; and Caggiula, A.R., *Sex differences in nicotine effects and self-administration: Review of human and animal evidence.* Nic and Tobacco Res 1:301–315, 1999.

Saccone, S.F., et al. *Cholinergic nicotinic receptor genes implicated in a nicotine dependence association study targeting 348 candidate genes with 3713 SNPs.* Hum Mol Genet 16(1):36–49, 2007.

Substance Abuse and Mental Health Services Administration, *Results from the 2010 National Survey on Drug Use and Health: Summary of National Findings*, NSDUH Series H-41, HHS Publication No. (SMA) 11-4658. Rockville, MD: Substance Abuse and Mental Health Services Administration, 2011. Available at http://oas.samhsa.gov/NSDUH/2k10NSDUH/2k10Results.htm

Thorgeirsson, T.E., et al. *A variant associated with nicotine dependence, lung cancer and peripheral arterial disease.* Nature 452(7187):638–642, 2008.

U.S. Department of Health and Human Services. *Reducing Tobacco Use: A Report of the Surgeon General.* Atlanta: Centers for Disease Control and Prevention, National Center for Chronic Disease Prevention and Health Promotion, Office on Smoking and Health, 2000. Available at: http://www.cdc.gov/tobacco/data_statistics/sgr/index.htm.

U.S. Department of Health and Human Services. *The Health Benefits of Smoking Cessation: A Report of the Surgeon General.* Atlanta: Centers for Disease Control and Prevention, National Center for Chronic Disease Prevention and Health Promotion, Office on Smoking and Health, 1990. Available at: http://www.cdc.gov/tobacco/data_statistics/sgr/index.htm.

U.S. Department of Health and Human Services. *The Health Consequences of Smoking: A Report of the Surgeon General.* Atlanta: Centers for Disease Control and Prevention, National Center for Chronic Disease Prevention and Health Promotion, Office on Smoking and Health, 2004. Available at: http://www.cdc.gov/tobacco/data_statistics/sgr/sgr_2004/index.htm.

Where can I get further information about tobacco/nicotine?

To learn more about tobacco/nicotine and other drugs of abuse, visit the NIDA Web site at **www.drugabuse.gov** or contact the DrugPubs Research Dissemination Center at 877-NIDA-NIH (877-643-2644; TTY/TDD: 240-645-0228).

What's on the NIDA Web Site

- Information on drugs of abuse and related health consequences
- NIDA publications, news, and events
- Resources for health care professionals
- Funding information (including program announcements and deadlines)
- International activities
- Links to related Web sites (access to Web sites of many other organizations in the field)

NIDA Web Sites

www.drugabuse.gov

www.drugabuse.gov/publications/term/160/DrugFacts

www.easyread.drugabuse.gov

www.drugabuse.gov/blending-initiative

For Physician Information

NIDAMED

www.drugabuse.gov/nidamed

Other Web Sites

Information on tobacco addiction is also available through these Web sites:

- Centers for Disease Control and Prevention: *www.cdc.gov/tobacco*
- National Cancer Institute: *www.cancer.gov*
- U.S. Department of Health and Human Services: *www.smokefree.gov*
- Substance Abuse and Mental Health Services Administration Health Information Network: *www.samhsa.gov/shin*
- Office of the Surgeon General: *www.surgeongeneral.gov/initiatives/tobacco*
- Society for Research on Nicotine and Tobacco: *www.srnt.org*
- The Robert Wood Johnson Foundation: *www.rwjf.org*
- Join Together Online: *www.quitnet.com*
- American Legacy Foundation: *www.americanlegacy.org*

NIH Publication Number 12-4342
Printed July 1998, Reprinted August 2001, Revised July 2006,
Revised June 2009, Revised July 2012.

Smokeless Tobacco Users

Check Monthly for Early Signs of Disease

The early signs of cancer in the mouth and tongue may be detected by self-examination. Dr. Elbert Glover, director of the Tobacco Research Center at West Virginia University, and the American Cancer Society recommend that the following self-check procedures be conducted every month.

- Check your face and neck for lumps on either side. Both sides of your face and neck should be the same shape.
- Look at your lips, cheeks, and gums. Look for sores, white or red patches, or changes in your gums by pulling down your lower lip. Check your inner cheeks, especially where you hold your tobacco. Gently squeeze your lip and cheeks to check for lumps or soreness.
- Put the tip of your tongue on the roof of your mouth. Place one finger on the floor of your mouth and press up under your chin with a finger from your other hand. Feel for bumps, soreness, or swelling. Check around the inside of your teeth from one side of your jaw to the other.
- Tilt your head back and open your mouth wide. Check for color changes or bumps or sores in the roof of your mouth.
- Stick out your tongue and look at the top. Gently grasp your tongue with a piece of cloth and pull it to each side. Look for color changes. Feel both sides of your tongue with your finger for bumps.

If you use smokeless tobacco and find anything that looks or feels unusual, see your dentist or physician as soon as possible.

From Decisions for Healthy Living by Pruitt, Stein and Pruitt, Addison Wesley Longman Educational Publishers, Inc.

When smoking tobacco through a water pipe, some nicotine is absorbed as it passes through a water bowl. Because most smokers stop when their nicotine craving has been satisfied, this may actually lead to longer smoking sessions, exposing the user to more smoke over a longer period of time (see Figure 9.5).

Some water pipe products and accessories are marketed and sold with claims of reducing the harmful effects of hookah, but according to the WHO, none have been shown to reduce the smoker's risk of exposure to toxins.

Smokeless tobacco comes in two forms: snuff and chewing tobacco. Snuff is a fine grain of tobacco, and chewing tobacco is shredded or bricked; either choice is placed in the mouth and the user sucks on the tobacco juices, spitting out the saliva. The sucking allows the nicotine to be absorbed in the bloodstream. It can be equally as dangerous and harmful as smoking. Smokeless tobacco is addictive. According to the Centers for Disease Control and Prevention (CDC) estimates, 6.1 percent of high school students are current

FIGURE 9.5 Comparison of Cigarette Smoking Session and Water Pipe Smoking Session

Cigarette Smoking Session	Water Pipe Smoking Session
8–12 puffs	50–100 puffs
5–7 minutes	20–80 minutes
0.5–0.6 liter of smoke inhaled	.015–1.0 liter of smoke inhaled

Note: Upon analysis, it is possible for a water pipe session to expose the smoker to the equivalent of up to one hundred cigarettes in a single session.

smokeless tobacco users. Nationally, an estimated 3.5 percent of adults are current smokeless tobacco users. (CDC, 2009).

The National Cancer Institute reports there are three thousand chemical compounds in smokeless tobacco. Nicotine is the addictive drug in all forms of tobacco. Holding one pinch of smokeless tobacco in your mouth for thirty minutes delivers as much nicotine as three to four cigarettes (National Cancer Institute, 2011). There have been at least twenty-eight cancer-causing agents found in smokeless tobacco:

- Nitrosamines—20 to 43,000 more nitrosamines are found in smokeless tobacco. Other consumer products like beer or bacon only contain five parts per billion
- Polonium 210—radioactive particles that turn into radon
- Formaldehyde—embalming fluid
- Cadmium—metallic element; its salts are poisonous
- Arsenic—poisonous element

Immediate effects from chewing tobacco are bad breath and stains on your teeth. Mouth sores also accompany smokeless tobacco users. The complications of long-term use can be very serious. These complications include increased gum and teeth problems, increased heart rate, irregular heartbeat, heart attacks, and cancer. Oral cancer can occur in the mouth, lips, tongue, cheeks, or gums. Other cancer possibilities resulting from smokeless tobacco can be stomach cancer, bladder cancer, and cancer of the esophagus.

Another major problem caused by smokeless tobacco is **leukoplakia**, a precancerous condition that produces thick, rough, white patches on the gums, tongue, and inner cheeks. A variety of cancers such as lip, pharynx, larynx, esophagus, and tongue can be attributed to smokeless tobacco. Dental and gum problems are major side effects as well. Smokeless tobacco used during pregnancy increases the risk for preeclampsia, a condition that includes high blood pressure, fluid retention, and swelling. It also puts the mother and newborn at higher risk for premature birth and low birth weight. In men, smokeless tobacco reduces sperm count and increases the liklihood of abnormal sperm cell. (CDC, 2010).

Environmental Tobacco Smoke

Environmental tobacco smoke (ETS), or secondhand smoke, contains more than 7,000 chemicals, including hundreds that are toxic and 70 carcinogens. Secondhand smoke exposure to non-smoking adults can cause heart disease, a 20–30 percent increased risk of lung cancer, and a 25–30 percent increased risk of heart attacks. Approximately 126 million non-smokers are exposed to secondhand smoke in homes, workplaces, and public places, resulting in an estimated 38,000 deaths and healthcare costs exceeding $10 billion annually. To those individuals with existing health issues, second-hand smoke exposure is an extremely high risk. There is no "risk-free" exposure to secondhand smoke; even brief exposure can be dangerous (CDC, 2006).

Secondhand smoke is especially dangerous to infants and children. In the United States, almost 22 million children are exposed to secondhand smoke. Globally almost half of the world's children breathe air polluted by tobacco smoke. This exposure can cause sudden infant death syndrome, acute respiratory infections, ear problems, slow lung growth, and severe asthma attacks. Each year in the United States, secondhand smoke is responsible for an estimated

There is no "risk-free" exposure to secondhand smoke.

FIGURE 9.6
When Smokers Quit

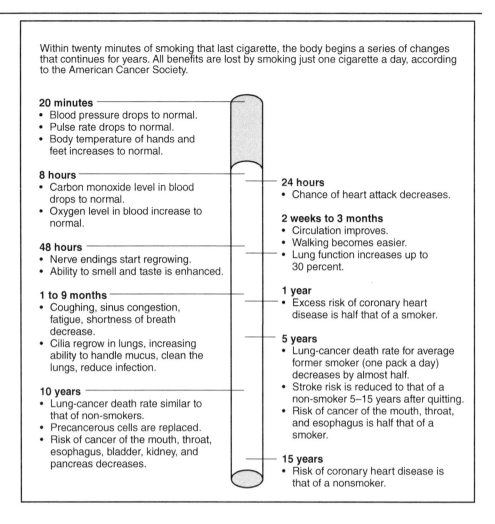

150,000–300,000 new cases of bronchitis and pneumonia in children less than 18 months, resulting in nearly 15,000 hospitalizations annually (CDC, 2012).

- American Heart Association (AHA)
  www.aha.org
  National Center
  7272 Greenville Avenue
  Dallas, TX 75231
  1-800-AHA-USA1
- American Lung Association (ALA)
  www.lungusa.org/
  1740 Broadway
  New York, NY 10019-4274
  (212) 315-8700
  1-800-LUNG-USA
- American Medical Association (AMA)
  www.ama-assn.org/
  515 North State Street
  Chicago, IL 60610
  (312) 464-5000
- Centers for Disease Control and Prevention
  www.cdc.gov/tobacco
- American Cancer Society (ACS)
  http://www.cancer.org
  1599 Clifton Road, N.E.
  Atlanta, GA 30329
  1-800-ACS-2345
- Cancer Research Foundation of America (CRFA)
  www.preventcancer.org
  1600 Duke Street
  Alexandria, VA 22314
  (703) 836-4412
- Doctors Ought to Care (DOC)
  www.bcm.tmc.edu/doc/
  5615 Kirby Drive
  Suite 440
  Houston, TX 77005
  (713) 528-1487

Cannabinoids

Marijuana refers to the dried leaves, flowers, stems, and seeds from the hemp plant *Cannabis sativa*, which contains the psychoactive (mind-altering) chemical delta-9-tetrahydrocannabinol (THC), as well as other related compounds. This plant material can also be concentrated in a resin called *hashish* or a sticky black liquid called *hash oil*.

Marijuana is the most common illicit drug used in the United States. After a period of decline in the last decade, its use has been increasing among young people since 2007, corresponding to a diminishing perception of the drug's risks that may be associated with increased public debate over the drug's legal status. Although the federal government considers marijuana a Schedule I substance (having no medicinal uses and high risk for abuse), two states have legalized marijuana for adult recreational use, and 20 states have passed laws allowing its use as a treatment for certain medical conditions (see "Is Marijuana Medicine?", below).

How Is Marijuana Used?

Marijuana is usually smoked in hand-rolled cigarettes (joints) or in pipes or water pipes (bongs). It is also smoked in blunts—cigars that have been emptied of tobacco and refilled with a mixture of marijuana and tobacco. Marijuana smoke has a pungent and distinctive, usually sweet-and-sour, odor. Marijuana can also be mixed in food or brewed as a tea.

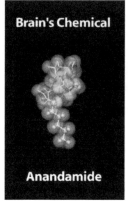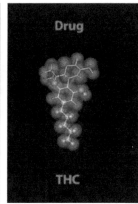

THC's chemical structure is similar to the brain chemical anandamide. Similarity in structure allows drugs to be recognized by the body and to alter normal brain communication.

How Does Marijuana Affect the Brain?

When marijuana is smoked, THC rapidly passes from the lungs into the bloodstream, which carries the chemical to the brain and other organs throughout the body. It is absorbed more slowly when ingested in food or drink.

However it is ingested, THC acts on specific molecular targets on brain cells, called cannabinoid receptors. These receptors are ordinarily activated by chemicals similar to THC that naturally occur in the body (such as anandamide; see picture, above) and are part of a neural

The behavioral effects on marijuana users include an impairment of shortterm memory, an overestimation of the passage of time, and loss of the ability to focus on a task.

communication network called the endocannabinoid system. This system plays an important role in normal brain development and function.

The highest density of cannabinoid receptors is found in parts of the brain that influence pleasure, memory, thinking, concentration, sensory and time perception, and coordinated movement. Marijuana overactivates the endocannabinoid system, causing the "high" and other effects that users experience. These effects include altered perceptions and mood, impaired coordination, difficulty with thinking and problem solving, and disrupted learning and memory.

Marijuana also affects brain development, and when it is used heavily by young people, its effects on thinking and memory may last a long time or even be permanent. A recent study of marijuana users who began using in adolescence revealed substantially reduced connectivity among brain areas responsible for learning and memory. And a large long-term study in New Zealand showed that people who began smoking marijuana heavily in their teens lost an average of 8 points in IQ between age 13 and age 38. Importantly, the lost cognitive abilities were not fully restored in those who quit smoking marijuana as adults. Those who started smoking marijuana in adulthood did not show significant IQ declines.

What Are the Other Health Effects of Marijuana?

Marijuana use may have a wide range of effects, particularly on cardiopulmonary and mental health.

Marijuana smoke is an irritant to the lungs, and frequent marijuana smokers can have many of the same respiratory problems experienced by tobacco smokers, such as daily cough and phlegm production, more frequent acute chest illness, and a heightened risk of lung infections. One study found that people who smoke marijuana frequently but do not smoke tobacco have more health problems and miss more days of work than those who don't smoke marijuana, mainly because of respiratory illnesses. It is not yet known whether marijuana smoking contributes to risk for lung cancer.

Marijuana also raises heart rate by 20-100 percent shortly after smoking; this

Is Marijuana Medicine?

Many have called for the legalization of marijuana to treat conditions including pain and nausea caused by HIV/AIDS, cancer, and other conditions, but clinical evidence has not shown that the therapeutic benefits of the marijuana plant outweigh its health risks. To be considered a legitimate medicine by the FDA, a substance must have well-defined and measureable ingredients that are consistent from one unit (such as a pill or injection) to the next. As the marijuana plant contains hundreds of chemical compounds that may have different effects and that vary from plant to plant, and because the plant is typically ingested via smoking, its use as a medicine is difficult to evaluate.

However, THC-based drugs to treat pain and nausea are already FDA approved and prescribed, and scientists continue to investigate the medicinal properties of other chemicals found in the cannabis plant—such as cannabidiol, a non-psychoactive cannabinoid compound that is being studied for its effects at treating pain, pediatric epilepsy, and other disorders. For more information, see
http://www.drugabuse.gov/publications/drugfacts/marijuana-medicine

effect can last up to 3 hours. In one study, it was estimated that marijuana users have a 4.8-fold increase in the risk of heart attack in the first hour after smoking the drug. This risk may be greater in older individuals or in those with cardiac vulnerabilities.

A number of studies have linked chronic marijuana use and mental illness. High doses of marijuana can produce a temporary psychotic reaction (involving hallucinations and paranoia) in some users, and using marijuana can worsen the course of illness in patients with schizophrenia. A series of large studies following users across time also showed a link between marijuana use and later development of psychosis. This relationship was influenced by genetic variables as well as the amount of drug used, drug potency, and the age at which it was first taken—those who start young are at increased risk for later problems.

Associations have also been found between marijuana use and other mental health problems, such as depression, anxiety, suicidal thoughts among adolescents, and personality disturbances, including a lack of motivation to engage in typically rewarding activities. More research is still needed to confirm and better understand these linkages.

Marijuana use during pregnancy is associated with increased risk of neurobehavioral problems in babies. Because THC and other compounds in marijuana mimic the body's own endocannabinoid chemicals, marijuana use by pregnant mothers may alter the developing endocannabinoid system in the brain of the fetus. Consequences for the child may include problems with attention, memory, and problem solving.

Additionally, because it seriously impairs judgment and motor coordination, marijuana contributes to risk of injury or death while driving a car. A recent analysis of data from several studies found that marijuana use more than doubles a driver's risk of being in an accident. The combination of marijuana and alcohol is worse than either substance alone with respect to driving impairment.

Rising Potency

The amount of THC in marijuana samples confiscated by police has been increasing steadily over the past few decades. In 2012, THC concentrations in marijuana averaged close to 15 percent, compared to around 4 percent in the 1980s. For a new user, this may mean exposure to higher concentrations of THC, with a greater chance of an adverse or unpredictable reaction. Increases in potency may account for the rise in emergency department visits involving marijuana use. For frequent users, it may mean a greater risk for addiction if they are exposing themselves to high doses on a regular basis. However, the full range of consequences associated with marijuana's higher potency is not well understood. For example, experienced users may adjust their intake in accordance with the potency or they may be exposing their brains to higher levels overall, or both.

Is Marijuana Addictive?

Contrary to common belief, marijuana is addictive. Estimates from research suggest that about 9 percent of users become addicted to marijuana; this number increases among those who start young (to about 17 percent, or 1 in 6) and among people who use marijuana daily (to 25-50 percent).

Long-term marijuana users trying to quit report withdrawal symptoms including irritability, sleeplessness, decreased appetite, anxiety, and drug craving, all of which can make it difficult to abstain. Behavioral interventions, including cognitive-behavioral therapy and motivational incentives (i.e., providing vouchers for goods or services to patients who remain abstinent) have proven to be effective in treating marijuana addiction. Although no medications are currently available, recent discoveries about the workings of the endocannabinoid system offer promise for the development of medications to ease withdrawal, block the intoxicating effects of marijuana, and prevent relapse.

How Does Marijuana Affect a User's Life?

Research shows marijuana may cause problems in daily life or make a person's existing problems worse. Heavy marijuana users generally report lower life satisfaction, poorer mental and physical health, more relationship problems, and less academic and career success compared to non-marijuana-using peers. For example, marijuana use is associated with a higher likelihood of dropping out of school. Several studies also associate workers' marijuana smoking with increased absences, tardiness, accidents, workers' compensation claims, and job turnover.

Learn More

For additional information on marijuana and marijuana abuse, please see http://www.drugabuse.gov/publications/research-reports/marijuana-abuse

Opioids

Derived from poppy seeds, **opium** is the base compound used for all narcotics. Opiates, which are narcotics, include opium and other drugs derived from opium, such as morphine, codeine, and heroin. Methadone is a synthetic chemical that has a morphine-like action, and also falls into this category of drugs.

Heroin

Heroin is considered a semi-synthetic narcotic because it is derived from a naturally occurring substance in the Oriental poppy plant called opium. It is a highly effective, fast-acting analgesic (painkiller) if injected when used medicinally; however, its benefits are outweighed by its risk of toxicity and high dependence rate. Heroin can be injected, snorted, or smoked. When heroin enters the brain it produces a dream-like euphoria (NIDA, 2010). Abuse is common because this drug creates a strong physical and psychological dependence and tolerance. Recently heroin has become more popular among young people. The risks of heroin use are increased due to the use of needles for injection. There is an increased likelihood of transmission of communicable diseases like HIV and hepatitis due to the practice of sharing needles. Although abrupt withdrawal from heroin is rarely fatal, the discomfort associated with going "cold turkey" is extremely intense.

Heroin users are at high risk for addiction. It is estimated that approximately 23% of heroin users become dependent. Anyone can become dependent, and life expectancy of the heroin addict who injects the drug intravenously is significantly lower than that of one who does not. Overdosing on heroin can result in death within minutes.

Stimulants

Caffeine

Caffeine is a stimulant as well as a psychotropic (mind affecting) drug. Caffeine is generally associated with coffee, tea, and cola, but can also be found in chocolate, cocoa, and other carbonated beverages (see Table 9.2), as well as some medications, both prescription and non-prescription, i.e., Excedrin®. Approximately 65–180 mg of caffeine are found in one cup of coffee, compared to tea, which contains 40–100 mg per cup, and cola, which contains 30–60 mg per twelve ounce serving. Caffeine is readily absorbed into the body and causes stimulation of the cerebral cortex and medullary centers in the brain, resulting in mental alertness.

Moderation is the key when using caffeine. Researchers agree that 300 mg of caffeine is considered moderate intake, which is equivalent to approximately three cups of coffee. Some individuals are more sensitive to caffeine than others and may feel the effects at smaller doses. According to research, caffeine in beverage form is not dehydrating, but if ingesting caffeine from food or tablets, be sure to rehydrate from the drug's diuretic action.

Excessive consumption of caffeine increases plasma levels of epinephrine, norepinephrine, and renin. It also can cause serious side effects, such as tremors, nervousness, irritability, headaches, hyperactivity, arrhythmia, dizziness, and insomnia. It can elevate the blood pressure and body temperature, increase the breathing rate, irritate the stomach and bowels, and dehydrate the body.

A study by researchers at Duke University Medical Center shows that caffeine taken in the morning has effects on the body that persist until bedtime and amplifies stress throughout the day. In addition to the body's physiological response in blood pressure elevations and stress

Moderation is the key when using caffeine.

TABLE 9.2 ♦ Caffeine Content of Selected Beverages

Drink	Ounces	Caffeine (mg)	mg/oz
5 Hour Energy	2	138	69.0
Amp	8.4	75	8.9
AriZona Green Tea Energy	16	200	12.5
Coca-Cola Classic	12	35	2.9
Coffee (Drip)	8	145	18.1
Coffee (Espresso)	1.5	77	51.3
Diet Coke	12	45	3.8
Diet Mountain Dew	12	55	4.6
Dr. Pepper	12	41	3.4
Full Throttle	16	144	9.0
Lipton Iced Teas	20	50	2.5
Mountain Dew	12	55	4.6
NOS	16	260	16.2
Red Bull	8.46	80	9.5
Rockstar	16	160	10.0
SoBe No Fear	16	174	10.9
Starbucks Double Shot	6.5	130	20.0
Starbucks Grande Coffee	16	330	20.6
Tea (Brewed)	8	47	5.9
VitaminWater Energy Citrus	20	42	2.1

Modified from Caffeine In Energy Drinks—How Much Caffeine Is In Your Energy Drink? www.ShapeFit.com

> **How can I reduce my caffeine consumption?**
> - Keep a log of how much caffeine you consume daily
> - Limit your consumption to 200–300 mg/daily
> - Substitute herbal tea or decaf coffee
> - Stop smoking—caffeine and smoking often go together
> - Remember coffee does not help sober up after drinking (McKinley, 2005)

hormone levels, it also magnifies a person's perception of stress. According to this study, caffeine enhances the effects of stress and can make stress even more unhealthy (Lane, Pieper, Phillips-Butte, Bryant, and Kuhn, 2002).

Excessive amounts of caffeine may increase the incidence of premenstrual syndrome (PMS) in some women and may increase fibrocystic breast disease (noncancerous breast lumps) as well. The U.S. Surgeon General recommends that women avoid or restrict caffeine intake during pregnancy. Withdrawal symptoms from caffeine may include headaches, depression, drowsiness, nervousness, and a feeling of lethargy.

In 2006, more than five hundred new energy drinks were released on the market worldwide. College students have been known to use caffeinated products like these for extra energy when studying, driving long distances, or needing more energy in general. A common practice of mixing energy drinks and alcohol is of special concern. Drinking large amounts of caffeine (a stimulant) combined with large amounts of alcohol (a depressant) can cause people to misjudge their level of intoxication. The combination of drugs may mask symptoms such as headache, weakness, and muscle coordination, but in reality visual reaction time and motor coordination are still negatively affected by alcohol. Driving or making any other important decisions under these circumstances can be extremely dangerous.

It is a naturally occurring psychoactive substance contained in the leaves of the South American coca plant. Crack cocaine, a rock-like crystalline form of cocaine made by combining cocaine hydrochloride with common baking soda, can be heated in the bowl of a pipe, enabling the vapors to be inhaled into the lungs. Cocaine is used occasionally as a topical anesthetic medicinally; however, more commonly it is inhaled (snorted), injected, or smoked illegally. The effects of cocaine use are rapid and short lived (from five to thirty minutes). Snorting enables only about 60 percent of the drug to be absorbed because the nasal vessels constrict immediately. Cocaine use causes dopamine and norepinephrine to be released into the brain, causing a feeling of euphoria and confidence; however, at the same time electrical impulses to the heart that regulate its rhythm are impaired. There is evidence that both psychological and physical dependence on cocaine occurs rapidly.

Today, the smoking of crack cocaine is more prevalent than inhalation. When smoked, the drug reaches the central nervous system immediately, affecting several neurotransmitters in the brain. The effects are short lived (usually around five to ten minutes), leaving the user with feelings of depression. Abuse of this drug can result in convulsions, seizures, respiratory distress, and sudden cardiac failure.

The relatively short-lived "high" from cocaine requires frequent use to maintain feelings of euphoria and is therefore quite costly for the addict. A single dose of crack sells for $30 or more, so to maintain a habit would cost hundreds of dollars a day. To pay for their habit, addicts will often turn to criminal activities such as dealing drugs, stealing, or prostitution. Crack houses are known for promoting the spread of HIV infection, and thousands of babies are born to crack-addicted mothers. These babies have severe physical and neurological problems, requiring significant medical attention. The cost of this drug is high, not just for the user, but for society as well.

Amphetamines are drugs that speed up the nervous system. They do not occur naturally and must be manufactured in a laboratory. When used in moderation, amphetamines stimulate receptor sites for two naturally occurring neurotransmitters, having the effect of elevated mood, increased alertness, and feelings of well-being. In addition, the activity of the stomach and intestines may be slowed and appetite suppressed. When amphetamines are eliminated from the body, the user becomes fatigued. With abuse, the user will experience rapid tolerance and a strong psychological dependence, along with the possibility of impotence and episodes of psychosis. When use stops, the abuser may experience periods of depression.

Methamphetamines

An extremely addictive and powerful drug that stimulates the central nervous system is commonly known as "meth." In its smoked form, it is called "crystal," "crank," or "ice." It is chemically similar to amphetamines but much stronger. The effects from methamphetamine can last up to eight hours or in some cases even longer. It comes in many forms and can be injected, inhaled, orally ingested, or snorted.

Methamphetamine is considered to be the fastest growing drug in the United States. According to the director of the Substance Abuse and Mental Health Services Administration (SAMHSA), the growth and popularity of this drug is because of its wide availability, easy production, low cost, and highly addictive nature.

Methamphetamine is a psychostimulant but different than others like cocaine or amphetamine. Methamphetamine, like cocaine, results in an accumulation of dopamine. Dopamine is a neurotransmitter in regions of the brain that deal with emotion, movement, motivation, and pleasure. The large release of dopamine is presumed to help the drug's toxic effects on the brain. However, unlike cocaine, which is removed and metabolized quickly from the body, methamphetamine has a longer duration of action, which stays in the body and brain longer, leading to prolonged stimulant effects. Chronic methamphetamine abuse significantly changes the way the brain functions (NIDA, 2010).

Methamphetamine abusers may display symptoms that include violent behavior, confusion, hallucinations, and possible paranoid or delusional feelings, also causing severe personality shifts. These feelings of paranoia can lead to homicidal or suicidal thoughts or tendencies.

Methamphetamines are highly addictive and can be fatal with a single use. Deadly ingredients include antifreeze, drain cleaner, fertilizer, battery acid, or lantern fuel. The results when overused can cause heart failure and death. Long-term physical effects can lead to strokes, liver, kidney, and lung damage. Abuse can also lead to permanent and severe brain and psychological damage.

Club Drugs

MDMA

MDMA, also known as **ecstasy**, has a chemical structure similar to methamphetamines and mescaline, causing hallucinogenic effects. As a result, it can produce both stimulant and psychedelic effects. In addition to its euphoric effects, MDMA can lead to disruptions in body temperature and cardiovascular regulation causing panic, anxiety, and rapid heart rate. It also damages nerves in the brain's serotonin system and possibly produces long-term damage to brain areas that are critical for thought and memory (NIDA, 2010). Physical effects can include muscle tension, teeth clenching, nausea, blurred vision, and faintness. The psychological effects can include confusion, depression, sleep disorders, anxiety, and paranoia that can last long after taking the drug. It is most often available in tablet form and usually taken orally. Occasionally it is found in powder form and can be snorted or smoked, but it is rarely injected. An overdose can be lethal, especially when taken with alcohol or other drugs, for instance heroin ("H-bomb").

Flunitrazepam

Flunitrazepam is an illegal drug in the United States. In other parts of the world, it is generally prescribed for sleep disorders. The commercial/street name is "Rohypnol."

A 2-mg tablet is equal to the potency of a six-pack of beer. Flunitrazepam is a tranquilizer, similar to Valium, but ten times more potent, producing sedative effects including muscle relaxation, dizziness, memory loss, and blackouts. The effects occur twenty to thirty minutes after use and lasts for up to eight hours.

Rohypnol, more commonly known as "roofies," is a small, white, tasteless, pill that dissolves in food or drinks. It is most commonly used with other drugs, such as alcohol, ecstasy, heroin, and marijuana to enhance the feeling of the other drug. Although Rohypnol alone can be very dangerous, as well as physically addicting, when mixed with other drugs it can be fatal. It is also referred to as the "date rape" drug because there have been many reported cases of individuals giving Rohypnol to someone without their knowledge. The effects incapacitate the victim, and therefore they are unable to resist a sexual assault. It produces an "anterograde amnesia," meaning they may not remember events experienced while under effects of the drug (NIDA, 2010).

Gamma-Hydroxybutyrate (GHB)

GHB is a fast-acting, powerful drug that depresses the nervous system. It occurs naturally in the body in small amounts.

Commonly taken with alcohol, it depresses the central nervous system and induces an intoxicated state. GHB is commonly consumed orally, usually as a clear liquid or a white powder. It is odorless, colorless, and slightly salty to taste. Effects from GHB can occur within fifteen to thirty minutes. Small doses (less than 1 g) of GHB act as a relaxant with larger doses causing strong feelings of relaxation, slowing heart rate, and respiration. There is a very fine line to cross to find a lethal dose, which can lead to seizures, respiratory distress, low blood pressure, and coma.

According to the Drug Abuse Warning Network, The Drug Induced Rape Prevention and Punishment Act of 1996 was enacted into federal law in response to the abuse of Rohypnol. This law makes it a crime to give someone a controlled substance without his/her knowledge and with the intent to commit a crime. The law also stiffens the penalties for possession and distribution of Rohypnol and GHB. Used in Europe as a general anesthetic and treatment for insomnia, GHB is growing in popularity and is widely available underground. Manufactured by non-professional "kitchen" chemists, concerns about quality and purity should be considered.

Dissociative Drugs

Ketamine Hydrochloride

"Special K" or "K" was originally created for use in a medical setting on humans and animals. Ninety percent is legally sold for veterinary use. Ketamine usually comes in liquid form and is cooked into a white powder for snorting. Higher doses produce a hallucinogenic effect and may cause the user to feel far away from their body. This is called a "K-hole" and has been compared to near-death experiences. Low doses can increase heart rate and numbness in the extremities with higher doses depressing consciousness and breathing. This makes it extremely dangerous if combined with other depressants such as alcohol or GHB.

Phencyclidine (phencyclidine)

Also known as PCP or angel dust, phencyclidine hydrochloride is sometimes considered a hallucinogen, although it does not easily fit into any category.

First synthesized in 1959, it is used intravenously and as an anesthetic that blocks pain without producing numbness. Taken in small doses, it causes feelings of euphoria. The harmful side effects include depression, anxiety, confusion, and delirium. High doses of PCP cause mental confusion, hallucinations, and can cause serious mental illness and extreme aggressive and violent behavior, including murder.

Hallucinogens

Hallucinogens, also called psychedelics, are drugs that affect perception, sensation, awareness, and emotion. Changes in time and space and hallucinations may be mild or extreme depending on the dose, and may vary on every occasion. There are many synthetic as well as natural hallucinogens in use. Synthetic groups include LSD, which is the most potent; mescaline, which is derived from the peyote cactus, and psilocybin, derived from mushrooms, have similar effects.

Lysergic Acid Diethylamide (LSD)

LSD is a colorless, odorless, and tasteless liquid that is made from lysergic acid, which comes from the ergot fungus. It was first converted to lysergic acid diethylamide (LSD) in 1938. In 1943, its psychoactive properties accidentally became known (NIDA, 2009).

Hallucinations and illusions often occur, and effects vary according to the dosage, personality of the user, and conditions under which the drug is used. A flashback is a recurrence of some hallucinations from a previous LSD experience days or months after the dose. Flashbacks can occur without reason, occurring to heavy users more frequently. After taking LSD, a person loses control over normal thought process. Street LSD is often mixed with other substances and its effects are quite uncertain (NIDA, 2009).

Other Compounds/Drugs

Anabolic Steroids

Anabolic androgenic steroids are man-made and very similar to male sex hormones. The word *anabolic* means "muscle building," and *androgenic* refers to masculine. Legally, steroids are prescribed to individuals to treat problems occurring when the body produces abnormally low amounts of testosterone and problems associated with delayed puberty or impotence. Other cases for prescribed steroids use would involve individuals with whom a disease has resulted in a loss of muscle mass (NIDA, 2005).

Although steroids are a banned substance in all professional and collegiate sports, most people who use steroids do so to enhance physical performance in sports or other activities. Some choose to use steroids to improve physical appearance or to increase muscle size and to reduce body fat. Some steroids can be taken orally or injected into the muscle. There are also some forms of steroid creams and gels that are to be rubbed into the skin. Most doses taken by abusers are ten to one hundred times the potency of normal doses used for medicinal purposes.

Consequences from steroid abuse can cause some serious health issues. There can be some problems with the normal hormone production in the individual, which can be very severe and irreversible. Major side effects of steroid abuse can lead to cardiovascular disease, high blood pressure, and stroke because it increases the LDL cholesterol levels while decreasing the HDL levels. There can also be liver damage, muscular and ligament damage, as well as stunted bone growth. In addition to these problems, the side effects for males can be shrinkage of the testes and a reduction in sperm count. For females, steroid use can cause facial hair growth and the cessation of the menstrual cycle.

Research also suggests some psychological and behavioral changes. Steroid abusers can become very aggressive and violent and have severe mood swings. Users are reported to have paranoid and jealous tendencies along with irritability and impaired judgment. Depression has also been linked to steroid use once the individual stops taking the drug, therefore leading to continued use. This depressed state can lead to serious consequences, and in some cases it has been reported to lead to suicidal thoughts (NIDA, 2005).

Inhalants

Inhalants are poisonous chemical gases, fumes, or vapors that produce psychoactive effects when sniffed. When inhaled, the fumes take away the body's ability to absorb oxygen. Inhalants are considered delerients, which can cause permanent damage to the heart, brain, lungs, and liver. Common inhalants include model glue, acetone, gasoline, kerosene, nail polish, aerosol sprays, Pam™ cooking spray, Scotchgard™ fabric protectant, lighter fluids, butane, and cleaning fluids, as well as nitrous oxide (laughing gas). These products were not created to be inhaled or ingested. They were designed to dissolve things or break things down, which is exactly what they do to the body.

Inhalants reach the lungs, bloodstream, and other parts of the body very quickly. Intoxication can occur in as little as five minutes and can last as long as nine hours. Inhaled lighter fluid/butane displaces the oxygen in the lungs, causing suffocation. Even a single episode can cause asphyxiation or cardiac arrhythmia and possibly lead to death.

The initial effects of inhalants are similar to those of alcohol, but they are very unpredictable. Some effects include dizziness and blurred vision, involuntary eye movement, poor coordination, involuntary extremity movement, slurred speech, euphoric feeling, nosebleeds, and possible coma.

Health risks involved with the use of inhalants may include hepatitis, liver and/or kidney failure, as well as the destruction of bone marrow and skeletal muscles. Respiratory impairment and blood abnormalities, along with irregular heartbeat and/or heart failure, are also serious side effects of inhalants. Regular use can lead to tolerance, the need for more powerful drugs, and addiction (NIDA, 2011).

Prescription Drugs

According to the National Institute on Drug Abuse (NIDA), approximately 16 million Americans misuse and abuse prescription drugs for non-medicinal purposes. To add to the severity of the problem, many of those who abuse prescription drugs abuse alcohol and other drugs as well (NIDA, 2009).

It is not that these drugs should not be used for the purpose intended, as "they have an important place in the treatment of debilitating conditions," says Richard Brown, M.D., associate professor at the University of Wisconsin (FDA, 2005). It is generally uncommon for addiction to occur to patients using the drug as prescribed. It is the individuals abusing and misusing these drugs that can lead to problems. Some of these problems occur because prescription drugs are received through false prescriptions, over-prescribing, and pharmacy theft. Sometimes problems arise because there is a lack of communication or information provided to the patient (FDA, 2005).

Some of the most commonly abused prescription drugs are pain killers such as morphine, codeine, Oxycodone, Vicodin, and Demerol.

Depressants

Depressants are sedatives or anxiolytic (anti-anxiety) drugs that depress the central nervous system. Benzodiazepines such as Valium and Xanax and barbituates like Nembutal, Secobarbital, and Phenobarbibtal can be prescribed to releive tension, induce relaxation and sleep, or treat panic attacks. All of these

Psychoactive Drug Types

Commonly Abused Drugs
Visit NIDA at www.drugabuse.gov

National Institutes of Health
U.S. Department of Health and Human Services
NIH… Turning Discovery Into Health

Substances: Category and Name	Examples of Commercial and Street Names	DEA Schedule*/ How Administered**	Acute Effects/Health Risks
Tobacco			
Nicotine	Found in cigarettes, cigars, bidis, and smokeless tobacco (snuff, spit tobacco, chew)	Not scheduled/smoked, snorted, chewed	Increased blood pressure and heart rate/chronic lung disease; cardiovascular disease; stroke; cancers of the mouth, pharynx, larynx, esophagus, stomach, pancreas, cervix, kidney, bladder, and acute myeloid leukemia; adverse pregnancy outcomes; addiction
Alcohol			
Alcohol (ethyl alcohol)	Found in liquor, beer, and wine	Not scheduled/swallowed	In low doses, euphoria, mild stimulation, relaxation, lowered inhibitions; in higher doses, drowsiness, slurred speech, nausea, emotional volatility, loss of coordination, visual distortions, impaired memory, sexual dysfunction, loss of consciousness/increased risk of injuries, violence, fetal damage (in pregnant women); depression; neurologic deficits; hypertension, liver and heart disease; addiction; fatal overdose
Cannabinoids			
Marijuana	Blunt, dope, ganja, grass, herb, joint, bud, Mary Jane, pot, reefer, green, trees, smoke, sinsemilla, skunk, weed	I/smoked, swallowed	Euphoria, relaxation, slowed reaction time, distorted sensory perception; impaired balance and coordination; increased heart rate and appetite; impaired learning, memory; anxiety; panic attacks; psychosis/cough; frequent respiratory infections; possible mental health decline; addiction
Hashish	Boom, gangster, hash, hash oil, hemp	I/smoked, swallowed	
Opioids			
Heroin	Diacetylmorphine: smack, horse, brown sugar, dope, H, junk, skag, skunk, white horse, China white; cheese (with OTC cold medicine and antihistamine)	I/injected, smoked, snorted	Euphoria, drowsiness, impaired coordination, dizziness, confusion, nausea, sedation; feeling of heaviness in the body; slowed or arrested breathing/constipation; endocarditis, hepatitis, HIV, addiction; fatal overdose
Opium	Laudanum, paregoric: big O, black stuff, block, gum, hop	II, III, V/swallowed, smoked	
Stimulants			
Cocaine	Cocaine hydrochloride: blow, bump, C, candy, Charlie, coke, crack, flake, rock, snow, toot	II/snorted, smoked, injected	Increased heart rate, blood pressure, body temperature, metabolism; feelings of exhilaration; increased energy, mental alertness; tremors; reduced appetite; irritability; anxiety; panic; paranoia; violent behavior; psychosis/weight loss; insomnia; cardiac or cardiovascular complications; stroke; seizures; addiction **Also, for cocaine**—nasal damage from snorting **Also, for methamphetamine**—severe dental problems
Amphetamine	Biphetamine, Dexedrine: bennies, black beauties, crosses, hearts, LA turnaround, speed, truck drivers, uppers	II/swallowed, snorted, smoked, injected	
Methamphetamine	Desoxyn: meth, ice, crank, chalk, crystal, fire, glass, go fast, speed	II/swallowed, snorted, smoked, injected	
Club Drugs			
MDMA (methylenedioxymethamphetamine)	Ecstasy, Adam, clarity, Eve, lover's speed, peace, uppers	I/swallowed, snorted, injected	**MDMA**—mild hallucinogenic effects; increased tactile sensitivity, empathic feelings; lowered inhibition; anxiety; chills; sweating; teeth clenching; muscle cramping/ sleep disturbances; depression; impaired memory; hyperthermia; addiction **Flunitrazepam**—sedation; muscle relaxation; confusion; memory loss; dizziness; impaired coordination/addiction **GHB**—drowsiness; nausea; headache; disorientation; loss of coordination; memory loss/ unconsciousness; seizures; coma
Flunitrazepam***	Rohypnol: forget-me pill, Mexican Valium, R2, roach, Roche, roofies, rofinol, rope, rophies	IV/swallowed, snorted	
GHB***	Gamma-hydroxybutyrate: G, Georgia home boy, grievous bodily harm, liquid ecstasy, soap, scoop, goop, liquid X	I/swallowed	
Dissociative Drugs			
Ketamine	Ketalar SV: cat Valium, K, Special K, vitamin K	III/injected, snorted, smoked	Feelings of being separate from one's body and environment; impaired motor function/anxiety; tremors; numbness; memory loss; nausea **Also, for ketamine**—analgesia; impaired memory; delirium; respiratory depression and arrest; death **Also, for PCP and analogs**—analgesia, psychosis; aggression; violence; slurred speech; loss of coordination; hallucinations **Also, for DXM**—euphoria; slurred speech; confusion; dizziness; distorted visual perceptions
PCP and analogs	Phencyclidine: angel dust, boat, hog, love boat, peace pill	I, II/swallowed, smoked, injected	
Salvia divinorum	Salvia, Shepherdess's Herb, Maria Pastora, magic mint, Sally-D	Not scheduled/chewed, swallowed, smoked	
Dextromethorphan (DXM)	Found in some cough and cold medications: Robotripping, Robo, Triple C	Not scheduled/swallowed	
Hallucinogens			
LSD	Lysergic acid diethylamide: acid, blotter, cubes, microdot, yellow sunshine, blue heaven	I/swallowed, absorbed through mouth tissues	Altered states of perception and feeling; hallucinations **Also, for LSD and mescaline**—increased body temperature, heart rate, blood pressure; loss of appetite; sweating; sleeplessness; numbness; dizziness; weakness; tremors; impulsive behavior; rapid shifts in emotion **Also, for LSD**—Flashbacks, Hallucinogen Persisting Perception Disorder **Also, for psilocybin**—nervousness; paranoia; panic
Mescaline	Buttons, cactus, mesc, peyote	I/swallowed, smoked	
Psilocybin	Magic mushrooms, purple passion, shrooms, little smoke	I/swallowed	
Other Compounds			
Anabolic steroids	Anadrol, Oxandrin, Durabolin, Depo-Testosterone, Equipoise: roids, juice, gym candy, pumpers	III/injected, swallowed, applied to skin	**Steroids**—no intoxication effects/hypertension; blood clotting and cholesterol changes; liver cysts; hostility and aggression; acne; in adolescents—premature stoppage of growth; in males—prostate cancer, reduced sperm production, shrunken testicles, breast enlargement, in females—menstrual irregularities, development of beard and other masculine characteristics **Inhalants** (varies by chemical)—stimulation; loss of inhibition; headache; nausea or vomiting; slurred speech; loss of motor coordination; wheezing/cramps; muscle weakness; depression; memory impairment; damage to cardiovascular and nervous systems; unconsciousness; sudden death
Inhalants	Solvents (paint thinners, gasoline, glues); gases (butane, propane, aerosol propellants, nitrous oxide); nitrites (isoamyl, isobutyl, cyclohexyl): laughing gas, poppers, snappers, whippets	Not scheduled/inhaled through nose or mouth	

276 CHAPTER 9

Substances: Category and Name	Examples of Commercial and Street Names	DEA Schedule*/ How Administered**	Acute Effects/Health Risks
Prescription Medications			
CNS Depressants	For more information on prescription medications, please visit http://www.nida.nih.gov/DrugPages/PrescripDrugsChart.html		
Stimulants			
Opioid Pain Relievers			

*Schedule I and II drugs have a high potential for abuse. They require greater storage security and have a quota on manufacturing, among other restrictions. Schedule I drugs are available only by prescription (unrefillable) and require a form for ordering. Schedule III and IV drugs are available by prescription, may have five refills in 6 months, and may be ordered orally. Some Schedule V drugs are available over the counter.
**Some of the health risks are directly related to the route of drug administration. For example, injection drug use can increase the risk of infection through needle contamination with staphylococci, HIV, hepatitis, and other organisms.
***Associated with sexual assaults.

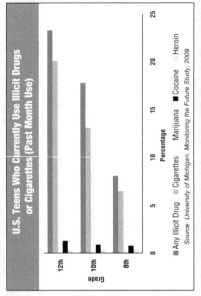

U.S. Teens Who Currently Use Illicit Drugs or Cigarettes (Past Month Use)

Any Illicit Drug • Cigarettes • Marijuana • Cocaine • Heroin

Source: University of Michigan, Monitoring the Future Study, 2009

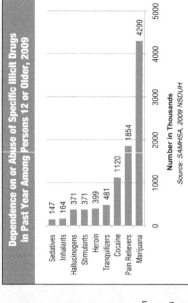

Dependence on or Abuse of Specific Illicit Drugs in Past Year Among Persons 12 or Older, 2009

- Sedatives 147
- Inhalants 164
- Hallucinogens 371
- Stimulants 371
- Heroin 399
- Tranquilizers 481
- Cocaine 1120
- Pain Relievers 1854
- Marijuana 4299

Number in Thousands
Source: SAMHSA, 2009 NSDUH

NATIONAL INSTITUTE ON DRUG ABUSE
NIDA DRUGPUBS
RESEARCH DISSEMINATION CENTER

Order NIDA publications from DrugPubs:
1-877-643-2644 or 1-240-645-0228 (TTY/TDD)

Revised March 2011
Reprinted April 2012

Principles of Drug Addiction Treatment

More than three decades of scientific research show that treatment can help drug-addicted individuals stop drug use, avoid relapse and successfully recover their lives. Based on this research, 13 fundamental principles that characterize effective drug abuse treatment have been developed. These principles are detailed in NIDA's *Principles of Drug Addiction Treatment: A Research-Based Guide*. The guide also describes different types of science-based treatments and provides answers to commonly asked questions.

1. **Addiction is a complex but treatable disease that affects brain function and behavior.** Drugs alter the brain's structure and how it functions, resulting in changes that persist long after drug use has ceased. This may help explain why abusers are at risk for relapse even after long periods of abstinence.

2. **No single treatment is appropriate for everyone.** Matching treatment settings, interventions, and services to an individual's particular problems and needs is critical to his or her ultimate success.

3. **Treatment needs to be readily available.** Because drug-addicted individuals may be uncertain about entering treatment, taking advantage of available services the moment people are ready for treatment is critical. Potential patients can be lost if treatment is not immediately available or readily accessible.

4. **Effective treatment attends to multiple needs of the individual, not just his or her drug abuse.** To be effective, treatment must address the individual's drug abuse and any associated medical, psychological, social, vocational, and legal problems.

5. **Remaining in treatment for an adequate period of time is critical.** The appropriate duration for an individual depends on the type and degree of his or her problems and needs. Research indicates that most addicted individuals need at least 3 months in treatment to significantly reduce or stop their drug use and that the best outcomes occur with longer durations of treatment.

6. **Counseling—individual and/or group—and other behavioral therapies are the most commonly used forms of drug abuse treatment.** Behavioral therapies vary in their focus and may involve addressing a patient's motivations to change, building skills to resist drug use, replacing drug-using activities with constructive and rewarding activities, improving problemsolving skills, and facilitating better interpersonal relationships.

7. **Medications are an important element of treatment for many patients, especially when combined with counseling and other behavioral therapies.** For example, methadone and buprenorphine are effective in helping individuals addicted to heroin or other opioids stabilize their lives and reduce their illicit drug use. Also, for persons addicted to nicotine, a nicotine replacement product (nicotine patches or gum) or an oral medication (bupropion or varenicline), can be an effective component of treatment when part of a comprehensive behavioral treatment program.

8. **An individual's treatment and services plan must be assessed continually and modified as necessary to ensure it meets his or her changing needs.** A patient may require varying combinations of services and treatment components during the course of treatment and recovery. In addition to counseling or psychotherapy, a patient may require medication, medical services, family therapy, parenting instruction, vocational rehabilitation and/or social and legal services. For many patients, a continuing care approach provides the best results, with treatment intensity varying according to a person's changing needs.

9. **Many drug-addicted individuals also have other mental disorders.** Because drug abuse and addiction—both of which are mental disorders—often co-occur with other mental illnesses, patients presenting with one condition should be assessed for the other(s). And when these problems co-occur, treatment should address both (or all), including the use of medications as appropriate.

10. **Medically assisted detoxification is only the first stage of addiction treatment and by itself does little to change long-term drug abuse.** Although medically assisted detoxification can safely manage the acute physical symptoms of withdrawal, detoxification alone is rarely sufficient to help addicted individuals achieve long-term abstinence. Thus, patients should be encouraged to continue drug treatment following detoxification.

11. **Treatment does not need to be voluntary to be effective.** Sanctions or enticements from family, employment settings, and/or the criminal justice system can significantly increase treatment entry, retention rates, and the ultimate success of drug treatment interventions.

12. **Drug use during treatment must be monitored continuously, as lapses during treatment do occur.** Knowing their drug use is being monitored can be a powerful incentive for patients and can help them withstand urges to use drugs. Monitoring also provides an early indication of a return to drug use, signaling a possible need to adjust an individual's treatment plan to better meet his or her needs.

13. **Treatment programs should assess patients for the presence of HIV/AIDS, hepatitis B and C, tuberculosis, and other infectious diseases, as well as provide targeted risk-reduction counseling to help patients modify or change behaviors that place them at risk of contracting or spreading infectious diseases.** Targeted counseling specifically focused on reducing infectious disease risk can help patients further reduce or avoid substance-related and other high-risk behaviors. Treatment providers should encourage and support HIV screening and inform patients that highly active antiretroviral therapy (HAART) has proven effective in combating HIV, including among drug-abusing populations.

differ in action, absorption, and metabolism, but all produce similar intoxication and withdrawal symptoms.

Depressants can produce both a physical and psychological dependence within two to four weeks. Those with a prior history of abuse are at greater risk of abusing sedatives, even if prescribed by a physician. If there is no previous substance abuse history, one rarely develops problems if prescribed and monitored by a physician. Depressants can be very dangerous, if not lethal, if used in combination with alcohol, leading to respiratory depression, respiratory arrest, and death.

Some of the physiological effects of depressants include drowsiness, impaired judgment, poor coordination, slowed breathing, confusion, weak and rapid heartbeat, relaxed muscles, and pain relief.

A major health risk associated with the use of depressants is the development of a dependence to the drug, leading to serious side effects, such as stupors, coma, and death.

Hydrocodone

Hydrocodone is a narcotic used to relieve pain and suppress cough. This drug, which can lead to both physiological and psychological dependence, saw a dramatic increase in legal sales between 1991 and 2010 with prescriptions topping 100 million (NIDA, 2011).

Codeine

A natural derivative of opium. Codeine is medically used as a mild painkiller or a cough suppressant. Although widely used, there is potential for physical dependence.

Morphine

This is the main alkaloid found in opium. It is ten times stronger than opium and brings quick relief from pain. It is most effectively used as an anesthetic during heart surgery, to relieve pain in post-operative patients, and sometimes used to relieve pain for cancer patients.

Oxycodone

Oxycodone, a drug used for moderate to severe pain relief, has a high potential for abuse. Tablets should be take orally, but when crushed and injected intravenously or snorted, a potentially lethal dose is released (FDA, 2004).

According to recent reports from the FDA, a highly abused stimulant among middle and high school students is methylphenidate, commonly known as **Ritalin.** This drug is more powerful than caffeine but not as potent as amphetamines and is prescribed for individuals with attention-deficit/hyperactivity disorders, ADHD, and sometimes to treat narcolepsy. Researchers speculate that Ritalin increases the slow and steady release of dopamine, therefore improving attention and focus for those in need of the increase. "Individuals abuse Ritalin to lose weight, increase alertness and experience the euphoric feelings resulting from high doses" (U.S. Dept. of Justice, 2006). When abused, the tablets are either taken orally or crushed and snorted; some even dissolve the tablets in water and inject the mixture. Addiction occurs when it induces large and fast increases of dopamine in the brain (DOJ, 2006).

Adderall is another stimulant used to treat ADHD as well as narcolepsy. Physical and psychological dependence may occur with this drug. Symptoms of Adderall overdose include dizziness, blurred vision, restlessness, rapid breathing, confusion, hallucinations, nausea, vomiting, irregular heartbeat, and seizures.

Stimulant ADHD Medications

Stimulant medications including amphetamines (e.g., Adderall) and methylphenidate (e.g., Ritalin and Concerta) are often prescribed to treat children, adolescents, or adults diagnosed with attention-deficit hyperactivity disorder (ADHD).

People with ADHD persistently have more difficulty paying attention or are more hyperactive or impulsive than other people the same age. This pattern of behavior usually becomes evident when a child is in preschool or the first grades of elementary school; the average age of onset of ADHD symptoms is 7 years. Many people's ADHD symptoms improve during adolescence or as they grow older, but the disorder can persist into adulthood.

ADHD diagnoses are increasing. According to the U.S. Centers for Disease Control and Prevention, as of 2011, 11 percent of people ages 4–17 have been diagnosed with ADHD.

How Are Prescription Stimulants Used?

Prescription stimulants have a paradoxically calming and "focusing" effect on individuals with ADHD. They are prescribed to patients for daily use, and come in the form of tablets or capsules of varying dosages. Treatment of ADHD with stimulants, often in conjunction with psychotherapy, helps to improve ADHD symptoms along with the patient's self-esteem, thinking ability, and social and family interactions.

Prescription stimulants are sometimes abused however—that is, taken in higher quantities or in a different manner than prescribed, or taken by those without a prescription. Because they suppress ap-

---

Do Prescription Stimulants Make You Smarter?

A growing number of teenagers and young adults are abusing prescription stimulants to boost their study performance in an effort to improve their grades in school, and there is a widespread belief that these drugs can improve a person's ability to learn ("cognitive enhancement").

Prescription stimulants do promote wakefulness, but studies have found that they do not enhance learning or thinking ability when taken by people who do not actually have ADHD. Also, research has also shown that students who abuse prescription stimulants actually have lower GPAs in high school and college than those who don't.

petite, increase wakefulness, and increase focus and attention, they are frequently abused for purposes of weight loss or performance enhancement (e.g., to help study or boost grades in school; see box, previous page). Because they may produce euphoria, these drugs are also frequently abuse for recreational purposes (i.e., to get high). Euphoria from stimulants is generally produced when pills are crushed and then snorted or mixed with water and injected.

How Do Prescription Stimulants Affect the Brain?

All stimulants work by increasing dopamine levels in the brain—dopamine is a neurotransmitter associated with pleasure, movement, and attention. The therapeutic effect of stimulants is achieved by slow and steady increases of dopamine, which are similar to the way dopamine is naturally produced in the brain. The doses prescribed by physicians start low and increase gradually until a therapeutic effect is reached.

When taken in doses and via routes other than those prescribed, prescription stimulants can increase brain dopamine in a rapid and highly amplified manner (similar to other drugs of abuse such as methamphetamine), thereby disrupting normal communication between brain cells and producing euphoria and, as a result, increasing the risk of addiction.

What Are the Other Health Effects of Prescription Stimulants?

Stimulants can increase blood pressure, heart rate, and body temperature and decrease sleep and appetite. When they are abused, they can lead to malnutrition and its consequences. Repeated abuse of stimulants can lead to feelings of hostility and paranoia. At high doses, they can lead to serious cardiovascular complications, including stroke.

Addiction to stimulants is also a very real consideration for anyone taking them without medical supervision. Addiction most likely occurs because stimulants, when taken in doses and routes other than those prescribed by a doctor, can induce a rapid rise in dopamine in the brain. Furthermore, if stimulants are abused chronically, withdrawal symptoms—including fatigue, depression, and disturbed sleep patterns—can result when a person stops taking them. Additional complications from abusing stimulants can arise when pills are crushed and injected: Insoluble fillers in the tablets can block small blood vessels.

> **Do Prescription Stimulants Affect a Patient's Risk of Substance Abuse?**
>
> Concerns have been raised that stimulants prescribed to treat a child's or adolescent's ADHD could affect an individual's vulnerability to developing later drug problems—either by increasing the risk or by providing a degree of protection. The studies conducted so far have found no differences in later substance use for children with ADHD who received treatment and those that did not. This suggests treatment with ADHD medication appears not to affect (either negatively or positively) an individual's risk for developing a substance use disorder.

Learn More

For additional information on prescription stimulants, see
http://www.drugabuse.gov/publications/research-reports/prescription-drugs

Commonly Abused Prescription Drugs

Visit NIDA at www.drugabuse.gov

National Institutes of Health
U.S. Department of Health and Human Services

Substances: Category and Name	Examples of Commercial and Street Names	DEA Schedule*/How Administered	Intoxication Effects/Health Risks
Depressants			Sedation/drowsiness, reduced anxiety, feelings of well-being, lowered inhibitions, slurred speech, poor concentration, confusion, dizziness, impaired coordination and memory/slowed pulse, lowered blood pressure, slowed breathing, tolerance, withdrawal, addiction; increased risk of respiratory distress and death when combined with alcohol
Barbiturates	Amytal, Nembutal, Seconal, Phenobarbital: barbs, reds, red birds, phennies, tooies, yellows, yellow jackets	II, III, IV/injected, swallowed	
Benzodiazepines	Ativan, Halcion, Librium, Valium, Xanax, Klonopin: candy, downers, sleeping pills, tranks	IV/swallowed	
Sleep Medications	Ambien (zolpidem), Sonata (zaleplon), Lunesta (eszopiclone)	IV/swallowed	for barbiturates—euphoria, unusual excitement, fever, irritability/life-threatening withdrawal in chronic users
Opioids and Morphine Derivatives**			Pain relief, euphoria, drowsiness, sedation, weakness, dizziness, nausea, impaired coordination, confusion, dry mouth, itching, sweating, clammy skin, constipation/slowed or arrested breathing, lowered pulse and blood pressure, tolerance, addiction, unconsciousness, coma, death; risk of death increased when combined with alcohol or other CNS depressants
Codeine	Empirin with Codeine, Fiorinal with Codeine, Robitussin A-C, Tylenol with Codeine: Captain Cody, Cody, schoolboy; (with glutethimide: doors & fours, loads, pancakes and syrup)	II, III, IV/injected, swallowed	
Morphine	Roxanol, Duramorph: M, Miss Emma, monkey, white stuff	II, III/injected, swallowed, smoked	for fentanyl—80–100 times more potent analgesic than morphine
Methadone	Methadose, Dolophine: fizzies, amidone, (with MDMA: chocolate chip cookies)	II/swallowed, injected	for oxycodone—muscle relaxation/twice as potent analgesic as morphine; high abuse potential
Fentanyl and analogs	Actiq, Duragesic, Sublimaze: Apache, China girl, dance fever, friend, goodfella, jackpot, murder 8, TNT, Tango and Cash	II/injected, smoked, snorted	for codeine—less analgesia, sedation, and respiratory depression than morphine
Other Opioid Pain Relievers: Oxycodone HCL, Hydrocodone Bitartrate, Hydromorphone, Oxymorphone, Meperidine, Propoxyphene	Tylox, Oxycontin, Percodan, Percocet: Oxy, O.C., oxycotton, oxycet, hillbilly heroin, percs; Vicodin, Lortab, Lorcet: vike, Watson-387; Dilaudid: juice, smack, D, footballs, dillies; Opana, Numorphan, Numorphone: biscuits, blue heaven, blues, Mrs. O, octagons, stop signs, O Bomb; Demerol, meperidine hydrochloride: demmies, pain killer; Darvon, Darvocet	II, III, IV/chewed, swallowed, snorted, injected, suppositories	for methadone—used to treat opioid addiction and pain; significant overdose risk when used improperly
Stimulants			Feelings of exhilaration, increased energy, mental alertness/increased heart rate, blood pressure, and metabolism, reduced appetite, weight loss, nervousness, insomnia, seizures, heart attack, stroke
Amphetamines	Biphetamine, Dexedrine, Adderall: bennies, black beauties, crosses, hearts, LA turnaround, speed, truck drivers, uppers	II/injected, swallowed, smoked, snorted	for amphetamines—rapid breathing, tremor, loss of coordination, irritability, anxiousness, restlessness/delirium, panic, paranoia, hallucinations, impulsive behavior, aggressiveness, tolerance, addiction
Methylphenidate	Concerta, Ritalin: JIF, MPH, R-ball, Skippy, the smart drug, vitamin R	II/injected, swallowed, snorted	for methylphenidate—increase or decrease in blood pressure, digestive problems, loss of appetite, weight loss
Other Compounds			
Dextromethorphan (DXM)	Found in some cough and cold medications: Robotripping, Robo, Triple C	not scheduled/swallowed	Euphoria, slurred speech/increased heart rate and blood pressure, dizziness, nausea, vomiting, confusion, paranoia, distorted visual perceptions, impaired motor function

* Schedule I and II drugs have a high potential for abuse. They require greater storage security and have a quota on manufacturing, among other restrictions. Schedule I drugs are available for research only and have no approved medical use. Schedule II drugs are available only by prescription and require a new prescription for each refill. Schedule III and IV drugs are available by prescription, may have five refills in 6 months, and may be ordered orally. Most Schedule V drugs are available over the counter.

** Taking drugs by injection can increase the risk of infection through needle contamination with staphylococci, HIV, hepatitis, and other organisms. Injection is a more common practice for opioids, but risks apply to any medication taken by injection.

Psychoactive Drug Types

Facts About Prescription Drug Abuse

Medications can be effective when they are used properly, but some can be addictive and dangerous when abused. This chart provides a brief look at some prescribed medications that—when used in ways or by people other than prescribed—have the potential for adverse medical consequences, including addiction.

In 2010, approximately 16 million Americans reported using a prescription drug for nonmedical reasons in the past year; 7 million in the past month.

What types of prescription drugs are abused?

Three types of drugs are abused most often:

- Opioids—prescribed for pain relief
- CNS depressants—barbiturates and benzodiazepines prescribed for anxiety or sleep problems (often referred to as sedatives or tranquilizers)
- Stimulants—prescribed for attention-deficit hyperactivity disorder (ADHD), the sleep disorder narcolepsy, or obesity.

How can you help prevent prescription drug abuse?

- Ask your doctor or pharmacist about your medication, especially if you are unsure about its effects.
- Keep your doctor informed about all medications you are taking, including over-the-counter medications.
- Read the information your pharmacist provides before starting to take medications.
- Take your medication(s) as prescribed.
- Keep all prescription medications secured at all times and properly dispose of any unused medications.

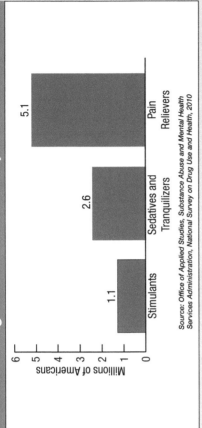

~7.0 Million Americans Reported Past-Month Use of Rx Drugs for Nonmedical Purposes in 2010

- Pain Relievers: 5.1
- Sedatives and Tranquilizers: 2.6
- Stimulants: 1.1

Source: Office of Applied Studies, Substance Abuse and Mental Health Services Administration, National Survey on Drug Use and Health, 2010

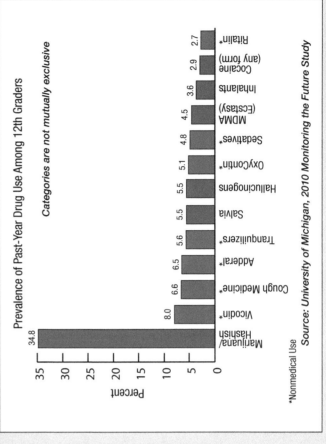

After Marijuana, Prescription and Over-the-Counter Medications* Account for Most of the Commonly Abused Drugs

Prevalence of Past-Year Drug Use Among 12th Graders
Categories are not mutually exclusive

- Marijuana/Hashish: 34.8
- Vicodin*: 8.0
- Cough Medicine: 6.6
- Adderall*: 6.5
- Tranquilizers*: 5.6
- Salvia: 5.5
- Hallucinogens: 5.5
- OxyContin*: 5.1
- Sedatives*: 4.8
- MDMA (Ecstasy): 4.5
- Inhalants: 3.6
- Cocaine (any form): 2.9
- Ritalin*: 2.7

*Nonmedical Use

Source: University of Michigan, 2010 Monitoring the Future Study

Revised October 2011

NIDA DRUGPUBS RESEARCH DISSEMINATION CENTER

Order NIDA publications from DrugPubs:
1-877-643-2644 or 1-240-645-0228 (TTY/TDD)

The danger from prescription and over-the-counter drugs is often underestimated by students. Many assume that if the drug is legal and prescribed by a physician, even if for someone else, it must be safe. However, what they fail to realize is that medications and dosages are tailored to each patient and may not be appropriate in the manner they intend to use them.

For more information:

National Institutes of Health
www.nih.gov

Web of Addictions
www.well.com/user/woa.

Substance Abuse and Mental Health Services Administration
www.samhsa.gov

National Institute on Drug Abuse
www.nida.gov

FDA Guidelines on How to Use Prescription Drugs Safely:
- Always follow medication directions
- Do not increase or decrease doses without consulting your physician
- Do not stop taking medication on your own
- Do not crush or break pills
- Be clear about the drug's effect on driving and other tasks
- Know the drug's potential interactions with alcohol and other drugs
- Inform your doctor if you have had past problems with substance abuse
- Do not use others' prescription medications, and do not share yours.

Ethical Considerations of Illicit Prescription Use

Gabriel Neal, MD

The rise in the diagnosis of cognitive deficits such as Attention Deficit and Hyperactivity Disorder over the past two decades has resulted in more students being treated with prescription nootropics (medications that improve cognitive function). Although safely treating deficient cognitive function does not raise ethical concern, the use of the same medications to improve normal cognitive function does, and the rise in use of prescription nootropics is getting the attention of high school and college administrators and faculty. Recent surveys suggest that 6–16% of college students are using prescription nootropics illicitly at some point during their studies.[i, ii] These studies demonstrated that while Ritalin and Adderall were the prescription drugs of choice, they were also used both for studying and partying. Students who used prescription medications illicitly were more likely to engage in illegal drugs such as marijuana and cocaine as well.

The illicit use of nootropic medications raises ethical concern in the areas of authenticity (the degree to which a student may take credit for work done) and fairness in the classroom. The argument for lesser authenticity is the claim that the more one is enhanced, the less credit one can receive. Could one use a calculator and still take credit for work? What if only a pencil and paper were used? The line between significant and insignificant enhancement is not perfectly clear. A response to the challenge of authenticity is that even with the use of nootropics, some level of study and hard work is still required, which could authenticate the achievement.

A student who uses Adderall to enhance her studying may be able to argue that her ideas are still authentically hers, but it remains to be answered whether it is fair that she use Adderall while other students are denied access to it. The magnitude of the enhancement becomes important. In competitive arenas such as the classroom, rules are created to keep out advantages that are of improper magnitude. Consider the use of caffeine. Although caffeine can heighten alertness, no one complains to the professor if they see someone drinking some coffee before an exam. The reason for this is that caffeine does not confer cognitive benefit of such magnitude that anyone thinks it is unfair to use it in class, and caffeine is accessible to any student. But as better, more potent enhancers such as Adderall emerge, the rules to ensure fairness may need to be updated. Some have suggested that those using prescription enhancers disclose it when submitting academic work.[iii] How professors would grade work differently remains to be seen.

While no one wishes to keep students with legitimate cognitive deficits from receiving proper treatment, students should be concerned about the unfair advantage that illicit use of prescription nootropics offers their classmates; and for those students engaged in illicit use of prescription nootropics, they would do well to consider whether they are cheating themselves out of authentic work and others out of a proper grade.

[i] McCabe, S.E. "Non-Medical Use of Prescription Stimulants among US College Students: Prevalence and Correlates from a National Survey." *Addiction* 100, January (2005): 96–106.

[ii] Prudhomme White, B., Becker-Blease, K., Grace-Bishop, K. "Stimulant Medication Use, Misuse, and Abuse in an Undergraduate and Graduate Student Sample." *Journal of American College Health* 54, no. 5 (2006): 261–268.

[iii] Farah, M.J., Illes, J., Cook-Deegan, R., Gardner, H., Kandel, E., King P. et al. "Neurocognitive Enhancement: what can we do and what should we do about it?" *Nature Reviews: Neuroscience* 5, May (2004): 421–425.

Source: *Gabriel A. Neal, M.D., Family Medical Practice: Obstestrics, St. Joseph Regional Health Center, Bryan, TX*

References

American Heart Association (AHA). *Annual Report.* 2006.

Broadhead, Raymond. (2005). *Synapses and Drugs.* http://outreach.mcb.harvard.edu/lessonplans_S05.htm; Retrieved April 19, 2014.

Center on Addiction and Substance Abuse at Columbia University. Commission on Substances Abuse at Colleges and Universities.

Centers for Disease Control and Prevention. *Behavioral Risk Factor Surveillance System,* 2010.

Centers for Disease Control and Prevention. *National Health Interview Survey,* 2010.

Centers for Disease Control and Prevention. *Smokeless Tobacco,* 2009 and 2010.

Centers for Disease Control and Prevention. *Second Hand Smoke,* 2012.

CollegeDrinkingPrevention.gov. *Interactive Body Content.* www.collegedrinkingprevention.gov; Retrieved April 19, 2014.

Department of Justice. National Drug Intelligence Center. Ritalin Fast Facts, 2006.

Everett, S. A., Lowry, R., Cohen, L. R., Dellinger, A. M. Unsafe motor vehicle practices among substance-using college students. *Accident Analysis,* 1999.

Ewing, J. Detecting Alcoholism: the CAGE Questionnaire. *Journal of the American Medical Association,* 1984.

Dennis, M. E. and the Texas Commission on Alcohol and Drug Abuse. *Instructor Manual, Alcohol Education Program for Minors.* Austin: TCADA. 2005.

Herman, A., et al. In an ongoing search to understand the mechanisms of fetal alcohol syndrome. National Institute on Alcohol Abuse and Alcoholism, 2003.

Hewlings, S. J., and Medeiros, D., M. *Nutrition: Real People, Real Choices* (2nd ed). Dubuque, IA:

Kendall/Hunt Publishing Company. 2011.

Hoeger, W. and Hoeger, S. *Principles and Labs for Fitness and Wellness* (5th ed). Englewood, CO: Morton Publishing Company. 1999.

Journal of the American Medical Association. Moderate alcohol intake and lower risk of coronary heart disease, 1994.

Journal of the American Medical Association. Lifetime alcohol consumption and breast cancer risk among postmenopausal women in Los Angeles, 1995.

Lane, J., Pieper, C., Phillips-Butte, B, Bryant, J., and Kuhn, C. Caffeine's Effects Are Long-Lasting and Compound Stress. *Psychosomatic Medicine.* National Institutes of Health, July /August, 2002.

McCusker, R., Goldberger, B., and Cone, E. The Content of Energy Drinks, Carbonated Sodas, and Other Beverages. *Journal of Analytical Toxicology,* Vol. 30, March 2006.

McKinley Health Center. University of Illinois at Urbana-Champaign, 2005.

Mick, Elizabeth. (2005). *The Nervous System Presentation.* http://outreach.mcb.harvard.edu/lessonplans_S05.htm; Retrieved April 19, 2014.

Miller, W., Tonigan, J., Longabaugh, R. Drinking Inventory of Consequences. An instrument for assessing adverse consequences of alcohol abuse. Test manual. Rockville, MD: National Institute on Alcohol Abuse and Alcoholism, 1995.

Miller, E. K., Erickson, C. A., & Desimone, R. Neural mechanisms of visual working memory in prefrontal cortex of the macaque. *Journal of Neuroscience,* 1996.

National Cancer Institute, U.S National Institutes of Health, 2008.

National Highway Traffic and Safety Administration (NHTSA). *Annual Report.* 2010.

National Institute on Alcohol Abuse and Alcoholism. National Institutes of Health. Statistic Snapshot of College Drinking, 2010.

National Institute on Alcohol Abuse and Alcoholism (NIAAA). College Drinking— Changing the Culture. "A Snapshot of Annual High-Risk College Drinking Consequences." 2010.

National Institute on Alcohol Abuse and Alcoholism. National Institutes of Health. Integrative Genetic Analysis Of Alcohol Dependence Using the Genenetwork Web Resources, 2008.

National Institute on Drug Abuse. *Alcohol: Drugs of Abuse.* www.drugabuse.gov/; Retrieved April 19, 2014.

National Institute on Drug Abuse. *Commonly Abused Drugs Chart.* Revised March 2011. www.drugabuse.gov/; Retrieved April 19, 2014.

National Institute on Drug Abuse. *Commonly Abused Prescription Drugs Chart.* Revised October 2011. www.drugabuse.gov/; Retrieved April 19, 2014.

National Institute on Drug Abuse. *Drugs, Brains, and Behavior: The Science of Addiction.* Revised January 2010. www.drugabuse.gov/; Retrieved April 19, 2014.

National Institute on Drug Abuse. *Drug Facts: Marijuana.* Revised January 2014. www.drugabuse.gov/; Retrieved April 19, 2014.

National Institute on Drug Abuse. *Drug Facts: Stimulant ADHD Medications – Methylphenidate and Amphetamines*. Revised January 2014. www.drugabuse.gov/; Retrieved April 19, 2014.

National Institute of Drug Abuse (NIDA). U.S. Dept. of Health and Human Services, 2008.

National Institute on Drug Abuse. Research Report Series: Tobacco/Nicotine. Revised July 2012. www.drugabuse.gov/; Retrieved April 25, 2014.

National Institute on Drug Abuse. The Science of Drug Abuse & Addiction. NIDA Info Facts: Hallucinogens, 2010.

National Institute on Drug Abuse. The Science of Drug Abuse & Addiction. NIDA Info Facts: Anabolic Steroids, 2010.

National Institute of Drug Abuse (NIDA). Update on Ecstacy. *NIDA Notes*, Volume 16, Number 5, Dec. 2001.

National Traffic Safety Administration. Traffic Safety Facts. Annual Assessment of Alcohol Related Fatalities, 2012.

Ray, O. and Ksir, C. *Drugs, Society, and Human Behavior* (8th ed). New York: WCB McGraw-Hill. 1999.

SAMHSA, National Clearinghouse for Alcohol and Drug Abuse; www.health.org

Centers for Disease Control and Prevention (CDC). Smoking and Tobacco Use. Fast Facts. Atlanta: Author. 2006.

Taber's Cyclopedic Medical Dictionary – 22nd Ed. (2013)

The International Center for Alcohol Policies. *Annex 1. The Basics about Alcohol*. www.icap.org/; Retrieved April 19, 2014.

U.S. Food and Drug Administration. Prescription Drug Use and Abuse: Complexities of Addiction, 2005.

U.S. Food and Drug Administration. Oxycodone. FDA Statement. Statement on Generic Oxycodone Hydrochloride Extended Release Tablets, 2004.

U.S. Food and Drug Administration, Department of Health and Human Services. FDA Issues Regulation Prohibiting Sale of Dietary Supplements Containing Ephedrine Alkaloids and Reiterates Its Advice That Consumers Stop Using These Products, 2005.

World Health Organization (WHO). *Waterpipe Tobacco Smoking: Health Effects, Research Needs and Recommended Actions by Regulators*. Geneva: Author. 2005.

Name _____ Section _____ Date _____

IN-CLASS ACTIVITY

The Physical Effects of Smoking

This test consists of twenty statements about the effects of smoking. Put a check to show whether you think each statement is true or false. If you don't know whether a statement is true or false, put a check under "Don't know."

		True	False	Don't Know
1.	Smoking low-tar and low-nicotine cigarettes reduces the risk of all smoking-related diseases.	____	____	____
2.	Carbon monoxide is inhaled when a person smokes.	____	____	____
3.	How deeply a smoker inhales is not related to his or her chance of developing lung cancer.	____	____	____
4.	Most experts agree that the harmful effects of smoking on health are not as great for women as for men.	____	____	____
5.	Cigarette smoking increases the risk of developing breathing problems.	____	____	____
6.	Cigarette smoke can increase air pollution in homes and offices.	____	____	____
7.	Cigarette smoking increases the health dangers associated with taking birth control pills.	____	____	____
8.	Frequent pipe and cigar smokers are more likely than nonsmokers to develop lung cancer.	____	____	____
9.	The average life expectancy of a smoker is the same as that of a non-smoker.	____	____	____
10.	People who smoke filter cigarettes inhale less carbon monoxide than do people who smoke nonfilter cigarettes.	____	____	____
11.	Most people gain weight when they quit smoking.	____	____	____
12.	Smokers have an increased risk of developing a lung infection after an operation.	____	____	____
13.	Smoking during pregnancy does not increase a baby's risk of death.	____	____	____
14.	Pipe smokers have a greater risk of developing cancer of the mouth than do cigarette smokers.	____	____	____
15.	Smoking causes the heart to beat more slowly.	____	____	____
16.	The health risks due to smoking do not change even after a person stops smoking.	____	____	____
17.	The more a person smokes, the greater the chance of developing heart disease.	____	____	____
18.	Cigarette smoke in the air can cause eye soreness in nonsmokers.	____	____	____
19.	On average, babies born to mothers who smoke during pregnancy are smaller than babies born to nonsmokers.	____	____	____
20.	Nicotine does not cause dependence similar to other addictive drugs.	____	____	____

Source: U.S. Department of Health and Human Services.

Answers
1 F; 2 T; 3 F; 4 F; 5 T; 6 T; 7 T; 8 T; 9 F; 10 F; 11 T; 12 T; 13 F; 14 T; 15 F; 16 F; 17 T; 18 T; 19 T; 20 F

Name _____ Section _____ Date _____

NOTEBOOK ACTIVITY

"Why Do You Smoke?" Test

		Always	Frequently	Occasionally	Seldom	Never
A.	I smoke cigarettes to keep myself from slowing down.	5	4	3	2	1
B.	Handling a cigarette is part of the enjoyment of smoking it.	5	4	3	2	1
C.	Smoking cigarettes is pleasant and relaxing.	5	4	3	2	1
D.	I light up a cigarette when I feel angry about something.	5	4	3	2	1
E.	When I have run out of cigarettes, I find it almost unbearable until I can get them.	5	4	3	2	1
F.	I smoke cigarettes automatically without even being aware of it.	5	4	3	2	1
G.	I smoke cigarettes for the stimulation, to perk myself up.	5	4	3	2	1
H.	Part of the enjoyment of smoking a cigarette comes from the steps I take to light up.	5	4	3	2	1
I.	I find cigarettes pleasurable.	5	4	3	2	1
J.	When I feel uncomfortable or upset about something, I light up a cigarette.	5	4	3	2	1
K.	I am very much aware of the fact when I am not smoking a cigarette.	5	4	3	2	1
L.	I light up a cigarette without realizing I still have one burning in the ashtray.	5	4	3	2	1
M.	I smoke cigarettes to give me a "lift."	5	4	3	2	1
N.	When I smoke a cigarette, part of the enjoyment is watching the smoke as I exhale it.	5	4	3	2	1
O.	I want a cigarette most when I am comfortable and relaxed.	5	4	3	2	1
P.	When I feel "blue" or want to take my mind off cares and worries, I smoke cigarettes.	5	4	3	2	1
Q.	I get a real gnawing hunger for a cigarette when I haven't smoked for a while.	5	4	3	2	1
R.	I've found a cigarette in my mouth and didn't remember putting it there.	5	4	3	2	1

Source: U.S. Department of Health and Human Services.

Scoring Your Test

Enter the numbers you have circled on the test questions in the spaces provided below, putting the number you circled for question A on line A, for question B on line B, etc. Add the three scores on each line to get a total for each factor. For example, the sum of your scores for lines A, G, and M gives you your score on "Stimulation," lines B, H, and N give the score on "Handling," etc. Scores can vary from 3 to 15. Any score 11 and above is high; any score 7 and below is low.

A _____ + G _____ + M _____ = _____ Stimulation

B _____ + H _____ + N _____ = _____ Handling

C _____ + I _____ + O _____ = _____ Pleasure/Relaxation

D _____ + J _____ + P _____ = _____ Crutch: Tension Reduction

E _____ + K _____ + Q _____ = _____ Craving: Psychological Addiction

F _____ + L _____ + R _____ = _____ Habit

Name _____ Section_____ Date_____

NOTEBOOK ACTIVITY

"Do You Want to Quit?" Test

		Strongly Agree	Mildly Agree	Mildly Disagree	Strongly Disagree
A.	Cigarette smoking might give me a serious illness.				
B.	My cigarette smoking sets a bad example for others.				
C.	I find cigarette smoking to be a messy kind of habit.				
D.	Controlling my cigarette smoking is a challenge to me.				
E.	Smoking causes shortness of breath.				
F.	If I quit smoking cigarettes, it might influence others to stop.				
G.	Cigarettes damage clothing and other personal property.				
H.	Quitting smoking would show that I have willpower.				
I.	My cigarette smoking will have a harmful effect on my health.				
J.	My cigarette smoking influences others close to me to take up or continue smoking.				
K.	If I quit smoking, my sense of taste or smell would improve.				
L.	I do not like the idea of feeling dependent on smoking.				

Scoring Your Test

Write the number you have circled after each statement on the test in the corresponding space to the right. Add the scores on each line to get your totals. For example, the sum of your scores A, E, I gives you your score for the Health factor. Scores can vary from 3 to 12. Any score of 9 or over is high, and a score of 6 or under is low.

A _____ + E _____ + I _____ = _____ Health

B _____ + F _____ + J _____ = _____ Example

C _____ + G _____ + K _____ = _____ Aesthetics

D _____ + H _____ + L _____ = _____ Mastery

Source: U.S. Department of Health and Human Services.

Name _____ Section _____ Date _____

NOTEBOOK ACTIVITY

Alcohol Screening Self-Assessment

This self-assessment tool is designed to assist you in understanding your use of alcohol.

The following ten questions pertain to your use of alcoholic beverages during the past year. Check your answers and record the score (the number next to each choice) for each question. In the questions, a "drink" is equal to 10 oz. of beer, 4 oz. of wine, or 1.25 oz. of 80 proof liquor.

1. How often do you have a drink containing alcohol?
   _____ Never (0)
   _____ Monthly or less (1)
   _____ 2 to 4 times a month (2)
   _____ 2 to 3 times a week (3)
   _____ 4 or more times a week (4)

2. How many drinks containing alcohol do you have on a typical day when you are drinking?
   _____ Never (0)
   _____ 1 or 2 (1)
   _____ 3 or 4 (2)
   _____ 5 or 6 (3)
   _____ 7 to 9 (4)
   _____ 10 or more (5)

3. How often do you have six or more drinks on one occasion?
   _____ Never (0)
   _____ Less than monthly (1)
   _____ Monthly (2)
   _____ Weekly (3)
   _____ Daily or almost daily (4)

4. How often during the last year have you found that you were unable to stop drinking once you had started?
   _____ Never (0)
   _____ Less than monthly (1)
   _____ Monthly (2)
   _____ Weekly (3)
   _____ Daily or almost daily (4)

5. How often during the last year have you failed to do what was normally expected from you because of drinking?
   _____ Never (0)
   _____ Less than monthly (1)
   _____ Monthly (2)
   _____ Weekly (3)
   _____ Daily or almost daily (4)

Source: Courtesy of the World Health Organization.

6. How often during the last year have you needed a drink first thing in the morning to get yourself going after a heavy drinking session?

   _____ Never (0)

   _____ Less than monthly (1)

   _____ Monthly (2)

   _____ Weekly (3)

   _____ Daily or almost daily (4)

7. How often during the last year have you had a feeling of guilt or remorse after drinking?

   _____ Never (0)

   _____ Less than monthly (1)

   _____ Monthly (2)

   _____ Weekly (3)

   _____ Daily or almost daily (4)

8. How often during the last year have you been unable to remember what happened the night before because you had been drinking?

   _____ Never (0)

   _____ Less than monthly (1)

   _____ Monthly (2)

   _____ Weekly (3)

   _____ Daily or almost daily (4)

9. Have you or someone else been injured as the result of your drinking?

   _____ Never (0)

   _____ Less than monthly (1)

   _____ Monthly (2)

   _____ Weekly (3)

   _____ Daily or almost daily (4)

10. Has a relative, friend, or a doctor or other health worker been concerned about your drinking or suggested you cut down?

    _____ Never (0)

    _____ Less than monthly (1)

    _____ Monthly (2)

    _____ Weekly (3)

    _____ Daily or almost daily (4)

Scoring and Interpretation

Determine your total score by adding up the scores for all ten questions. A score of eight or more indicates that a harmful level of alcohol consumption is likely and that you should seek help.

About This Instrument

This self-assessment tool is the Alcohol Use Disorders Identification Test (AUDIT) developed by the World Health Organization and tested in a world-wide trial.

Name _____ Section _____ Date _____

NOTEBOOK ACTIVITY

Making Changes

Breaking the Habit

Here's a six-point program to help you or someone you love quit smoking. (Caution: Don't undertake the quit-smoking program until you have a two- to four-week period of relatively unstressful work and study schedules or social commitments.)

1. *Identify your smoking habits.* Keep a daily diary (a piece of paper wrapped around your cigarette pack with a rubber band will do) and record the time you smoke, the activity associated with smoking (after breakfast, in the car), and your urge for a cigarette (desperate, pleasant, or automatic). For the first week or two, don't bother trying to cut down; just use the diary to learn the conditions under which you smoke.

2. *Get support.* It can be tough to go it alone. Phone your local chapter of the American Cancer Society, or otherwise get the names of some ex-smokers who can give you support.

3. *Begin by tapering off.* For a period of one to four weeks, aim at cutting down to, say, twelve or fifteen cigarettes a day; or change to a lower-nicotine brand, and concentrate on not increasing the number of cigarettes you smoke. As indicated by your diary, begin by cutting out those cigarettes you smoke automatically. In addition, restrict the times you allow yourself to smoke. Throughout this period, stay in touch, once a day or every few days, with your ex-smoker friend(s) to discuss your problems.

4. *Set a quit date.* At some point during the tapering-off period, announce to everyone—friends, family, and ex-smokers—when you're going to quit. Do it with flair. Announce it to coincide with a significant date, such as your birthday or anniversary.

5. *Stop.* A week before Q-day, smoke only five cigarettes a day. Begin late in the day, say after 4:00 p.m. Smoke the first two cigarettes in close succession. Then, in the evening, smoke the last three, also in close succession, about fifteen minutes apart. Focus on the negative aspects of cigarettes, such as the rawness in your throat and lungs. After seven days, quit and give yourself a big reward on that day, such as a movie or a fantastic meal or new clothes.

6. *Follow-up.* Stay in touch with your ex-smoker friend(s) during the following two weeks, particularly if anything stressful or tense occurs that might trigger a return to smoking. Think of the person you're becoming—the very person cigarette ads would have you believe smoking makes you. Now that you're quitting smoking, you're becoming healthier, sexier, more sophisticated, more mature, and better looking—and you've earned it!

Source: Courtesy of the World Health Organization.

Chapter 10
The Reproductive System

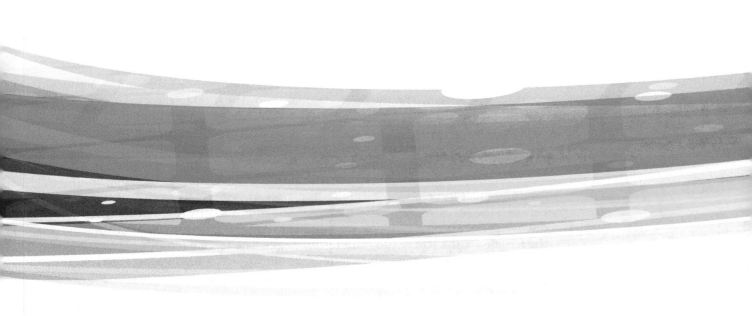

OBJECTIVES

- Explain the structures and function of the male and female reproductive tissues.
- Analyze the role of the endocrine system in each stage of reproduction.
- Explain the steps in a normal menstrual cycle.
- Differentiate methods of pregnancy prevention.

From *Human Anatomy and Physiology, 2nd Edition* by Joseph F. Crivello. Copyright © 2013 by Kendall Hunt Publishing Company. Reprinted by permission.

KEY TERMS

acrosome	interstitial cells	ovulation	spermatogenesis
bulbourethral	luteal phase	penis	spermatogonia
cervix	mammary gland	perineum	spermatozoa
corpus luteum	menopause	prostate	sustenacular cells
ductus deferens	menstrual cycle	puberty	testes
ejaculatory duct	menstruation	scrotum	thecal cells
epididymis	oocyte	semen	uterine tubes
follicular phase	oogenesis	seminal vesicles	uterus
gametes	oogonia	seminiferous tubules	vagina
granulosa cells	ovaries	spermatocytes	zygote
inguinal canal			

Introduction

Most organ systems are crucial for the survival of an individual but not the reproductive system. The overall continuation of the human race, however, depends on the reproductive system functioning properly in a sufficient number of people. The reproductive system is not essential for the survival of an individual and can be nonfunctional in an otherwise healthy individual; in fact, it normally becomes less functional as people age. Hormones produced by the reproductive system are responsible for sex-specific developmental patterns and for differences in behavior. Although most organ systems show little difference between the sexes, reproductive tissues vary significantly by sex. The male reproductive system produces sperm cells and the tissues required to transfer them to the female vagina where they can fertilize eggs. The female reproductive system produces oocytes, or eggs, and tissues that allow for fertilization, fetal development, parturition, and nutrition for the baby after birth.

Even though the male and female reproductive systems show striking differences in structure and function, they also share a number of similarities. Reproductive tissues in both sexes are derived from the same embryonic structures. Reproductive hormones are the same in both sexes, even though they can have different actions, and both have stem cells that are capable of dividing and differentiating into sperm or oocytes.

This chapter begins with a discussion of the male reproductive anatomy and its physiology. This is followed by a discussion of the female reproductive anatomy and its physiology. Chapter 11 discusses the steps involved in fertilization, fetal development, parturition, and the genetic basis of inheritance.

Anatomy of the Male Reproductive System

The structure of the male reproductive system clearly shows its intended function: to produce and transport sperm to where they can successfully fertilize an egg (Figure 10.1). The male reproductive system is composed of a pair of **testes**, found within the perineum, which produce sperm; ducts, found within the abdominal and pelvic region, which carry sperm out of the body;

accessory glands, within the pelvic region, which produce essential fluids; and supporting structures. The ducts start with the **epididymis** (L. *epi*, on; *didymos*, twins), are followed by the **ductus** (L. *ductus*, to lead) **deferens**, and then the urethra. Three pelvic accessory glands: the **seminal** vesicles, **prostate** (G. *prostates*, one standing before) gland, and **bulbourethral** gland, produce fluids and substances that support sperm and aid in the fertilization process. The external male genitalia are composed of the **scrotum**, urethra, and **penis** (Figure 10.2).

Scrotum

The testes and epididymis, areas in which sperm cells develop, are found outside of the pelvic region within the scrotum. The scrotum is about 3°C cooler than the internal pelvic region. Sperm maturation is very temperature sensitive, and sperm would not form properly within the body at 37°C. Another reason why sperm develop at cooler temperatures is that once they enter the woman's body at 37°C, they are maximally active for 4-6 hours, after which they become inactive. Storing them at a cooler temperature keeps them inactive until they are released.

Sperm production within the testes and epididymis requires that sperm reenter the body through the ductus deferens into the pelvis. The ductus deferens joins with the seminal vesicle ducts to form the ampullae. Extensions of the ampullae, the **ejaculatory ducts**, pass through the prostate and empty into the urethra within the prostate. The urethra, in turn, exits from the pelvis and passes through the penis to the outside of the body.

In addition to the testes, epididymis, and start of the ductus deferens, the scrotum also contains blood vessels and nerves. The septum, an incomplete connective tissue membrane, divides the scrotum into two internal compartments. The *raphe* (L. *raphe*, seam), an irregular ridge, externally marks the external scrotum midline. The raphe extends in a posterior direction to

thought question
Why is the production of sperm so temperature sensitive? Doesn't having the testes outside of the body make them more vulnerable to physical damage?

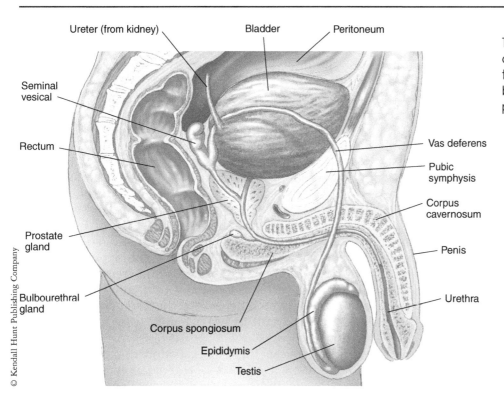

FIGURE 10.1

The male reproductive organs include the penis, testes, seminal vesicles, bulbourethral glands, and prostate gland

FIGURE 10.2

The penis, testes, epidiymus, and accessory glands essential for gamete function. The testes reside within the scrotum and are the site of spermatozoa production

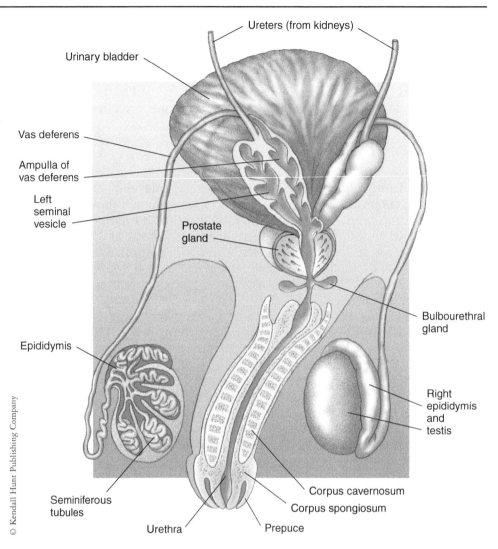

the anus and in an anterior direction to the penis inferior surface. The skin covers the outer layer of the scrotum. Beneath the skin are loose connective tissue and the *dartos* (L. *dartos*, to the skin) muscle, a smooth muscle layer.

Exposure to cold temperatures causes the dartos to contract, making the scrotum firm, wrinkled, and reduced in size. Contraction of the dartos muscle decreases the scrotum volume and the rate of heat loss. Heat loss is further reduced by contraction of the cremaster muscles. The cremaster muscles are extensions of abdominal muscles that extend to the scrotum. Cremaster muscle contraction pulls the testes toward the body. The body is 3°C warmer than the testes, and moving the testes counters the heat loss to the environment. Exposure to warm temperatures or after exercise when the body is warmer, the dartos and cremaster muscles relax, dropping the testes away from the body. After relaxation of the dartos and cremaster muscles, the skin covering the scrotum becomes loose and thin, allowing the testes to descend away from the body, which helps keep the testes cool. The dartos and cremaster muscles are important in testes temperature regulation and normal sperm maturation.

A diamond-shaped area between the thighs, bounded anteriorly by the symphysis pubis, posteriorly by the coccyx, and laterally by the ischial tuberosities, is known as the **perineum** (L. *peri*, between). The perineum is divided into two triangles: The anterior urogenital triangle is associated with penis base and scrotum, and the posterior anal triangle contains the anus. The perineum is rich in sensory nerve endings, and tactile stimulation of the perineum during sexual activities is important for the maintenance of an erection and ejaculation (see later discussion).

Testes

The testes have an exocrine and endocrine function. The exocrine function is associated with the production and release of sperm, whereas the endocrine function is associated with steroid production. Superficially, the testes are small (4 to 5 cm in length), ellipsoid-shaped organs found entirely within the scrotum (Figure 10.3).

A connective tissue capsule known as the tunica albuginea covers the outer testis. The thick, white, mostly fibrous tunica albuginea forms incomplete septa (L. *septa*, fence) after it enters the testes. The testis is divided into about 300 to 400 cone-shaped lobules by the septa. Within the lobules are **seminiferous** (L. *semen*, seed; *fero,* to carry) **tubules** and **Leydig cells** or *interstitial cells*, and **Sertoli cells** or *sustenacular cells*. The seminiferous tubules are the site of sperm cell development. The Leydig cells, endocrine cells that secrete testosterone, are found within a loose connective tissue that surrounds the tubules (Figure 10.4 on page 978).

If laid end to end, the overall length of all of the seminiferous tubules in both testes is about a kilometer. Each seminiferous tubule empties into a set of short, straight tubules and then into the *rete testis*, a tubular network. The rete testis empties into efferent ductules lined with ciliated pseudostratified columnar epithelium. The beating of these cilia propels sperm out of the testis and into the epididymis. The efferent ductule goes through the tunica albuginea to exit the testis and join with the epididymis.

Clinical–In Context: Inguinal Hernias

A *hernia* (L., rupture) is described as the protrusion (bulging) of a tissue part (like the intestine) through the tissue that contains it (like the abdomen). An *inguinal* (L., groin) hernia occurs in the groin, and parts of the intestine bulge out through the inguinal canal. The inguinal canal is a small opening between layers of abdominal muscle through which the testes descend in male individuals. Any action that increases intra-abdominal pressure, like obesity, pregnancy, heavy lifting, or straining to defecate, pushes the intestine against the inguinal canal and a hernia can occur if the inguinal canal is weakened or not sealed properly.

An inguinal hernia is quite painful, and a lump is usually found in the groin near the thigh. The intestine may be damaged if abdominal muscles block the flow of food or blood. The common treatment for an inguinal hernia is returning the intestine to the abdominal cavity and surgical repair of the weakened inguinal canal.

Duct System

After birth, the testes do not change significantly until a boy reaches puberty. The prepubescent (before puberty) testes do not produce sperm, interstitial cells are sparse and widely separated, and the seminiferous tubules are not functional. Puberty (see later discussion) initiates several changes (mostly hormonal) that stimulate the growth and hypertrophy of interstitial cells, development of the seminiferous lumen, and production of sperm cells.

FIGURE 10.3

The testes are composed of lobules that contain seminiferous tubules

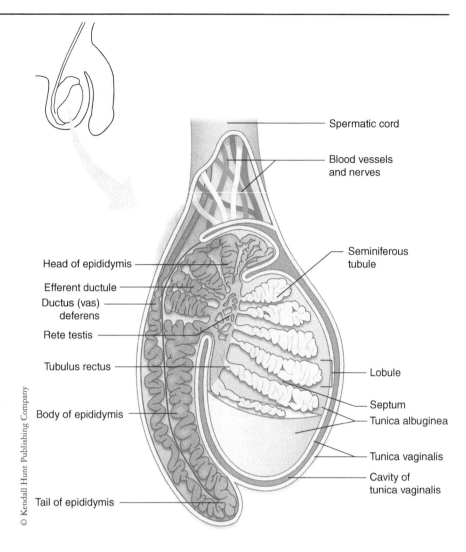

thought question
Besides preventing the immune system from attacking sperm, what other effects could the testes-blood barrier have?

thought question
Why does the immune system recognize sperm cells as nonself?

thought question
Why isn't the ampulla surrounded with skeletal muscle?

The seminiferous tubules contain stem cells (see chapter 3) that create sperm cells. The process of cell division and differentiation is known as **spermatogenesis** (Figure 10.5 on page 380). **Spermatogonia,** or *stem cells*, are found within the seminiferous tubule lumen, together with Sertoli, cells (Enrico Sertoli, 1842–1910, Italian histologist). The Sertoli cell does not differentiate to form sperm, but maintains the seminiferous tubule luminal environment to allow spermatozoa to develop. Sertoli cells are found adjacent to spermatogonia at the outermost tubule margin. Tight junctions between Sertoli cells prevent seminiferous tubule lumen contents from entering the body (and vice versa) and form a testes-blood barrier that isolates the developing sperm from the rest of the body. This barrier prevents the immune system from attacking and destroying sperm cells.

Epididymis

The epididymis is a narrow, tightly coiled duct that joins with efferent duct near each testis and then to the ductus deferens (Figure 10.3). The epididymis can be divided anatomically into three different regions: caput (head), corpus (body), and cauda (tail). The cauda arises from the efferent ductules emptying into a single convoluted tubule, the epididymis duct. This duct is several meters long. The cauda is lined with pseudostratified columnar epithelium with stereocilia (elongated microvilli). The stereocilia increase epithelial surface area and increase the absorption of luminal fluid.

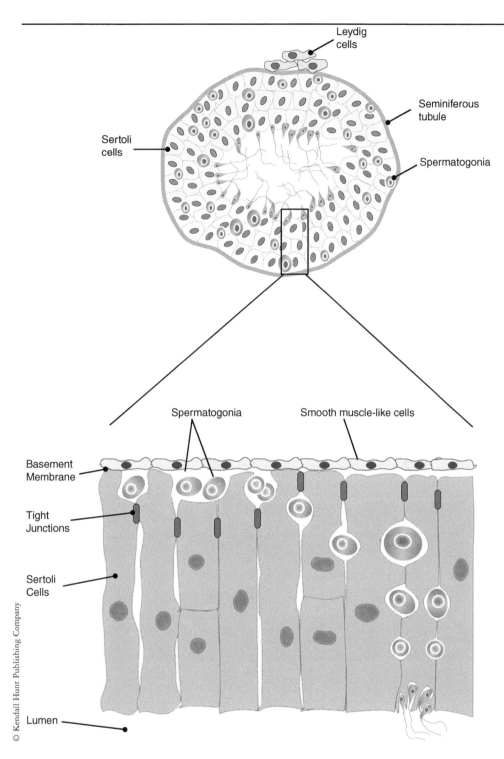

FIGURE 10.4
Sertoli cells are connected by tight junctions that help separate sperm development from immune surveillance. Leydig cells are found between seminiferous tubules

Sperm enter the caput, move into the corpus, and then into the cauda of the epididymis, where they are stored. Sperm that enter the epididymis are not completely functional; they lack the ability to swim and are incapable of egg fertilization. The length of the epididymis allows for continued sperm maturation, because of the time required for sperm to move through the long (several meters) epididymis. Sperm maturation is completed after ejaculation in the female reproductive tract. During male ejaculation, sperm are pushed by peristaltic smooth muscle contractions from the cauda into the ductus

FIGURE 10.5

Mitosis and meiosis during spermatozoa development

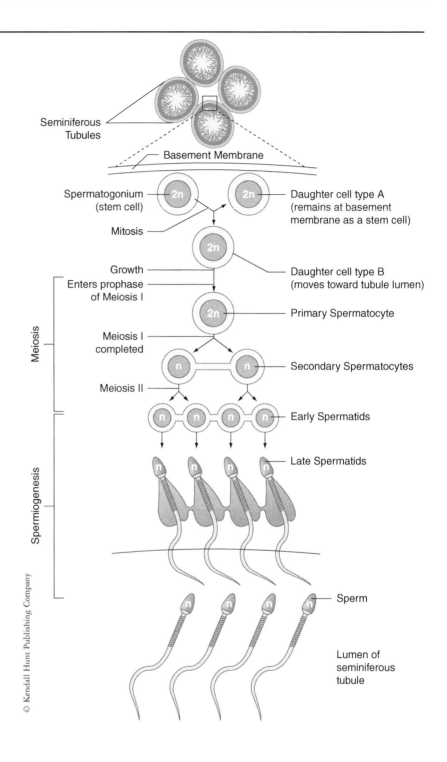

deferens. The sperm are packed so tightly in the cauda region of the epididymis that they cannot swim.

Ductus Deferens

There are two ductus deferens, or *vas deferens* (L., carrying-away vessel), which join the epididymis on one end and the ejaculatory duct on the other end (Figure 10.3). The ductus deferens begins at the epididymis and ascends along the back of the testes with blood vessels and nerves. The ductus deferens, blood vessels (testicular artery and venous sinus), lymphatic vessels, and nerves make up the spermatic cord. The outside of the spermatic cord is covered by connective tissue (external and internal spermatic fascia), and the cremaster muscle (which extends from the abdomen). Ductus deferens smooth muscle contracts and provides the peristaltic motion that propels sperm toward the ejaculatory duct during ejaculation. The lumen of the ductus deferens is lined with pseudostratified columnar epithelium. The ductus deferens is cut and tied off during a vasectomy.

The spermatic cord ascends and passes into the pelvic cavity through the inguinal canal bringing with it the ductus deferens. Within the pelvic cavity, the ductus deferens crosses the lateral and posterior pelvic cavity walls. The ductus deferens then passes over the ureter and moves over the posterior surface of the urinary bladder as it nears the prostate gland. As the ductus deferens nears the prostate, its end swells to form an ampulla.

Ejaculatory Duct

The ejaculatory duct (Figure 10.2) is a small duct (about 2 to 3 cm in length) that connects the ductus deferens ampulla and the seminal vesicle. The seminal vesicle is a small, sac-shaped gland that produces fluid that is part of **semen**. Semen is a fluid that contains materials to nurture and support sperm, proteins, and enzymes. Sperm move into the ejaculatory duct during ejaculation where they are mixed with seminal vesicle fluid. The sperm then move through the prostate and empty into the urethra.

Urethra

The urethra is a tube that connects the urinary bladder to the outside of the body at the tip of the penis (Figure 10.3). The urethra has an excretory function in men and women, to allow urine to leave the body. In men, the urethra also acts as a tube that allows sperm to leave the body during ejaculation.

The male urethra is longer than the female urethra (20 vs. 4 cm). This has the effect of making women more susceptible to bladder infections than men (bacterial infections have a smaller distance to travel). The male urethra also contains a number of bends, making it difficult to catheterize.

The male urethra is divided into three parts: prostatic, membranous, and spongy. The prostatic urethra extends from the urinary bladder through the prostate. Near the urinary bladder, the prostatic urethra is lined with transitional epithelium. As it moves through the prostate, the lumen becomes lined with stratified columnar epithelia. Inside the prostate, up to 30 small glands empty into the urethra. These small glands produce fluids that make up part of semen. The two ejaculatory ducts also empty into the prostatic urethra. The narrowest and shortest membranous urethra extends from the prostate through the external urethral sphincter.

The spongy urethra extends from the external urethral sphincter to the penis tip. The spongy urethra passes through the corpus spongiosum (see discussion in the next section). The lumen of the spongy urethra is lined with stratified squamous epithelium and many small mucous secreting glands. The

Clinical–In Context: Bicycle Seat Neuropathy

Bicycle seat neuropathy is one of the more common injuries reported by cyclists. The injuries and symptoms are due to the cyclist supporting his or her body weight on a narrow seat and is thought to be due to either vascular or neurologic injury to the pudendal nerve. Symptoms include numbness or pain in the pudendal area, penile numbness, and impotence lasting longer than one month.

The cause of bicycle seat neuropathy has been attributed to severe compression of the pudendal nerve as it passes through Alcock's canal. Alcock's canal is enclosed laterally by the ischial bone and medially by the obturator internus muscle. The pudendal nerve exits the canal ventrally, below the symphysis pubis, and innervates the genital and perineal regions. This has led to the redesign of bicycle seats to reduce the pressure on Alcock's canal.

glands secrete a mucus that protects the epithelium from the caustic effects of acidic urine (urine is usually acidic; see chapter 25).

Penis

The penis (L., tail) is the external male copulatory organ and the external organ of urination (Figure 10.2). The penis is capable of erection (stiffening as a result of fluid pressure), as a result of engorgement with blood, allowing it to be used for copulation. The penis relies on engorgement with blood to become fully erect; it also cannot be withdrawn back into the pelvic area (as it can in other mammals). The penis is made up of three columns of erectile tissue: two *corpora cavernosa* and one *corpus spongiosum*. The corpora cavernosa form the dorsal (upper) sides of the penis, and the corpus spongiosum forms the ventral (lower) side of the penis. The corpus spongiosum swells at its distal end to form the *glans penis* and at its proximal end to form the bulb of the penis. The glans penis supports the *prepuce*, or *foreskin*, a loose fold of skin that can be pulled back to show the glans penis. The frenulum is the dorsal area of the penis where the prepuce is attached. *Circumcision* (L. *circum*, around; *caedo*, to cut) is the surgical removal of the prepuce after birth. Circumcisions have been justified on religious and medical grounds, but there is no apparent medical reason for circumcisions.

The corpora cavernosa expands at its proximal end to form the crus of the penis. The bulb and crus of the penis come together to form the root of the penis and the cruca. The cruca attaches the penis to the coxae. The urethra passes through the corpus spongiosum to its external opening, the meatus or *external urethral orifice*, at the tip of the glans penis. The skin that covers the penis, except for the glans penis, is loosely attached to the underlying connective tissue. This allows the penis to increase in size during an erection. Skin is firmly attached to the glans penis base and thins as it covers the glans penis. The skin covering the penis, and especially the glans penis, is highly innervated with sensory receptors.

The penis is supplied arterial blood through the penile artery. Within the penis, the penile artery branches into the bulbourethral, dorsal, and cavernosa arteries. The penile artery, penile veins, and nerves pass along the dorsal surface of the penis.

Accessory Glands

Several accessory glands contribute fluid to the ejaculate. They include the seminal vesicles, bulbourethral gland, and prostate.

Seminal Vesicles

The seminal vesicles are a pair of sac-shaped glands found on the posterior surface of the urinary bladder (Figure 10.2). The glands are adjacent to the ductus deferens ampulla, and a short seminal vesicle duct joins with the ductus deferens to form the ejaculatory duct. The seminal vesicles produce about 60% of the fluid (seminal fluid) that becomes semen. Seminal vesicle secretions contain proteins (fibrinogen to cause semen to clot within the vagina), enzymes, fructose (to provide energy for sperm), phosphorylcholine, and prostaglandins (to cause uterine contractions and help propel the sperm toward the uterine tubes). Seminal fluid enters the ejaculatory duct as a consequence of smooth muscle contractions.

Prostate Gland

The prostate gland is a small walnut-sized gland (4 cm long and 2 cm wide) normally weighing 25 g (Figure 10.2). The main function of the prostate gland is to secrete a clear, slightly basic fluid (about 30% of the total) that becomes part of semen. The prostate surrounds the prostatic urethra. The prostate lies superior to the rectum and can be felt during a digital rectal examination. The prostate outer connective tissue capsule is fibrous and extends into the prostate forming numerous lobules (partitions). The fibrous partitions are filled with smooth muscle covered with columnar epithelia. The columnar epithelia are secretory cells that produce the prostatic fluid.

There are 15 to 30 prostate excretory ducts that enter the urethra as it passes through the prostate. Each of these excretory ducts receives secretions from 4 to 6 prostatic lobules. Under the regulation of testosterone, columnar epithelial cells produce secretory proteins that are stored as viscous secretions. During ejaculation, sympathetic nerves stimulate prostate smooth muscle contraction and excretion of prostate fluid into the ducts and into the urethra to help form the ejaculate. The milky prostatic secretion has a high pH (8), as do the secretions of the bulbourethral glands and urethral mucous glands, to help neutralize the acidic urethra. These secretions also help to neutralize the acidic vaginal secretions after ejaculation. Prostatic secretions are also important in stimulating semen coagulation because they contain proteases that convert fibrinogen to insoluble fibrin. The coagulation reaction keeps the semen as a single mass for a few minutes after ejaculation; then the clot is dissolved, releasing sperm cells to make their way up the female reproductive tract.

Bulbourethral Glands

The *bulbourethral glands*, or *Cowper's glands*, are two small glands (about the size of peas) lateral and posterior to the membranous urethra (Figure 10.2). The bulbourethral glands decrease in size with age. The bulbourethral glands are compound mucous glands (see chapter 3 for a discussion of mucous glands). Each gland has an excretory duct that opens into the spongy urethra at the base of the penis. The bulbourethral glands secrete a clear fluid before ejaculation but after sexual arousal. The clear fluid removes traces of urine from the urethra before ejaculation, neutralizes the spongy urethra's acidic pH, and helps reduce vaginal acidity.

In Review...

10.1	What is the role of the scrotum?	
10.2	How do the cremaster and dartos muscles affect sperm development?	
10.3	How does the testis descend into the scrotum?	
10.4	What tissues are found within the testes?	
10.5	What are sustenacular cells? What are their functions?	
10.6	What is the role of the epididymis? The ductus deferens? The ejaculatory duct?	
10.7	What is the function of the urethra in male reproduction? In the urinary system?	
10.8	What internal structures make up the penis?	
10.9	What is secreted by the seminal vesicles? By the bulbourethral gland? By the prostate gland?	
10.10	What are the roles of the accessory gland secretions in the fertilization process?	

Male Reproductive System Physiologic Functions

The male and female reproductive systems share some physiologic functions, but the female reproductive system has additional physiologic functions. The overall physiologic function of the male reproductive system is to:

- **Gametes:** Produce gametes (cells with 23 chromosomes) from stem cells;
- **Transport:** Transport gametes to a suitable place where fertilization can occur.

The male reproductive system provides for the production of gametes and the tissues and organs required for the movement of gametes to a place where fertilization can occur inside a woman's body. The female reproductive system carries out those physiologic functions but in addition, carries out fertilization, embryogenesis, fetal development, parturition and post-natal (after birth) support. The ability of both the male and female reproductive systems to carry out these physiologic functions depends on specific tissues, cells, and hormones.

Embryonic Development of the Male Reproductive System

The *SRY* locus on the Y chromosome in male babies produces a transcription factor that stimulates the developing reproductive tissue to produce males (Figure 10.6a). Embryos have the potential to be either male or female, and in the absence of the *SRY* transcription factor, all babies are female, even if the embryo is XY or XXY (the *SRY* gene can be non-functional or partially missing). In the presence of the *SRY* transcription factor all babies are phenotypically male, even if the embryo is XX or XXY (the *SRY* gene can be translocated to the X chromosome during gamete formation). The *SRY* transcription factor induces the formation of *Sox9*, another transcription factor, which increases the expression of fibroblast growth factor 9 (FGF9). Elevated levels of FGF9 increase *Sox9* levels in an autocrine positive-feedback manner, driving cell growth and differentiation into a male specific pathway. This leads to the growth and development of the *Wolffian ducts*, the precursors of the male reproductive organs, and the degeneration of the *Mullerian system*, precursors of the female internal sex organs (Figure 10.6b). During fetal development, the placenta secretes human chorionic gonadotropin (hCG), which stimulates fetal testes to produce and secrete testosterone. After birth, the testes are not stimulated to produce testosterone until puberty. As a result, the testis of a newborn male baby atrophies slightly and secretes very low amounts of testosterone.

FIGURE 10.6

(a) SRY gene locus and (b) precursors of male and female reproductive organs

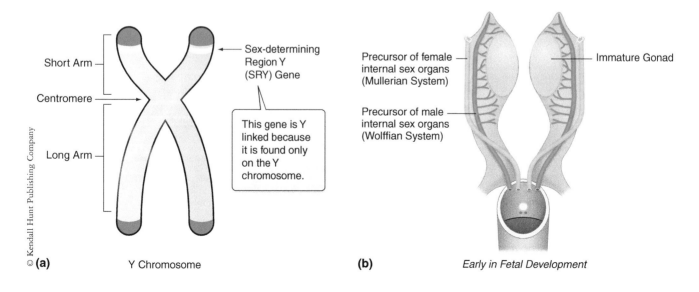

(a) Y Chromosome

(b) *Early in Fetal Development*

Testicular Descent

The testes develop within the body as retroperitoneal organs in the abdominopelvic cavity. During development, the testes move from the abdominopelvic cavity through the inguinal canals to the scrotum. The inguinal canals are bilateral oblique (angled downward) passageways in the anterior abdominal wall. In females, the inguinal canals also develop, but they are smaller than those in male individuals and the ovaries do not descend through them. *Cryptorchidism* (L. *crypto*, hidden; *chidism*, testicle) is a disorder characterized by the failure to find one or both testes in the scrotum after 1 year of age. The higher abdominal temperature prevents the normal formation of sperm (discussed later) and can lead to infertility or tumors. Normally, the testis descends through the inguinal canal and into the scrotum during the 4th month of fetal development. If untreated, cryptorchidism can lead to infertility.

Puberty

Males

Puberty (L. *pubertas*, grown up) is characterized by significant changes in the male and female bodies and the development of functional sperm and eggs (Figure 10.7). At puberty in males there are changes in hypothalamic secretion of gonadotropin-releasing hormone (GnRH). Before puberty low blood levels of sex-steroids inhibit hypothalamic GnRH release in a classic negative feedback manner. At puberty, the hypothalamic neurons become less sensitive to the blood levels of sex-steroids reducing inhibition of GnRH release. Increased GnRH results in increased LH (leutenizing hormone) and FSH (follicle stimulating hormone) release, and increasing blood testosterone levels and sperm cell formation. Testosterone still has a negative feedback effect on GnRH, but this effect requires much greater levels than before puberty.

The development of puberty is related in a complex way to male body size and body fat content. In a period of about 2 years before the onset of puberty,

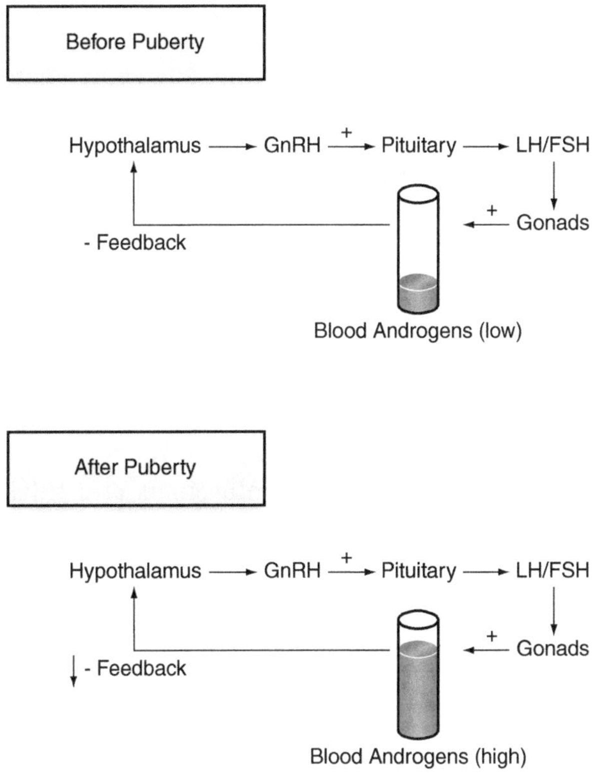

thought question
How could these neurons become less sensitive to testosterone?

FIGURE 10.7

Puberty is triggered by a change in the sensitivity of hypothalamic neurons to blood levels of androgens. The decrease in a negative feedback loop leads to an increased secretion of gonadotropin-releasing hormone (GnRH) and ultimately blood androgens. FSH, follicle-stimulating hormone; LSH, leutenizing hormone

there are two important processes that contribute to the physical changes during this period: adrenarche and gonadarche. *Adrenarche*, increased adrenal androgen secretion, normally occurs at 12 years of age as a result of increased pituitary adrenocorticotropin hormone (ACTH) secretion and reduced sensitivity of hypothalamic neurons to androgens (as described above). The exact age depends on heredity and nutrition. In most industrialized societies, this process normally starts at 12 years of age. However, in developing countries where more than one third of the total children are malnourished or undernourished, the onset of this process is usually delayed. Adrenarche causes the formation of axillary (underarm) and pubic hair, and an increase in body height and bone growth.

On average, boys enter puberty 2 years later than girls. Puberty is initiated by *gonadarche*, changes in the hypothalamus-pituitary axis that regulate the secretion of LH and FSH. At the onset of puberty, GnRH-mediated regulation of LH and FSH secretion changes to one in which hormones are released in a pulsatile manner. These hormonal changes trigger a set of classic changes in boys: sudden increases in height and weight, development of secondary sex characteristics, and increased sexual interest.

Several physiologic changes take place during this period; the most prominent is nocturnal emissions or wet dreams (*spermarche*). These involuntary ejaculations are the result of sexually stimulating dreams. Puberty usually occurs in boys between 13 and 15 years of age. Unlike girls, there are no visible signs that tell a boy he has entered puberty. Hence, it becomes difficult to know exactly when puberty occurs. The changes occur gradually over a period of time rather than as a single event. The attainment of puberty varies widely in boys in rural and urban areas, and depends on factors like heredity and nutrition. There are five stages of puberty in boys.

Stage 1 commonly occurs by age 12 and is characterized by the beginning of adrenarche. There are no outward signs of sexual development, but increased androgen production prepares boys for undergoing pubertal changes that become apparent in the second stage. Many boys in the transition phases between stage 1 and stage 2 may experience an enlargement in the size of the testicles and scrotum. At this point, the penis does not enlarge.

Stage 2 usually occurs at 13 to 14 years and is characterized by the first appearance of long, soft pubic hair (fine and straight) that covers only a small area around the genitals. The scrotum and testes start enlarging, accompanied by reddening and folding of the skin. Penis enlargement begins approximately 1 year after the testicles begin enlarging.

Stage 3 usually occurs at 14 to 15 years and is characterized by penis enlargement, mostly in length, although there may be small changes in penis diameter. Pubic hair coarsens and becomes darker as it continues to spread. Body size increases. Some boys may get some temporary breast swelling as a result of hormonal changes. Most boys have the capacity to ejaculate at this stage. They also might experience nocturnal emissions.

Stage 4 often occurs at age 16 to 17 years and is characterized by larynx changes that cause the voice to deepen. Further enlargement and development of the scrotum and testes take place and pubic hair thickens. Other changes involving the pubic hair include curling and coarsening in texture. These changes continue throughout the fourth stage, accompanied by enlargement of the penile glans and thickening of hair in the pubic area. There may be an increase in hair growth on the face, under the arms, and on the legs.

Stage 5 typically begins at age 17 to 18 years and completes sexual maturation. The penis, scrotum, and testes are fully matured and are adult-sized. Hair fills the pubic area and extends onto the surface of the thighs and up the abdomen. Pubic hair growth is complete and full height is attained. There is an increase in oily skin, pimples, and sweating.

Secondary Sex Characteristics

The male secondary sex characteristics are hormone-induced bodily changes that exclude changes to the testes. Secondary sex characteristics begin to develop during puberty and last through adult life. In the male body, these include changes in the deposition and characteristics of body hair, in the growth and development of skeletal muscle mass, in fat deposition, changes to the larynx that cause the voice to deepen, changes to the integumentary system, and changes in behavior. Sweat glands become more active and the sweat develops an odor. Oil glands become more active and acne may appear. In boys, the shoulders broaden, the length of bones in arms and legs increase, and there is an increase in the amount of body and facial hair. Testosterone also controls prostate growth and formation of the prostatic secretions.

Menopause

Men do not experience the equivalent of the female **menopause** or *climacteric* (a natural loss of fertility in response to hormonal changes or tissue changes). Nonetheless, there are many common age-related changes that occur in the male reproductive system. The testes can decrease in size (atrophy) with age, the number interstitial cells can decrease, and the wall of the seminiferous tubule can thin. These changes may be caused by a decrease in testes blood flow or a by decrease in testosterone production. The decrease in testosterone production may be a result of a decrease in testes blood flow. The testicular change results in a decrease in sperm cell production (decreasing the amount of sperm in the ejaculate) and an increase in the number of abnormal sperm cells. However, spermatogenesis does not stop and men remain fertile well into their eighties.

The prostate gland also has several age-related changes: a decreased prostate blood flow, increased epithelial cell lining, and decreased functional smooth muscle cells. These prostatic changes do not decrease fertility. There is an increase in the incidence of benign prostatic hypertrophy (BPH), a noncancerous prostate growth, which interferes with the flow of urine through the prostatic urethra. A significant number of men older than 60 (approximately 15%) are treated for BPH; by the age of 80, more than 50% of men need treatment. Before the age of 50, prostate cancer is extremely rare, but by the age of 55, it is the third leading cause of cancer deaths. Enlargement of the prostate by cancer commonly compresses the urethra and makes urination difficult.

Sexual impotence (inability to form and maintain an erection) also increases with age. At the age of 60, approximately 15% of all men suffer from impotence, and by the age of 80, the rate is 50%. There is an increase in fibrous connective tissue within the penis erectile tissue starting at the age of 60, decreasing the speed of an erection because of the loss of erectile tissue.

There is greater variability in the age-related affects on male reproduction than there is in female reproduction, but most men experience a decrease in sex drive, sexual activity, and sexual performance.

Spermatogenesis

Meiosis and Mitosis

Spermatogenesis requires meiosis and mitosis (discussed in detail in chapter 2). All human somatic cells (all cells except the **gametes**, or sex cells) contain 46 chromosomes and are diploid cells (they have 2 copies of each chromosome). Gametes (egg and sperm) are haploid and have one copy of each chromosome. During mitosis, daughter cells contain 46 chromosomes. During meiosis, daughter cells contain 23 chromosomes. Spermatogenesis requires mitosis and meiosis of spermatogonia to produce spermatocytes (Figure 10.5).

> **thought question**
> Could the increase in abnormal spermatocyte production be due to changes in testes temperature?

Clinical–In Context Prostate-Specific Antigen Test

Prostate-specific antigen (PSA) is an enzyme released by prostate epithelial cells in response to cell proliferation or damage. Each year, more than 30 million American men have their blood PSA levels checked as a cancer screening tool at a cost exceeding $3 billion a year. Recent medical opinion has begun to question the efficacy and usefulness of the PSA test to screen for prostate cancer. Men have a 15% to 20% chance of developing prostate cancer during their lives, but only a 3% chance of dying of it. Critics of the PSA points to the fact that PSA does not detect the presence of cancer specifically, ibuprofen and benign prostatic hypertrophy also increase PSA, and cannot differentiate between slow-growing cancers that don't kill men and rapid-growing cancers that do. A recent study reported in the New England Journal of Medicine showed that PSA screening did not reduce the prostate cancer death rate of men older than 55. The PSA test does have its place, though: a rapid increase in PSA levels in men who have been diagnosed for prostate cancer likely indicates its return, and men who have a genetic predisposition for prostate cancer can benefit by testing. Physicians question whether the benefits of wide spread testing justify its cost.

FIGURE 10.8

Hormonal regulation of spermatogenesis and oogenesis is through the same hormone pathways

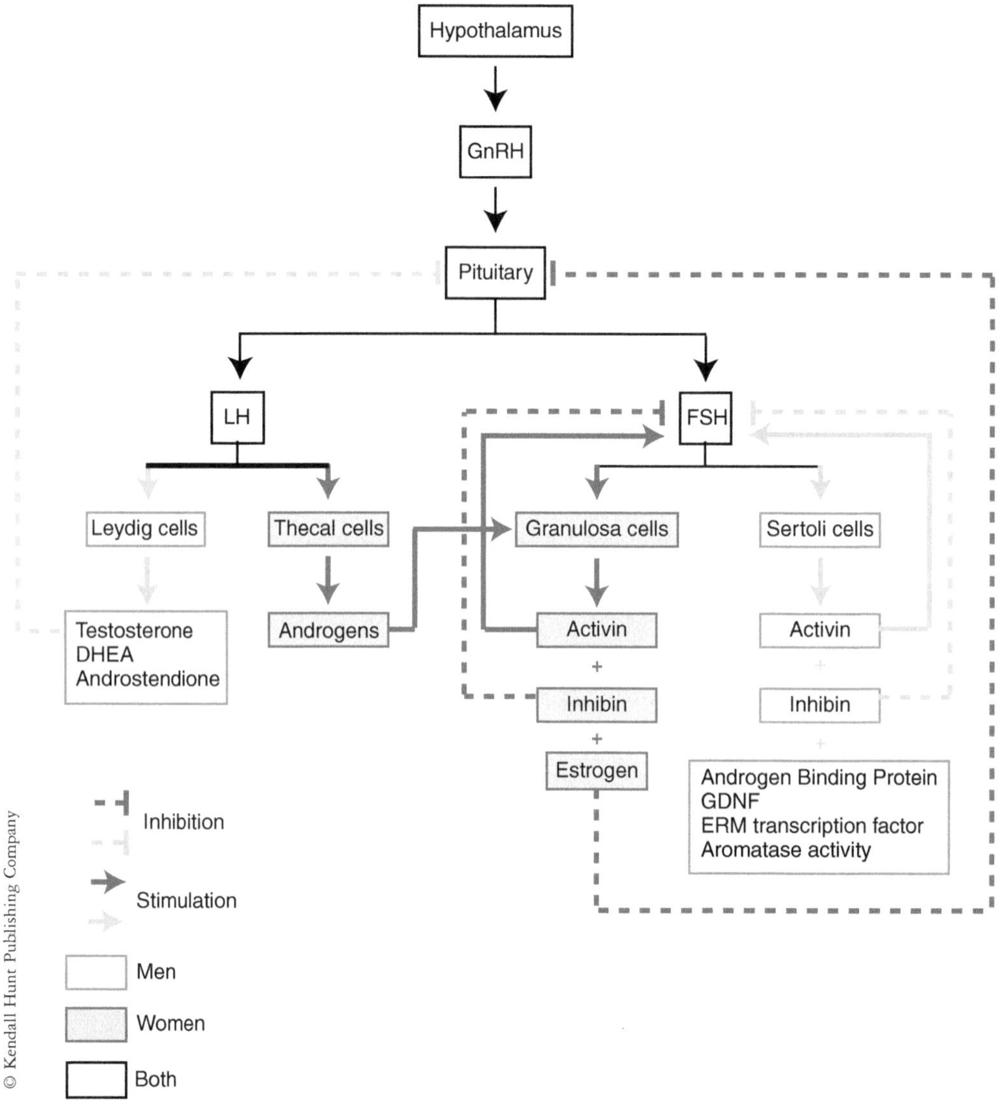

Spermatogenesis

The hypothalamus secretes GnRH, which in turn stimulates the secretion of LH and FSH that travels to the testes (Figure 10.8). Within the testes, LH binds to its receptor on interstitial cells and stimulates testosterone production. FSH binds to Sertoli cell membrane receptors and promotes sperm cell development by the secretion of proteins and nutrients into the seminiferous tubule. Sertoli cells also secrete inhibin and activin, polypeptide hormones. Inhibin inhibits FSH release from the anterior pituitary by a negative feedback mechanism, while activin stimulates the synthesis and release of FSH from the pituitary. FSH and LH work through a G-protein coupled receptor (GPCR) and stimulate the formation of cAMP within the cell (see chapter 17). In male individuals, LH and FSH release is fairly constant, unlike in female individuals, who also regulate their reproductive tracts with LH and FSH (see later).

Testosterone is not solely a male sex steroid. Female ovaries also produce testosterone, though at much lower levels, and testes produce testosterone and estrogen together with other steroids. The typical male individual has a 50:1 or 100:1 ratio of blood testosterone to estrogen; in female individuals, this ratio is reversed. Testosterone passes into the seminiferous tubule and binds to sustenacular cell receptors, which require testosterone for normal functioning. Sertoli cells convert testosterone into dihydrotestosterone (DHT) and estrogen. The normal ratio of testosterone and DHT to estrogen in adult men is 50:1 to 100:1. The sustenacular cells also produce and secrete androgen-binding protein (ABP) into seminiferous tubules. ABP binds testosterone and DHT, and is carried along with other secretions of the seminiferous tubules to the epididymis. ABP keeps testosterone and DHT within the seminiferous tubule and epididymis; the concentration of these hormones is much greater in the seminiferous tubules and epididymis than in the rest of the body.

thought question
Could the increase in abnormal sperm cell production be due to changes in testes temperature?

As described in chapter 17, steroid hormones exist in a bound and free form. The bound form is the percentage of steroid hormone bound to a carrier protein (in this case ABP), and the free form is the percentage found in solution (usually less than 1% of the total). The high concentrations of dihydrotestosterone bound to ABP within the seminiferous tubules ensure that the free form stays maximally increased (even though it is less than 1% of the total). Free DHT is essential for normal sperm development.

As a steroid, testosterone affects gene expression (see chapter 17) and increases or decreases the production of proteins critical for male reproductive function. During development, testosterone is required to masculinize the fetus and cause the regression of female reproductive organs (see chapter 11). Testosterone is critically important during puberty for development of male fertility and for male behavior.

Between the sustenacular cells are spermatogonia; stem cells that create sperm (see chapter 3 for a discussion of stem cells). Spermatogonia are found at the periphery of the seminiferous tubule, while dividing and developing cells move toward the seminiferous tubule lumen. The maintenance of spermatogonia depends, in part, on the interaction of cells with the basement membrane. Spermatogonia divide by mitosis to form daughter cells and spermatogonia. Daughter cells, under the influence of testosterone, DHT, and possibly estrogen, divide and differentiate to become primary **spermatocytes**.

Primary **spermatocytes** undergo meiosis to form two secondary spermatocytes (Figure 10.4). Each secondary spermatocyte undergoes a second round of meiosis to form *spermatids*. Spermatids undergo the last phase of spermatogenesis to form a mature sperm cell, or spermatozoa. A spermatid differs from a secondary spermatocyte in the formation of a head, midpiece, and flagellum. The spermatid head contains only 24 chromosomes because it was formed by meiosis, not mitosis. The cell membrane surrounding the spermatid head is known as the **acrosome** (L., cap) (Figure 10.9). The acrosome is modified by interaction with the egg membrane to release enzymes

FIGURE 10.9

A typical sperm can be divided into a head, midpiece, and tail

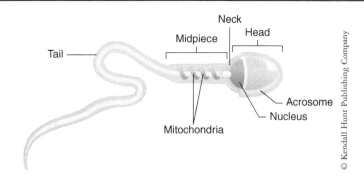

that help the sperm penetrate the egg (see chapter 11). The midpiece contains mitochondria that provide ATP to drive the microtubule motors within the flagellum (see chapter 2 for a discussion of molecular motors). The spermatid is the only human cell with a functional flagellum used for swimming. The flagellum moves side to side and pushes the spermatocyte through fluid.

Spermatogenesis begins near the testes blood barrier, and as sperm develop they move toward the seminiferous tubule lumen. At the end of spermatogenesis, sperm cells are found within the seminiferous tubule lumen with their heads orientated toward the spermatogonia and their tails orientated toward the lumen. When development is almost completed, spermatocytes are released into the seminiferous tubule lumen. Final stages of development occur while spermatocytes move through the epididymis.

Sperm cells must travel back into the body before they can be ejaculated through the penis. The sperm take the following path: They first pass through the tubule recti into the rete testes; they then pass through the efferent ductile, exit the testes, and join with the epididymis; sperm cells then move through the epididymis, ductus epididymis, ductus deferens, ejaculatory duct, and urethra to reach the body exterior.

Other Testosterone Effects

In the male adult, testosterone and DHT are the major androgens produced. Testes Leydig cells produce nearly all androgens, with small amounts produced by adrenal gland zona reticularis and sustenacular cells. Testosterone and DHT have a wide range of effects in the male body that include fetal and adult effects. Fetal effects include enlargement and differentiation of the testes and reproductive ducts, and stimulation of testes descent near the end of fetal development. Adult effects include stimulation of spermatogenesis, stimulation of pubic hair growth extending up the linea alba (L., white cord or line), and hair growth on the legs, chest, axillary region, face, and less commonly, the back. It also causes vellus hair to be converted to terminal hair, which is more pigmented and coarser. In the adult, testosterone causes the skin to become coarser, and because of an increase in melanin, darker; it also increases sebaceous gland secretions, especially in the face, increasing the likelihood of acne. Testosterone causes hypertrophy of the larynx, causing the voice to become deeper and increases the basal metabolic rate, red blood cell production, body fluids through a mineralocorticoid-like property, protein synthesis and skeletal muscle mass, bone growth, and the closure of the epiphyseal plates. Testosterone also initiates and maintains male sexual behavior.

Male Sexual Response

Testosterone and DHT are required to initiate and maintain male sexual behavior. Testosterone affects male sexual behavior through its direct and indirect actions on the hypothalamus and surrounding brain areas. Male sex behavior also depends on the conversion of testosterone to estrogen and other steroids in the brain.

Blood testosterone levels remain relatively constant from puberty until the age of 40. Thereafter, it decreases slowly to 20% of the original level by the age of 80, causing a slow decline in sex drive and fertility. The male sex act has five steps: penis erection (sexual arousal), secretion of mucus into the urethra, emission, ejaculation (and orgasm), and resolution.

Initiation

Male sexual arousal depends on memory, visual, auditory, and olfactory sensations that reach the brain. Tactile stimulation of sensory neurons in the perineum transmits afferent inputs to the sacral region of the spinal cord via the pudendal nerve, which integrates many of the reflexes of the male sex act. Afferent information from the sacral region is sent to the cerebrum, where the conscious sensation of sexual arousal is initiated.

Erection

Sexual arousal causes the penis to fill with arterial blood. The corpus cavernosa is an erectile tissue, filled with cavernous sinuses (tiny blood vessels). Each of the cavernous sinuses are surrounded by smooth muscles and supported by elastic fibrous tissue. Sexual arousal stimulates pudendal nerve parasympathetic efferent fibers to release acetylcholine (ACh) and nitric oxide (NO) within the penis. The main initiator of penis erection is now thought to be NO. ACh binds to muscarinic receptors on penis smooth muscle cells and causes cavernous sinus smooth muscle relaxation through a GPCR pathway (see chapter 17). NO travel within cavernous sinus smooth muscle cells and activates the enzyme guanylate cyclase. Guanylate cyclase then converts guanosine triphosphate (GTP) into cyclic guanosine monophosphate (cGMP). Cyclic GMP causes cavernous sinus smooth muscles to also relax. As the cavernous sinuses fill with blood (as a result of ACh and NO) from the penis artery, the corpus cavernosa fills with blood and the adjacent penis veins are compressed. Venous compression slows the loss of blood from the penis, accelerating blood accumulation. As the penis fills with blood, fluid pressure within the penis causes it to elongate and swell (or to become erect). NO is continuously released by the penis artery during sexual stimulation to keep the penis erect until ejaculation. This continuous release is required because NO lasts for only a few seconds after release.

Erection is normally stimulated by parasympathetic efferents from the S2 through S4 regions of the spinal cord, but can also be stimulated by sympathetic efferents from the T2 through L1 regions. Normally, the parasympathetic inputs are more important, but in cases of spinal cord damage in the sacral region, it is possible for an erection to occur because of sympathetic inputs. Parasympathetic inputs also stimulate urethra and bulbourethral mucous glands to secrete mucus.

Emission

Emission is the urethral accumulation of sperm cells, and prostate gland and seminal vesicle secretions. Emission is regulated by sympathetic inputs from the T12 through L1 regions of the spinal cord. Sexual stimulation results in the activation of sympathetic fibers that cause peristaltic contractions of the reproductive ducts and stimulate secretions from seminal vesicles and the prostate gland. As a result, semen accumulates in the prostatic urethra. The accumulation of semen within the prostatic urethra activates sensory fibers within the pudendal nerve. This sensory information is integrated within the spinal cord with sympathetic and motor neuron outputs. Sympathetic fibers constrict the urinary bladder internal sphincter, preventing the mixture of semen and urine. Motor fibers stimulate the rhythmic contraction of the urogenital diaphragm and base of the penis, forcing semen out of the urethra.

Ejaculation

Accessory gland secretions combine with sperm to form semen, which is composed of approximately 60% seminal vesicle fluid, 30% prostate gland fluid, 5% bulbourethral gland secretions, and 5% seminiferous tubule fluids. The normal volume of semen in each ejaculation is between 5 and 10 ml, with up to 4 billion sperm (from 75 to 400 million sperm/mL semen). Many of the sperm cells are abnormal and millions die as they move their way through vaginal and uterine mucous secretions.

Erection and ejaculation are dependent on the continued activation of sensory receptors in the glans penis and skin surrounding the penis. Activation of sensory receptors in the scrotum, anal, perineal, pubic regions, and other areas of the body reinforce sexual stimulation and help to maintain the erection and initiate ejaculation. Sexual stimulation can also be provided by the prostate and seminal vesicles.

Sexual stimulation is more than a physical stimulation of the penis and surrounding tissues; male sexual stimulation can also be increased by visual, auditory, and olfactory input, as well as thinking of previous encounters. Dreaming of erotic events can lead to ejaculation during sleep (wet dream), which is common in young male individuals just after puberty. Sexual stimulation can also be decreased by auditory, olfactory, and visual sensory inputs, as well as nonsexual thoughts. The maintenance of an erection, emission, and ejaculation rely on the male concentrating on sexual stimulations and ignoring nonsexual stimuli.

Pleasurable sensations that occur during sexual intercourse result in a climatic sensation, known as an *orgasm*, associated with ejaculation of semen. After ejaculation, a phase called *resolution* occurs, in which the penis becomes flaccid and a man feels an overall sense of satisfaction. Typically, a man is unable to achieve another erection and second ejaculation for many minutes to many hours or longer.

Disorders of the Male Reproductive System

Erectile Dysfunction

Men can suffer sometimes during their lives with *impotence* (L. *impotentia*, inability), the inability to achieve or maintain an erection and to accomplish the sexual act. Impotence can also be caused by damage to the pudendal nerve from cycling, prostate surgery, or restricted circulation.

Erectile dysfunction, the repeated inability to get or keep an erection firm enough for sexual intercourse, is a major source of frustration to some men and can lead to problems with personal relationships. It has been estimated that 15 to 30 million men in the United States have suffered from erectile dysfunction at one point in their lives. Visits to a physician's office for treatment of erectile dysfunction have tripled since 1990 as more effective treatments have been become available. The most publicized advance in the treatment of erectile dysfunction was the introduction of the oral medication sildenafil citrate (Viagra) in March 1998. In August 2003, the Food and Drug Administration (FDA) gave approval to a second oral medicine, vardenafil hydrochloride (Levitra). Additional oral medicines for treating impotence are now available.

Because an erection requires a precise sequence of events, erectile dysfunction can occur when any of the events is disrupted. In older men, erectile dysfunction usually has a physical cause, such as disease, injury, or side effects of drugs. Any disorder that causes injury to the nerves or impairs blood flow in the penis has the potential to cause erectile dysfunction. Diseases such as diabetes, kidney disease, chronic alcoholism, multiple sclerosis, atherosclerosis, vascular disease, and neurologic disease account for about 70% of erectile dysfunction cases. Between 35% and 50% of men with diabetes experience erectile dysfunction. In addition, surgery (especially radical prostate and bladder surgery for cancer) can injure nerves and arteries near the penis, causing erectile dysfunction, as can injury to the penis, spinal cord, prostate, bladder, and pelvis. Many common medicines (blood pressure drugs, antihistamines, antidepressants, tranquilizers, appetite suppressants, and cimetidine, an ulcer drug) can produce erectile dysfunction as an adverse side effect. Psychological factors such as anxiety, guilt, depression, and low self-esteem are thought to be responsible for 10% to 20% of all erectile dysfunction cases. Incidence increases with age: About 5% of 40-year-old men and between 15% and 25% of 65-year-old men experience erectile dysfunction, but it is not an inevitable part of aging.

Erectile dysfunction is treated by identifying the cause. If it is due to a drug, treatment is to reduce the dose of the drug. Other treatments are psychotherapy and behavior modification, oral or locally injected drugs, vacuum devices, and surgically implanted devices. In rare cases, surgery involving veins or arteries may be considered.

Taken an hour before sexual activity, Viagra and Levitra work by blocking the activity of a phosphodiesterase that converts cGMP to GMP. As a result, cGMP accumulates within arterial smooth muscle in erectile tissue and causes vasodilation, increasing blood flow to cavernosa sinuses and causing erections. Although these drugs enhance erections in male individuals, they also cause vasodilation of other tissues and increase the workload on the heart.

Prostate Cancer

Prostate cancer is the most common cancer diagnosed in male Americans, and the second most common cause of male cancer deaths. One of four American men will be operated on some time in his lifetime to surgically relieve BPH, and this is the second leading cause of male surgery in the United States, second only to cataract operations. Prostate cancer in men older than 50 is common enough that an annual or biannual test is highly recommended. Prostate cancer was thought to be detected by the presence of PSA protein in serum, but now that is being disputed (see Clinical | In Context Prostate-Specific Antigen Test). PSA is normally found in high concentrations in the ejaculate (1-2 mg/mL) where it nourishes sperm. When the prostate is damaged by abnormal growth or cancer, PSA protein enters the serum. The prostate grows normally with age, and normal PSA blood levels also increase. PSA levels greater than 4 ng/mL in the past were cause for concern and further testing. Many physicians believe that the rate at which PSA increases in blood is the best indicator of prostate cancer.

Prostate cancer is much more dangerous in younger men (<50 years of age) than in older men (>65 years of age). Prostate cancer in younger men tends to be more aggressive and must be treated more aggressively with surgery, radiation, and chemotherapy. Prostate cancer in older men tends to be much less aggressive and usually is not the eventual cause of death. The treatment for prostate cancer in older men is less invasive and usually involves placing a radioactive needle into the prostate tumor. The radioactive material kills cancer cells without significantly damaging surrounding tissues. Surgery is not usually performed in older men because the surgery can cause incontinence (inability to regulate urination) and impotence.

Hormonal Imbalances

Inadequate blood LH and FSH levels can reduce testes testosterone production and sperm counts. Reduced blood LH and FSH levels can result from hypothyroidism, hypothalamic trauma or clots, and tumors. Decreased testosterone levels, caused by a defect in sustenacular cells, can also reduce sperm cell counts.

Decreased sperm counts can be caused by administration of large amounts of GnRH analogues, leading to chemical castration, through a downregulation of testes LH receptors. Chemical castration is used clinically in men with prostate cancer, and is sometimes offered to sex offenders. Depo Provera is a synthetic form of progesterone that suppresses testosterone production. Estrogen inhibits some enzymes in the testosterone synthetic pathway and directly inhibits testosterone production.

Causes of Male Infertility

Approximately 15% of couples attempting their first pregnancy meet with failure. Conception (fertilization) is usually achieved within 12 months in 80% to 85% of couples who use no contraceptive measures. Male infertility, which is responsible for 50% of infertile couples, is commonly caused by low sperm counts. Low sperm counts can be caused by damage to the testes, infections (mumps), radiation exposure, failure of testes descent, or other factors.

It has been suggested that the average male sperm count has been decreased 1% a year since 1945, although there is a great deal of controversy about the accuracy of these reports. Scientists speculate that synthetic estrogen-like compounds (known as *xenoestrogens*), pollution, radiation exposure, and other factors are responsible for a reduction in sperm counts.

Fertility is not only affected by sperm count but by the structure and activity of sperm. Abnormal sperm are incapable of successfully fertilizing eggs. Abnormal sperm can be a result of chromosomal abnormalities, increased testes temperatures, or exposure to environmental chemicals. Abnormal sperm usually have reduced motility, resulting in infertility.

Another major cause of male infertility is autoimmune disorder, in which anti-sperm antibodies produced by the immune system bind to sperm cells. Fertility can be increased by collecting several ejaculations and concentrating sperm cells, followed by their introduction into the vagina with a syringe, a process known as *artificial insemination*.

Semen analysis is commonly conducted to assess male fertility. A man donates (by masturbation) a semen sample after a 36-hour period of sexual abstinence. The entire ejaculate volume is measured together with its appearance, color, thickness, and pH. Semen is allowed to liquefy after coagulation and drops are placed on microscope slides. The World Health Organization (WHO) standards for sperm analysis require the examination of at least 100 sperm. Sperm are graded for shape and motility. Normal semen has more than 20 million sperm/mL with less than 14% sperm with abnormal shapes. Common abnormalities are malformed sperm heads and sperm that did not separate during spermatogenesis. Normal semen also has at least 50% of the sperm showing swimming ability in a forward direction. Semen is also commonly examined for white blood cell counts to determine whether there is a reproductive tract infection. Biochemical tests can also examine other components (e.g., fructose levels).

An abnormally low semen volume is usually an indication of a prostate or seminal vesicle problem, because they contribute 90% of the semen volume. If sperm counts decrease to less than 20 million/mL, a male individual is usually infertile because too few sperm survive to reach the uterine tubes. If the semen appears to be normal, then a sperm sample is mixed with hamster oocytes on a microscope slide (the hamster test). Normal sperm can fertilize hamster oocytes, but the resulting zygote is unable to undergo cell division. If fertilization does not occur, the problem is usually found in the acrosomal cap (fertilization is discussed in chapter 11).

In Review...

10.10 What is the effect of testosterone on the body?
10.11 How is spermatogenesis regulated?
10.12 What is androgen-binding protein and what is its function?
10.13 What does testosterone and dihydrotestosterone do within the seminiferous tubule?
10.14 What is an acrosome?
10.15 What path do sperm take to leave the body?
10.16 What processes regulate the male sexual act?
10.17 What occurs during emission? During ejaculation?

Anatomy of the Female Reproductive System

The female reproductive system, like the male reproductive system, clearly shows its intended function: to produce and transport ova to where they can be fertilized, and then provide for fetal development, and childbirth (parturition) (Table 10.1). The female equivalents of the testes are paired **ovaries** (Figure 10.10). In addition, the female reproductive system has **uterine tubes** (or *Fallopian tube*s), a **uterus**, **vagina**, external genital organs, and **mammary** glands. The male testes are outside of the pelvic region, but all female reproductive organs are found within the pelvic region between the urinary bladder and the rectum. The uterus and vagina are on the midline, with the ovaries to each side of the uterus. The internal reproductive tissues are held in place by a group of ligaments that are peritoneal extensions. The main ligament, the broad ligament, is a sheet-like peritoneal extension that attaches to the uterine wall and encloses the uterine tube.

The female body has two small almond-shaped ovaries (L. *ovum*, eggs) about 3 cm in length and 1 to 1.5 cm in width (Figure 10.10). Like testes, the ovaries develop high up on the abdominal wall and descend before birth to their final resting area on the lateral wall of the pelvic cavity. The mesovarium, a peritoneal extension, supports and attaches each ovary to the posterior surface of the broad ligament. The ovaries are also supported by the suspensory ligament extending from the mesovarium to the body wall, and the ovarian ligament that attaches the ovary to the superior margin of the uterus. The ovaries are supplied by arteries, veins, and nerves that travel through the suspensory ligament and enter the ovary through the mesovarium.

The outer surface of each ovary is covered with a visceral peritoneum known as the *ovarian epithelium*. The tunica albuginea, a layer of dense fibrous connective tissue lies below the epithelial layer. The ovary outer cortex is composed of dense tissue that surrounds the medulla, which is filled with looser tissue. Ovarian **follicles** are distributed throughout the cortex. Each of these small vesicles contains an **oocyte** (G. *oon*, egg; *kytos*, a hollow)

The uterine tubes are found on each side of the uterus associated with an ovary (Figure 10.11). Each uterine tube is located along the superior margin of the broad ligament, in an area known as the *mesosalpinx* (G, *mesos*, middle; L. *salpinx*, trumpet-shaped). The uterine tube is connected to the uterus on the medial end and opens at its lateral end directly into the peritoneal cavity.

TABLE 10.1 ♦ Summary of Homologous Reproductive Structures in Men and Women

Female Structures	Male Structures
Ovaries	Testes
Oogonia	Spermatogonia
Labia majora	Scrotum
Labia minora	Spongy (penile) urethra
Vestibule	Membranous urethra
Vestibule bulb	Penis corpus spongiosum and penis bulb
Clitoris	Glans penis
Para urethral glands	Prostate
Greater vestibular glands	Bulbourethral glands
Nipples and areola	Nipples and areola

FIGURE 10.10

Female reproductive tissues. (a) The female reproductive tissues include the vagina, cervix, uterus, uterine tubes, and the ovaries. (b) A midsagittal view

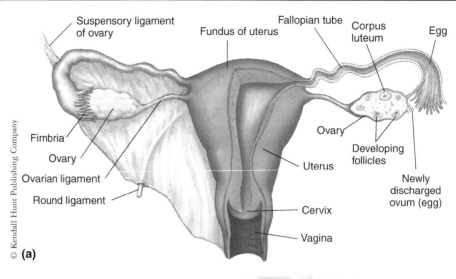

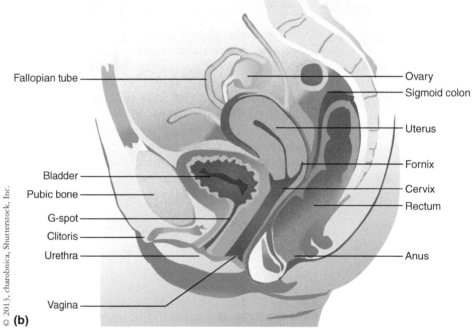

FIGURE 10.11

Ovaries are the site of oogenesis and supported by suspensory ligaments.

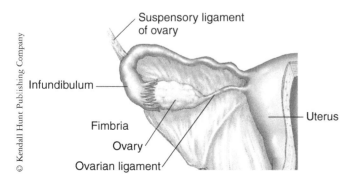

The lateral end of the uterine tube forms the *infundibulum* (L. *infundibulum*, funnel) with long, thin *fimbriae* (L. *fimbriae*, fringe) surrounding the infundibulum opening. The inner surfaces of the fimbriae are lined with ciliated mucous epithelium.

Medial to the infundibulum is the *ampulla*, the widest and longest part of the uterine tube representing approximately 80% of the overall 10 cm length. Adjacent to the uterus is the isthmus, the last 20% of the uterine tube, which is much narrower and with thicker walls than the ampulla. The portion of the tube that passes through the uterine wall, known as the uterine or *intramural part*, ends in a very small uterine opening.

The uterine tube wall consists of three layers throughout its length: the outer serosal layer formed by the peritoneum, the middle muscular layer composed of longitudinal and circular smooth muscle, and the inner mucosal layer composed of simple ciliated columnar epithelium. The mucosal is arranged into numerous longitudinal folds to increase its surface area.

Through the beating of ciliated cells, the uterine tube propels the follicle from the peritoneal space into the uterine tube. The mucosal layer provides nutrients for the oocyte as it moves through the uterine tube (chapter 11). The ciliated epithelia beat back and forth, pushing a small amount of fluid and the oocyte toward the uterus.

Uterus

The uterus (L. *uterus*, womb) is the size and shape of a medium-sized pear, about 7.5 cm in length and 5 cm in width (Figure 10.10). One end of the uterus, the **cervix** (L., neck) opens into the vagina, whereas the superior end is connected to the uterine tubes. The uterus is found within the pelvic cavity, immediately dorsal to the urinary bladder and ventral to the rectum. The uterus is slightly flattened anterioposteriorly and is orientated in the pelvic cavity, with the larger, rounded fundus (L., bottom) directed superiorly and the cervix is directed inferiorly. The main part of the uterus, the body, lies between the fundus and the cervix. A slight constriction called the *isthmus* marks the junction of the cervix and body. Internally, the uterine cavity continues as the cervical canal, which opens through the ostium (L., door) into the vagina.

thought question
Why isn't the myometrium made of skeletal muscle?

The uterus is held in place by several ligaments: the broad, round, and uterosacral ligaments. The broad ligament, which also holds the uterine tubes and ovaries in place, is an extension of the peritoneal membrane. It extends from the lateral margin of the uterus to the pelvic wall on each side. The round ligaments extend from the uterus through the inguinal canal to the *labia* (L., lip) majora of the external genitalia. The uterosacral ligaments attach the lateral uterine wall to the sacrum. The pelvic floor skeletal muscles also provide inferior support to the uterus. These skeletal muscles can be weakened by childbirth, causing a *prolapsed* (L. *prolapsus*, collapsing) uterus when the uterus extends inferiorly into the vagina.

The uterine wall is composed of three layers: the perimetrium, myometrium, and endometrium (Figure 10.12). The outer fibrous connective tissue

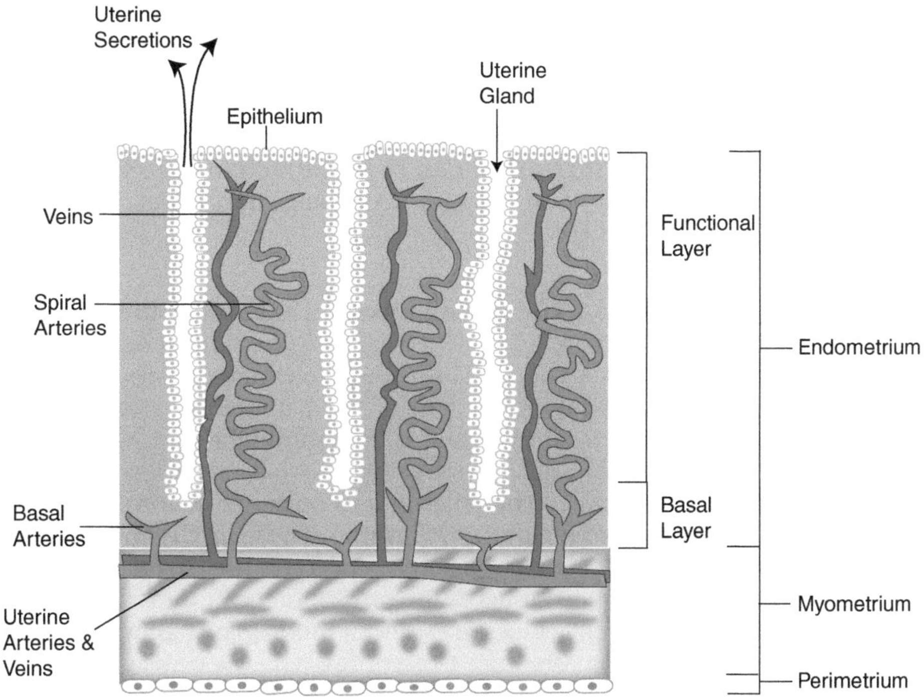

FIGURE 10.12
The uterine wall

perimetrium covers the uterus surface. The medial myometrium consists of a thick layer of smooth muscle accounting for most of the uterine wall mass and the thickest smooth muscle layer in the body. The cervix is less flexible and more rigid than the rest of the uterus because it contains less smooth muscle with denser connective tissue. The inner endometrium is a mucosa membrane made up of simple columnar epithelium and the lamina propria, a connective tissue layer. The lamina propria contains many simple tubular glands that open through the epithelial lining into the uterine cavity. The endometrium consists of two layers: the basal and functional layers. The basal layer is a thin layer continuous with the myometrium. The thicker functional layer lines the uterine cavity and is made up of the lamina propria and endothelium. The functional layer undergoes a monthly cycle of growth and sloughing during the menstrual cycle (see later discussion).

The cervical canal is lined with columnar epithelium that contains numerous mucous glands. Cervical mucous glands produce mucus that plugs the cervical canal. The cervical mucus plug prevents substances and cells (bacteria) from passing from the vagina into the uterus. Near the time of ovulation, hormonal changes make the cervical mucous plug thin, making it easier for sperm to enter the uterus from the vagina.

Common Uterine Disorders

Endometrial, or *uterine*, polyps originate from the uterus endometrium. They are usually noncancerous. The cause of endometrial polyps is not clear, but some are associated with excess estrogen.

Cervical cancer is one of the most common cancers that affect a woman's reproductive organs. Various strains of the human papillomavirus, a sexually transmitted disease, are responsible for the majority of cervical cancer cases. When exposed to human papillomavirus, the immune system response in most women destroys the virus or prevents it from infecting cells (see chapter 21). In a small group of women, however, the virus survives for years within cervical cells before it eventually converts some cells on the surface of the cervix into cancer cells. Older women and sexually active young women are most at risk for development of cervical cancer. This has lead to the recent development of a vaccine (Gardasil) for the papillomavirus.

Thanks largely to Pap test screening (named after the U.S. physician Dr. Papanicolaou), the mortality rate from cervical cancer has decreased greatly since 1970. Pap smears, a scrapping of the cervical surface to examine for the presence of abnormal cells, have a reliability of greater than 90% in detecting cervical cancer.

As a woman ages, female pelvic organs change, and many women report a feeling of pelvic pressure or heaviness. These symptoms may be caused by pelvic support problems that sometimes begin with childbirth. Other tissues; bladder, intestine, rectum, or uterus, sometimes bulge into or out of the vagina.

The vagina (L., sheath) is a 10-cm long tube that extends from the cervix to the external genitalia (Figure 10.9). The vagina is the female organ of copulation. The vagina receives the penis during intercourse, allows for the flow of menstrual blood, and the movement of the baby out of the uterus. The vagina lumen is characterized by columns, longitudinal ridges, which extend the length of the anterior and posterior vaginal walls and rugae, transverse ridges, which extend between the anterior and posterior columns. The *fornix* (L., domed) is the superior region of the vagina. The fornix is attached to the sides of the cervix so that a part of the cervix extends into the vagina.

Like the uterus wall, the vaginal wall consists of an outer smooth muscle layer and inner mucosal epithelia. The smooth muscle layer allows the vagina to increase in size during intercourse to accommodate the penis and to stretch to allow a baby to pass through during childbirth. A moist stratified squamous epithelium forms a protective mucosal surface in the vaginal lumen. The vaginal mucus membrane releases lubricating secretions during sexual intercourse, and the acidic mucus prevents bacterial infections in the vagina.

External Genitalia

The external female genitalia, the *vulva* (L., covering), or *pudendum*, is made up of a vestibule and surrounding structures (Figure 10.13). The vestibule, bordered on each side by the labia minora, a pair of thin, longitudinal skin folds, is the external space with the vaginal and urethral openings. The *clitoris* (G. *kleitoris*, body), a small erectile tissue, is located in the vestibule anterior margin. The clitoris is covered by the prepuce (L., before) formed by a combination of the two labia minora into a skin fold.

The *hymen* (G. *hymenaeus*, god of marriage) is part of the vulva. It is a thin tissue layer that partly covers the vaginal opening, or *orifice*. If this mucous membrane completely closes the vaginal opening, it is known as the imperforate hymen and must be surgically removed to allow menstrual blood flow. More commonly, the hymen is perforated by several small holes. The openings through the hymen are greatly enlarged during the first sexual intercourse, and can be torn or perforated before sexual intercourse by such things as strenuous physical exercise. The absence of an intact hymen is not necessarily a result of having had sexual intercourse, as once thought.

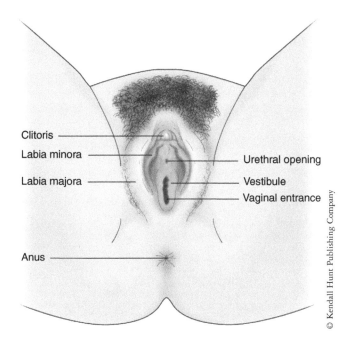

FIGURE 10.13

External female genitalia

The small clitoris (<2 cm in length) consists of a shaft and distal glans. Like the male penis, the clitoris is richly innervated with sensory receptors. Manual stimulation of the clitoris results in its erection by increased blood flow. The clitoris contains erectile tissue, corpora cavernosa, which extends at the clitoris base to form the crus of the clitoris and attaches the clitoris to the coxae. The corpora cavernosa is equivalent to the penis corpora cavernosa and becomes engorged with blood when a woman is sexually excited. In most women, sexual excitement results in an increase of the width, but not the length, of the clitoris. With an increased diameter, the clitoris makes better contact with the penis and surrounding tissues, and is more easily stimulated during sexual intercourse.

The vestibule also contains erectile tissue equivalent to the penis corpus spongiosum. This additional erectile tissue, known as the *vestibular bulb*, is found on either side of the vaginal orifice on the lateral margins of the vestibular floor. The vestibular bulb becomes engorged with blood during sexual activity, which increases the sensitivity of vestibule. As the vestibular bulbs fill with blood, the vaginal orifice is narrowed, increasing the contact with the penis during sexual intercourse.

On the lateral margins of the vestibule, between the vaginal opening and the labia minora, are the openings of the greater vestibular gland, a mucous gland. Additional small mucous glands, known as the *lesser vestibular glands*, are found near the clitoris and urethral opening. These mucous glands produce a lubricating fluid that helps to maintain vestibular moistness during sexual intercourse.

The vestibule is covered externally by labia majora, two prominent, rounded skin folds lateral to the labia minora. The external appearance of the labia majora is due primarily to the presence of subcutaneous fat. At the upper margin, the labia majora form the mons (L., mound) pubis. The mons pubis and labia majora lateral surfaces are covered with coarse pubic hair. The mons pubis and labia majora medial surfaces are covered with numerous sebaceous and sweat glands; the pudendal cleft is the space between the labia majora.

FIGURE 10.14

Anatomy of the breast. Lobes of the mammary gland produce milk, which flows to the areola, where it is released through the nipple

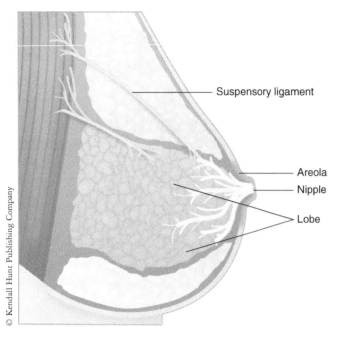

Normally, the pudendal cleft is closed by the contact of the labia majora across the midline, concealing the deeper structures within the vestibule.

The female perineum is separated into two triangles, like the male perineum, by the superficial and deep transverse perineal muscles. The external genitalia are found within the anterior urogenital triangle, whereas the anal opening is found within the posterior anal triangle. The region between the vagina and anus is known as the *clinical perineum* because the skin and muscle can tear during childbirth. To prevent tearing, an incision called an *episiotomy* is commonly made in the clinical perineum. The rationale behind an episiotomy is that the clean, straight incision is easier to repair than a ragged tear. Recently, though, the number of women receiving episiotomies in the United States has decreased dramatically because clinical research has shown that episiotomies are no better than natural tearing of the perineum.

Within the breasts are mammary (L. *mamma*, mother) glands, modified sweat glands, which produce milk after a pregnancy (Figure 10.14). The male and female breasts have raised nipples surrounded by a circular, pigmented *areola* (L., area). Rudimentary mammary glands, known as *areolar glands*, are found below of the areolar surface, giving it a bumpy surface. Secretions from these glands protect the nipple and areola during breast-feeding. Sexual stimulation can cause smooth muscle beneath nipples to contract, causing the nipple to become erect. These smooth muscle cells become activated during sexual arousal.

General breast structure is similar in prepubescent boys and girls. Prepubescent males and females possess a crude series of glandular lobes, consisting of ducts and alveoli. During puberty, increasing blood concentrations of estrogen and progesterone stimulate female breasts to grow. Estrogen and progesterone stimulate the growth of breasts, deposition of adipose within the breast, and development of the glandular lobes. During this growth period, the breasts can become sensitive and painful. During puberty in male individuals, there is commonly a brief period of breast enlargement, but this change is not permanent. On rare occasions, male breasts become enlarged, known as *gynecomastia*, because of hormonal imbalance or environmental factors.

The ability of a breast to provide milk to a baby is not dependent on its size but on the number of glandular lobes. After puberty, most breasts contain 15 to 20 glandular lobes encased in a significant amount of adipose tissue. The external appearance of a breast is primarily due to adipose tissue and not the number or structure of the glandular lobes. Milk produced by the glandular lobes is carried to the nipple surface by lactiferous ducts. Each glandular lobe has a single lactiferous duct, and they open on the nipple surface independently of each other. Each lactiferous duct expands at its distal end to form a small lactiferous sinus, which accumulates milk during lactation. Milk is produced with alveoli, *secretory sacs*, within the glandular lobes. The milk is transported through smaller branches that form into larger and larger ducts that finally become the lactiferous duct.

The breasts are held in place by mammary, or *Cooper's*, ligaments. Mammary ligaments extend from the pectoralis major muscles to the skin over the mammary glands and support the weight of the breasts. Mammary ligaments weaken and elongate over age because of weight gains, pregnancies, and the effects of gravity, allowing the breasts to sag and lose their conical shape.

In Review...

10.18 What role do the ovaries play in female reproduction?
10.19 What is the role of the uterine tube? The uterus? The vagina?
10.20 What structures within the breasts are responsible for milk production?

Physiology of the Female Reproductive System

In both male and female bodies, reproduction is under the control of hormones and the nervous system. Development and the normal functioning of the female reproductive system depend on a number of hormones.

Development of the Female Reproductive System

Embryonic Origin of Reproductive Tissues

Like the male testes, the female ovaries and reproductive tissues arise from ridges along mesonephros ventral borders during embryonic development (see chapter 11). Müllerian ducts, or *paramesonephric ducts*, develop lateral to mesonephric ducts, and create uterine tubes, the uterus, and part of the vagina.

Puberty

The development of puberty in female individuals is related in a complex way to body size and body fat content. Adrenarche, an early sign of puberty marked by the appearance of pubic hair, hormonal changes, and ovarian enlargement, normally occurs between 8 and 10 years of age; the exact age depends on heredity and nutrition. In developing countries where more than one third of the total children are malnourished or undernourished, the onset of this process is usually delayed.

As in boys, female puberty is initiated by changes in the hypothalamus-pituitary axis that regulates the secretion of LH/FSH. Before puberty, the secretion of LH/FSH is very low and regulated in a classic negative feedback manner. At the onset of puberty, the regulation of LH/FSH secretion changes to one in which the hormones are released in a pulsatile manner. The beginning of puberty has advanced 4 months per decade in the past century and appears to be due to hormonal changes rather than a nutritional change.

Several physiologic changes take place during this period, the most prominent in girls being the onset of menstruation (*menarche*). The onset of menarche is one of the most visible signs that a girl is entering puberty. The rate at which breasts grow and develop varies greatly and depends of the deposition of fat pads beneath the skin. The development of fat pads is different for each young woman and depends on many factors like heredity and nutrition.

Every girl enters into puberty at her own time. There is a great deal of variation in the average age of menarche between rural and urban girls, as well as across various socioeconomic and ethnic groups. This variation depends on factors such as body girths, weight, and nutrition. Five stages are associated with female puberty.

Stage 1 usually occurs at 10 to 11 years. There are no outward signs of development, but adrenarche has begun.

Stage 2 usually occurs at 11 to 12 years. The first outward pubertal change is breast development, including "breast buds," in which a small mound is formed by the elevation of the breast and papilla. Hair under the arms appears around 12 years of age. The areola (the circle of different-colored skin around the nipple) increases in size at this time. A girl may also gain considerable height and weight. Pubic hair, long, soft hair (fine and straight) in a small area around the genitals, follows shortly after breast development.

Stage 3 typically occurs at 12 to 13 years. The breasts continue to enlarge. Eventually, the nipples and the areolas will elevate, forming another projection on the breasts. At the adult state, only the nipple remains erect. Pubic hair coarsens and becomes darker as it continues to spread. The inner length of the vagina starts enlarging and may begin to produce a clear or whitish discharge, which is a normal self-cleansing process. Most girls experience their first menstrual period late in this stage, but menstrual periods remain irregular.

Stage 4 typically occurs at 14 to 15 years. There may be an increase in hair growth, not only in the pubic area, but also under the arms and on the legs. The pubic hair growth now takes the triangular shape of adulthood, but in a smaller area. It may spread to the thighs and sometimes up the stomach. By this time, ovulation begins in some girls, although it does not typically happen in a regular monthly routine until stage 5. Female body shape will also begin to change, not only an increase in height and weight, but wider hips and smaller waists. There may also be an increase in fat in the buttocks, legs, and stomach. Body size will increase, with the feet, arms, legs, and hands sometimes growing faster than the rest of the body. Adolescents may experience an increase in oily skin and sweating; acne may develop.

Stage 5 typically occurs at 16 to 17 years. This is the final stage of sexual development, when a girl is physically an adult. Breast and pubic hair growth are complete, and full height is attained. Menstrual periods are well established, and ovulation occurs monthly.

There are no standard norms for girls to attain each milestone of puberty by a specific age, especially in Third World countries. The age at which pubertal milestones are attained varies greatly across class, caste, race and ethnicity, and is influenced by activity level and nutritional status. Girls with low body fat, for example, competitive athletes, may face a significant delay in menarche.

Secondary Sex Characteristics

Female secondary sex characteristics are developed in response to increasing levels of estrogen during puberty. These changes include increased fat deposition especially around the hips and breasts, increase in height and weight, widening of the hips, growth of pubic hair and axillary hair, changes in the hypothalamic area of the brain, changes in the olfactory system (increasing its sensitivity compared to males), and increased sexual interest.

Menopause (Climacteric)

When a woman reaches her 40s, her menstrual cycles become less regular and ovulation occurs less frequently, until they eventually stop completely. Menopause, or *climacteric,* occurs when the menstrual cycles have completely stopped. The period from the onset of irregular menstrual cycles to menopause, which can last 10 years, is known as the female climacteric, or *perimenopause.*

Menopause is caused by ovarian changes. Previously, it had been believed that women are born with a set number of follicles, and only 400 or so actually undergo the process of follicular development and ovulation. As a woman reaches middle age, the number of follicles capable of undergoing growth and differentiation decreases to a small number. More recent work has shown that mammals (mice) can produce new oocytes in postnatal life. It is unclear whether female humans produce new oocytes in postnatal life, but this is an area of intense research.

Even if female humans produce new oocytes during their lives, as women, age follicles become less sensitive to LH and FSH stimulation because of a downregulation of LH and FSH receptors. Blood levels of LH and FSH are commonly increased as a woman ages, but the lack of receptors prevents them from stimulating follicular development. When the follicles become less responsive to these hormones, fewer mature follicles and corpora lutea are produced. This results in a gradual morphologic change in the female body because estrogen and progesterone are reduced.

A variety of symptoms occur during the female climacteric, including hot flashes, irritability, fatigue, anxiety, and sometimes severe emotional disturbances. Research suggests that the decrease in estrogen increases the risk for heart disease for the first few years after menopause. This beneficial effect of estrogen is one of the reasons many women are treated with small amounts of estrogen after menopause. This postmenopausal hormone treatment has been shown to increase the risk for cancer and other health problems.

In addition to the ovarian changes in climacteric women, there are many uterine changes: a decrease in uterine size, a decrease in endometrial thickness, and the tipping of the uterus into a more posterior position. If the supporting uterine ligaments weaken, the uterus can descend and protrude into the vagina, which is known as *uterine prolapse*. Vaginal changes include a decrease in mucous secretion, a decrease in vaginal wall thickness, a decrease in vaginal wall elasticity, a thinning of the protective epithelial layer, a decrease in smooth muscle activity, and an increase in the frequency of vaginal infections. Sexual intercourse can become painful because of a narrowing of the vagina and loss of elasticity. Because of these physical changes, sexual excitement takes longer time to develop, the peak levels of sexual excitement are lower, and the return to the resting state is more rapid.

Approximately 10% of all women will have breast cancer. The increase in breast cancer is most rapid in women between the ages of 45 and 65, and the incidence is greater for women with a history of breast cancer in their families, suggesting a genetic component. Cervical and endometrial cancers also increase between the ages of 50 and 65. Ovarian cancer is also more frequent in older women, and is the second most common reproductive cancer in women.

Oogenesis

Oogenesis requires the growth and development of ovarian stem cells, the oogonia, to form oocytes. Oogenesis requires meiosis (see chapter 2) to produce daughter cells with 23 chromosomes. Oogenesis, the female equivalent of spermatogenesis, is the process by which oocytes (eggs) develop within the ovary (Figure 10.15). Oocytes are formed during fetal development, and in the fourth month of prenatal life, the ovaries may contain as many as 5 million stem cell **oogonia** (L. *oon*, egg; *goné*, generation), from which oocytes develop. During the last 5 months of fetal development, almost two thirds of the oogonia have degenerated, and the remaining oogonia have begun meiosis. Oogonia meiosis is arrested at prophase I (see chapter 2). After meiosis arrest, the oogonia are known as *primary oocytes*. There are 1 to 2 million primary oocytes at birth. In contrast with the spermatocyte, the primary oocyte is covered with a single layer of flat **granulosa cells**. The granulosa cells and primary oocyte form a primordial follicle (Figure 10.16). By puberty, the number of primordial follicles declines to 300,000 to 400,000. Of the remaining primordial follicles, only 0.1%, or 400, become mature follicles that release an oocyte from the ovary.

Follicular Maturation

After puberty, FSH stimulates development of a small number of primordial follicles each month (approximately 20 in each ovary). Each month a woman can produce 0, 1, or 2 primary follicles. It is a common misconception that each month the ovaries switch with each other to produce a primary follicle.

thought question
In contrast with male individuals, why doesn't the female immune system recognize oocytes as nonself?

thought question
Why aren't the neural pathways regulating sexual activity in men and women different?

FIGURE 10.15

Oogenesis is the process of oocyte and follicle development within the ovary

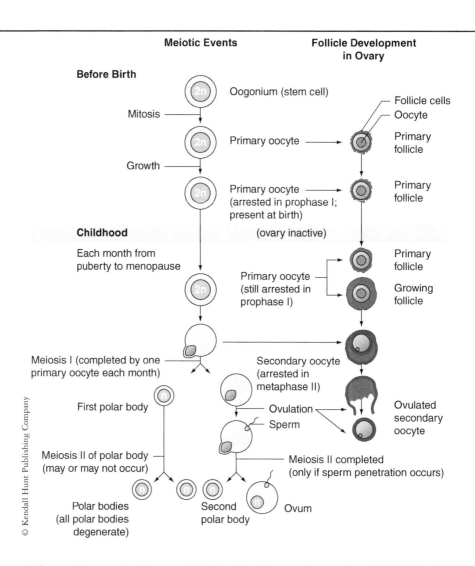

The pattern of primary follicle development can be left-right, right-right, right-left, left-left, or one ovary may produce many primary follicles in a row.

Not all follicles respond in the same way to FSH secretion because not all follicular cells have the same numbers of FSH receptors on their surface, and some follicles are better than others in stimulating angiogenesis (growth of new blood vessels). Increased blood flow to a follicle increases delivery of blood-borne FSH, allowing FSH to have a greater effect on the follicular cells. This results in the rapid development of a single follicle known as the *primary follicle*. The other primordial follicles that also started development stop doing so and degenerate. In the remaining primary follicle, the oocyte enlarges and the single layer of granulosa cells first becomes enlarged and then cuboidal. During this development, several layers of granulosa cells form, and a clear layer of material known as the *zona pellucida* (L. *zona*, area; *pellucid*, clear) (Figure 10.17), is deposited around the primary oocyte.

Under the influence of FSH, the primary follicle continues to develop and becomes a secondary follicle. FSH stimulates granulosa cells to divide and form an increasing number of layers around the oocyte. As the granulosa cells divide, a number of small, fluid-filled spaces (vesicles) form. The small, fluid-filled vesicles eventually fuse to form a single fluid-filled

FIGURE 10.16

An internal view of the ovary with the different stages of follicular development

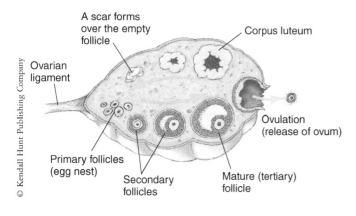

chamber, the antrum (G. *antron*, cave). As the secondary follicle enlarges, the surrounding cells form the **thecal** (G. *theke*, box) layer, or the *thecal capsule*. The thecal layer is made up of two layers: inner vascular theca interna and outer fibrous theca externa. Once the antrum has fully formed, the secondary follicle is known as a mature, or a *graafian*, follicle. As the follicle continues to develop under the influence of FSH, the antrum continues to fill with fluid and increases in size. The internal wall of the vagina is separated by 4 to 5 mm of tissue from the ovary, making it possible to feel a mature follicle on the ovarian surface during an internal vaginal examination.

Granulosa cells secrete the fluid that fills the antrum. Fluid pressure keeps the oocyte to one side of the follicle in a mass of follicular cells called the *cumulus* (L., heap) *mass*. The innermost cells of this mass resemble a crown radiating from the oocyte and are thus called the *corona radiata* (Figure 10.17).

In a mature follicle, just before ovulation, the primary oocyte completes the first meiotic division to produce a secondary oocyte and a polar body. The division of cytoplasmic contents is unequal, with the secondary oocyte receiving almost all of the cytoplasm and the polar body almost none. The secondary oocyte then begins a second round of meiosis and stops in metaphase II.

Figure 10.17
A primary follicle, with a zona pellucid and secondary oocyte

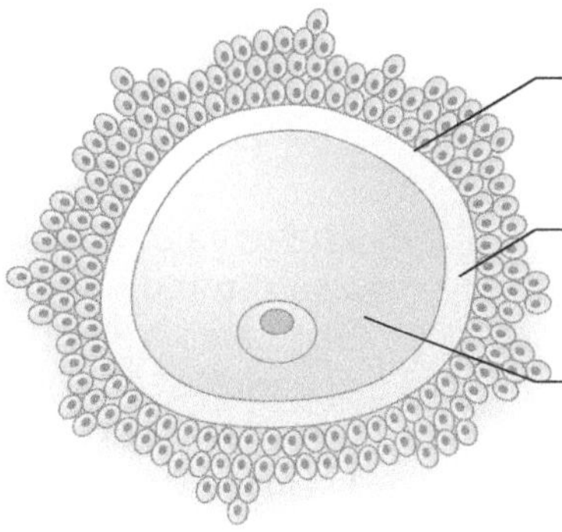

- Corona Radiata
- Zona Pellucida
- Ovulated Secondary Oocyte

Ovulation

As the mature follicle continues to swell, it can be seen on the surface of the ovary as a tight, translucent blister. The swelling is caused by follicular cells secreting fluid at a faster rate than can be accommodated by follicular growth. As a result, the outer layer of granulosa cells, the thecal layer, becomes very thin over the exposed region of the ovarian surface. The thecal cells also secrete several different enzymes, like hyaluronidase, which weaken the ovarian outer connective tissue capsule.

Eventually, the rapidly expanding follicle ruptures, forcing a small amount of fluid, blood, and the secondary oocyte (still surrounded by granulosa cells and the zona pellucida) from the follicle and out of the ovary. The force of the rupture also breaches the connective tissue capsule of the ovary. The release of the secondary oocyte, in this manner, is known as **ovulation**.

During ovulation, the development of the secondary oocyte stops at metaphase II. The secondary oocyte must be fertilized by a sperm to finish the second round of meiosis. If it is not fertilized, the second round of meiosis is never completed and the oocyte degenerates and passes out of the body. Once the secondary oocyte is fertilized, the secondary oocyte completes the second phase of meiosis, producing another polar body and a **zygote** (G. *zygotes*, yolked), a fertilized oocyte.

Corpus Luteum

After ovulation, the remaining follicle becomes transformed into the **corpus luteum** (L. *corpa*, body; *lutea*, yellow), a glandular structure. The corpus luteum has a convoluted appearance as the result of the collapse of the remaining follicle after ovulation. The remaining granulosa and thecal cells, now known as **luteal cells**, enlarge and begin to secrete progesterone and estrogen.

If the oocyte is fertilized and implanted into the uterus, the corpus luteum enlarges and remains throughout the pregnancy. The corpus luteum acts to produce progesterone and estrogen. If oocyte is not fertilized, the corpus luteum remains functional for an additional 10 to 12 days and then degenerates.

The blood levels of progesterone and estrogen decrease as the corpus luteum degenerates. The remaining cells with the corpus luteum become enlarged and clear, and are known as the corpus albicans (L. *albica*, white). The corpus albicans continues to degenerate and may completely disappear after several months, or even years.

Female Sex Hormones and the Reproductive Cycle

Female Sex Hormones

The female reproductive cycle is under the control of the same hypothalamic and pituitary hormones as the male reproductive cycle: GnRH, FSH, and LH. In female individuals, GnRH secretion stimulates the pituitary secretion of LH and FSH. Increasing LH and FSH levels then inhibit GnRH secretion in a negative feedback manner except during the middle of the menstrual cycle. LH and FSH stimulate ovarian cells to produce estrogen and progesterone. The female reproductive system also produces a small amount of testosterone. The normal ratio of estrogen to testosterone in women is 50 or 100:1, opposite of what it is in male individuals.

Menstrual Cycle

Each month, a sexually mature, nonpregnant woman undergoes cyclic changes in blood hormone levels, coupled with ovarian and uterine changes in a process known as the **menstrual** (L. *menses*, month) **cycle** (Figure 10.18). The typical menstrual cycle lasts 28 days, but can be as short as 18 days or as long as 40 days. The menstrual cycle ends with a period of mild hemorrhage during which tissue is sloughed off and expelled from the uterus through the vagina. **Menstruation** refers to the discharge of blood and tissues. The menstrual cycle is best understood by examining the ovarian changes that occur during the cycle. The ovarian changes are classified as the follicular, ovulatory, and luteal phases.

The hypothalamic neurons that secrete GnRH undergo three changes during the menstrual cycle: Initially, they are inhibited by blood estrogen; then they are stimulated by increasing blood estrogen levels; and finally, they are inhibited by increasing blood progesterone and estrogen.

Follicular Phase (Days 1 to 14)

The follicular phase begins at day 1 of the menstrual cycle and last 14 days. The follicular phase is characterized by increasing blood levels of estrogen, and rapid growth and development of follicular cells (Figure 10.18). During the follicular phase, the hypothalamus secretes GnRH, which, in turn, stimulates the release of LH and FSH. LH and FSH bind to receptors on cells within the ovary and stimulate the production of estrogen, among other things. Increasing blood estrogen levels inhibit the release of GnRH (a classic negative feedback loop). At day 1, a mature follicle begins to grow in response to FSH. LH primarily affects thecal cells, but later affects granulosa cells. FSH primarily affects granulosa cells.

In a mature, developing follicle, FSH upregulates LH receptors on granulosa cells. In thecal cells, LH stimulates the production of estrogen, which, in turn, increases LH receptors in thecal cells (a positive feedback loop). The overall secretion of LH and FSH from the pituitary does not increase markedly during the follicular phase, but the upregulation of follicular cell LH receptors increases LH binding and increases the effect of the LH that is present, thereby increasing estrogen production. Increasing blood levels of estrogen stimulate follicular growth and development (oogenesis). LH binding to granulosa cells also stimulates progesterone production, which diffuses to thecal cells, where it is converted to androgens. The androgens, in turn, are converted by granulosa cells to estrogen, further increasing blood levels of estrogen. The net result is

FIGURE 10.18
The menstrual cycle

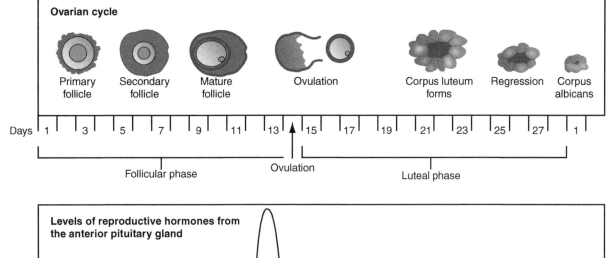

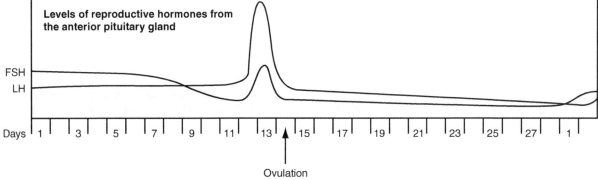

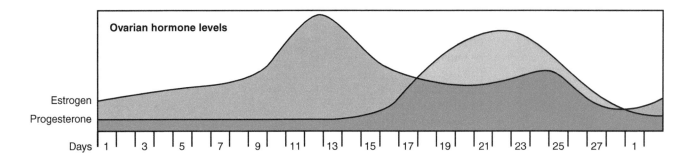

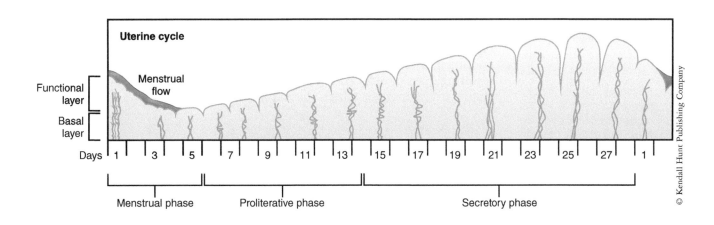

that even though there is only a small increase in LH and FSH during the follicular phase, there is a marked increase in estrogen production. FSH levels actually decrease during the follicular phase because follicular cells start to produce inhibin, which acts in a negative feedback manner to decrease FSH release.

Ovulatory Phase (Day 15)

The ovulatory phase is marked by the release of a secondary oocyte from the ovary. By day 14, increased estrogen alters hypothalamic neurons from responding in a negative feedback loop to a positive feedback loop. By day 15, the rapidly increasing estrogen blood levels increase GnRH release (a positive feedback effect). This rapidly increases estrogen production, which, in turn, increases LH and FSH secretion. The increasing LH stimulates follicular cells to secrete enzymes (like hyaluronidase) that weaken ovarian connective tissue, although at the same time stimulating follicular cells to increase the secretion of fluid into the maturing follicle (Figure 11.17). LH and FSH levels peak at day 15 and initiate ovulation. Although LH and FSH levels are similar from days 1 through 14, LH peaks earlier and to a significantly greater level than FSH. The LH surge also stimulates the oocyte to complete the first meiotic division just before or during ovulation.

Luteal Phase (Days 16 to 28)

The luteal phase is characterized by the formation of the corpus luteum and the preparation of the uterus for a fertilized oocyte. Shortly after ovulation, there is a rapid decrease in LH and FSH levels that, in turn, decreases estrogen levels. This occurs because the hypothalamic neurons now respond in a negative feedback manner to estrogen once again. Progesterone levels increase as follicular cells are transformed into the corpus luteum. The corpus luteum secretes both progesterone and estrogen, although estrogen levels are lower than in the follicular phase. Progesterone and estrogen act in a negative feedback manner on the hypothalamus to inhibit GnRH release. The decrease in GnRH release during the luteal phase ensures that no other follicles undergo development because of decreased levels of LH and FSH.

During the luteal phase, increasing blood estrogen and progesterone act on the uterus to cause the growth of the functional endometrium. The functional endometrium grows rapidly and prepares for the possibility of a fertilized egg. In addition, there is a rapid growth of blood vessels (angiogenesis) in the functional endometrium.

If fertilization occurs, the developing embryo secretes human chorionic gonadotropin (hCG), which prevents the corpus luteum from degenerating. As a result, blood estrogen and progesterone levels do not decrease, and menses does not occur. hCG has a nearly identical structure to LH and can bind to the LH receptor.

If fertilization does not occur, the corpus luteum degenerates. On day 25 or 26 of the menstrual cycle, the degeneration of the corpus luteum causes blood estrogen and progesterone levels to decline, and causes the functional endometrial cells to die and be sloughed off, together with blood vessels. The resulting mass of dead endometrial cells and clotted blood is lost through the vagina in the menses, which starts the cycle over again.

Uterine Cycle

The hormonal changes that occur during the menstrual cycle also affect the uterus. The changes that occur in the uterus during the menstrual cycle are known as the *uterine cycle*. Hormonal changes also affect the breasts and vagina but not to the extent that they affect the uterus (Figure 10.18).

During the follicular phase of the menstrual cycle, increasing blood concentrations of estrogen stimulate the growth of the functional endometrium, stimulate the growth of the myometrium (but to a lesser extent), stimulate growth of new blood vessels within the functional endometrium, upregulate progesterone receptors within the myometrium, and increase the irritability of uterine smooth muscle cells. After the previous menses, estrogen stimulates the remaining epithelial cells within the functional endometrium to divide and form cuboidal epithelial cells. The cuboidal epithelium differentiates into columnar epithelium, eventually folding to form tubular spiral glands. The dividing epithelium secretes angiogenesis factors that stimulate blood vessels, the spiral arteries, to grow and supply nutrients to the endometrial cells.

After ovulation, increasing progesterone levels also stimulate the growth of the functional endometrium and it thickens; the spiral glands develop further and begin to secrete small amounts of a fluid rich in glycogen. Approximately 7 days after ovulation, or day 21 of the menstrual cycle, the functional endometrium is ready to receive the fertilized oocyte. If the fertilized oocyte arrives in the uterus too early, or too late, the endometrium cannot provide a suitable environment.

During the luteal phase, increasing progesterone levels act to inhibit uterine smooth muscle contractions. If fertilization does not occur by day 25 or 26 of the menstrual cycle, the corpus luteum degenerates and blood estrogen and progesterone levels fall. As a result, the functional endometrium begins to degenerate, smooth muscle within the walls of the spiral arteries begin to constrict, reducing blood flow to the functional endometrium. The resulting ischemia causes the destruction of all but the basal functional endometrium. As the cells become necrotic, they are sloughed off. The menstrual fluid is made up of sloughed cells, mucous secretions, and a small amount of clotted blood released from the spiral arteries. The myometrium contracts in response to progesterone and prostaglandins, and some women experience cramps. Menstrual fluid is expelled through the vagina.

It is normal for menstrual bleeding to last up to 7 days. Abnormal bleeding occurs when bleeding lasts longer than normal, is heavier than normal, or when bleeding patterns change. Abnormal bleeding can occur when the menstrual period is not regular. There are many causes of abnormal bleeding: past or present illnesses, certain medications, birth control pills, exercise, hormonal imbalances, endometrial hyperplasia, and stress. Endometrial hyperplasia is caused by accelerated growth of the endometrial layer but is not cancerous.

Summary of Hormone Interactions and the Female Reproductive Cycle

In summary, the female reproductive cycle is dependent on cyclic changes in blood levels of LH, FSH, estrogen, and progesterone. These cyclic changes are dependent on the hypothalamic secretion of GnRH and stimulation of the pituitary to secrete LH and FSH. LH and FSH stimulate the ovaries to produce estrogen and progesterone, and oogenesis.

Female Sexual Response

The female sex drive, like the male sex drive, depends, in part, on hormone secretion. Blood levels of estrogen, and androgens (produced by the liver) affect brain neurons, especially in the hypothalamus, and influence sexual behavior. The female sex drive is also influenced by psychological factors, emotions, memory, and other sensory input (odors, sounds, etc.). Postmenopausal women (with the ovaries removed) often experience an increase in sex drive because they can no longer become pregnant.

The neural pathways that control sexual behavior are the same in both men and women. Female genital sensory information is carried to the spinal cord sacral region, where sexual reflexes are integrated. Ascending afferent pathways are carried by the spinothalamic tracts to the brain, and descending pathways pass through the sacral region of the spinal cord. Motoneurons innervate reproductive organs through both parasympathetic and sympathetic nerves.

During sexual excitement, parasympathetic stimulation causes erectile tissue within the clitoris, nipples, and around the vaginal opening to become engorged with blood, making them more sensitive to physical contact. The vestibule mucous glands secrete small amounts of mucous during sexual excitement, and the vaginal walls secrete large amounts of a mucous-like fluid, even though the vaginal wall does not have well-defined mucous glands. These secretions allow for easy penis penetration and movement during sexual intercourse. During sexual intercourse, tactile stimulation of the female's genitals together with psychological stimuli can trigger an orgasm. During an orgasm, the vaginal, uterine, and perineal muscles contract rhythmically, and muscle tension increases throughout the body. After orgasm, women experience an overall sensation of relaxation and satisfaction. In contrast with men, women can be continuously stimulated and achieve multiple orgasms. Although an orgasm is a pleasure component of the sexual act, it is not necessary for a female to achieve orgasm for fertilization to occur.

Disorders of the Female Reproductive System

The female reproductive cycle can be disrupted by weight loss, low body weight, anorexia and bulimia, and vigorous exercise. These factors alter the normal menstrual cycle or prevent its occurrence.

Dysmenorrhea is painful menstruation caused by strong myometrial contractions that occur before and during menstruation. Most uterine contractions are never noticed, but strong ones are painful. During strong contractions, the uterus may contract too strongly or too frequently, causing the blood supply to the uterus to be temporarily cut off. This deprives the muscle of O_2, causing pain. In addition to painful uterine cramping with menses, women with dysmenorrhea may experience nausea, vomiting, diarrhea, headaches, weakness, and/or fainting. Dysmenorrhea can be an incapacitating problem, causing significant disruption in a woman's life each month.

Therapies for dysmenorrhea include rest, applying a heating pad to the lower abdomen or back, nutrition, aerobic exercise, and medication. Nutrition therapy includes a well-balanced diet with an adequate intake of Ca and a fluid intake of 2 quarts of water each day. Medication for dysmenorrhea usually involves over-the-counter, nonsteroidal anti-inflammatory drugs such as aspirin, ibuprofen, or naproxen. Hormonal alteration of the menstrual cycle is usually accomplished by taking oral contraceptives to prevent ovulation, decrease endometrium thickness, and prevent prostaglandin production.

Amenorrhea occurs when a woman of childbearing age fails to menstruate. The normal menstrual cycle ranges from 23 to 35 days. There are two types of amenorrhea: primary amenorrhea, also called *delayed menarche*; and secondary amenorrhea.

Primary amenorrhea occurs when a woman has not had her first menstrual period (menarche) by age 16. Primary amenorrhea is most often due to late puberty, which is common in teenage girls who are very thin or very athletic. These young women are typically underweight and their bodies do not have the normal puberty-related increase in body fat that is required to trigger the beginning of menstrual cycling.

Secondary amenorrhea happens when a woman who has menstruated previously fails to menstruate for 3 months. Secondary amenorrhea can be caused by pregnancy (the most common cause), breast-feeding, menopause, menopause before age 40, hysterectomy (surgical removal of the uterus), and stopping birth-control pills. Other factors include emotional or physical stress, rapid weight loss, obesity, frequent strenuous exercise, chronic illness, cancer chemotherapy, and ovarian tumors or cysts.

Premenstrual syndrome (PMS) is a disorder characterized by a set of hormonal changes that trigger disruptive symptoms in women for up to 2 weeks before menstruation. Of the estimated 40 million suffers, more than 5 million require medical treatment for marked mood and behavioral changes. Often symptoms tend to taper off with menstruation and women remain symptom-free until the 2 weeks or so before the next menstrual period. These regularly recurring symptoms from ovulation until menses typify PMS.

More than 150 symptoms have been attributed to PMS. Symptoms can vary from month to month, and some women may even have symptom-free months. Symptoms can be both physical and emotional. Physical symptoms can include headache, migraine, fluid retention (bloating), fatigue, constipation, painful joints, backache, abdominal cramping, heart palpitations, and weight gain. Emotional and behavioral changes can include anxiety, depression, irritability, panic attacks, tension, lack of co-ordination, decreased work or social performance, and altered libido.

Depending on individual symptoms and their severity, patients can manage PMS by eating six small meals, high in complex carbohydrates and low in simple sugars, at regular, 3-hour intervals. This helps to maintain a steady blood glucose level and avoids energy highs and lows. Substantially reducing and eliminating use of caffeine, alcohol, salt, fats, and simple sugars can reduce bloating, fatigue, tension, and depression. Taking daily supplemental vitamins and minerals including B_6, B-complex, Mg^{2+}, vitamin E, and vitamin C have been shown to reduce symptoms. Exercising can reduce stress and tension, elevate mood, provide a sense of well-being, and improve blood circulation by increasing natural production of β-endorphins.

Breast Cancer

Breast cancer is the second most common type of cancer among women in the United States after skin cancer and is the second leading cause of cancer deaths after lung cancer (10.19). A female born today has a 1 in 8 chance of being diagnosed with breast cancer sometime during her life.

A large number of risk factors are associated with breast cancer. A woman older than age 60 is at greatest risk, and the disease is uncommon before menopause. Other risk factors include previous cancer in one breast; breast cancer in a mother, sister, or daughter before the age of 40; the presence of abnormal cells within the breasts; genetic mutations in certain genes (BRAC1, BRAC2, and others); a history of abnormal reproductive and menstrual cycles; becoming pregnant late in life or never being pregnant; the age of the first menstruation; beginning menopause after 55; estrogen-replacement therapy after menopause; racial differences; radiation exposure; the amount of dense (not fatty) tissue within the breast; lack of physical activity; and/or obesity.

Breast cancer is detected by mammograms; breast self-examinations, and biopsies. Mammograms are considered to the best tool for early detection of breast cancer. Screening mammograms can often show a breast lump before it can be felt. They also can show a cluster of very tiny specks of calcium called *microcalcifications*. Lumps or specks can be signs of cancer. However, mammograms may miss existing cancers (false negative) or detect tissue that is not cancerous (false positive), and some fast-growing tumors may already have spread to other parts of the body before a mammogram detects them.

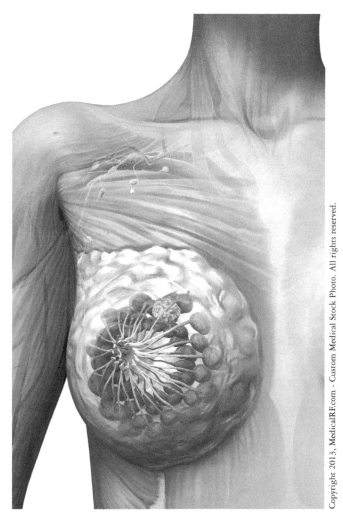

FIGURE 10.19

Human breast cancer

During a clinical breast examination, the breasts are examined although a woman is standing, sitting up, or lying down. Physicians look for differences between breasts, including unusual differences in size or shape. The skin of each breast is checked for a rash, dimpling, or other abnormal signs. During a self-breast examination, women should examine their breasts in a mirror to look for differences in size, shape, and color with their arms at their sides. Then they should examine their breasts with their arm raised above their head. The nipples should also be examined for the presence of a discharge or blood. Woman should also examine their breasts for lumps although lying down and sitting up. Routine self-breast examinations and clinical breast examinations by a doctor have been shown to be an effective way to detect breast cancer. Breast cancer can change how the breast or nipple feels; cause the formation of a lump or thickening in or near the breast or in the underarm area; change breast or nipple shape; change breast size; cause formation of a rash on the breast skin, areola, or nipple; or cause a nipple discharge.

If breast cancer is suspected, definitive proof is provided by biopsy. During a fine-needle biopsy, a small needle is inserted into the suspected tumor, and a few cells are removed and examined histologically for presence of cancerous cells. Use of a larger needle in a core biopsy allows the physician to remove more breast tissues. In an incisional biopsy, a surgeon removes a sample of a lump or abnormal area. In an excisional biopsy, the surgeon removes the entire lump or abnormal area. After a definitive diagnosis of breast cancer has been made, additional tests are often performed. Many breast cancer cells require estrogen or progesterone to grow, and the presence of a receptor for hEGF (human epidermal growth factor) increases the chance that the cancer will come back.

Women with breast cancer have many treatment options, including surgery, chemotherapy, radiation therapy, hormonal therapy, and biological therapy. Surgery is the most common treatment for breast cancer. An operation to remove the cancer but not the breast is known as a *partial mastectomy*. The underarm lymph nodes are often also removed because of the high risk that breast cancer cells have migrated there. A complete mastectomy removes the entire breast.

Radiation therapy uses high-energy rays to kill cancer cells. It generally follows breast-sparing surgery. Sometimes, depending on the size of the tumor and other factors, radiation therapy also is used after mastectomy. The radiation destroys breast cancer cells that may remain in the area. Chemotherapy for breast cancer is usually a combination of drugs. Hormonal therapy, such as tamoxifen, an antagonist of the estrogen receptor, keeps cancer cells from getting the natural hormones (estrogen and progesterone) they need to grow. Biological therapy uses the body's natural ability (immune system) to fight cancer.

Causes of Female Infertility

There are many causes of female infertility, including but not limited to uterine tube malformation, decreased pituitary or ovary hormone secretion, and problems with zygote implantation in the uterus. One possible consequence of pelvic inflammatory disease, caused by many different infections, is the formation of adhesions that can block one or both of the uterine tubes. The formation of these adhesions is a common cause of female infertility.

Decreased ovulation is commonly a result of decreased pituitary secretion of LH and FSH. Decreased LH and FSH secretion can occur as a result of hypothyroidism, hypothalamic trauma, infarctions of hypothalamic or pituitary blood vessels, and tumors. Scarring of the uterine tube by pelvic inflammatory disease can reduce the transport of oocyte to the uterus. Physicians can use gamete intrauterine tube transfer to move oocytes from the ovary to the uterine tubes in a position where they can be fertilized.

Zygote implantation can be affected by uterine tumors or abnormal hormone secretion. Premature corpus luteum degradation will cause progesterone levels to decrease and the functional endometrium will degenerate. The loss of the functional endometrium will cause the developing zygote to degenerate. Secondary amenorrhea will also cause implantation problems.

Endometriosis, the presence of endometrium in the ovaries, uterine tubes, and abdominal cavity, is a common cause of infertility. Endometriosis is thought to be caused by the movement of uterine endometrial cells through the uterine tube to the abdominal cavity. The endometrial cells then invade the peritoneum and grow in the presence of estrogen and progesterone outside the uterus. The result is a monthly inflammation of the areas outside the uterus where endometrial cells are found. Endometriosis is very painful and can cause infertility.

In Review...

10.21	What are the typical changes that occur in girls during puberty?
10.22	What is a menstrual cycle? What hormonal changes are responsible for the control of the menstrual cycle?
10.23	What events occur during the follicular phase of the menstrual cycle? During the ovulatory phase? During the luteal phase?
10.24	What is the role of hCG in the menstrual cycle?
10.25	What changes occur during the uterine cycle?
10.26	What events occur during female sexual excitement?
10.27	What factors influence the formation of a female orgasm?
10.28	What is oogenesis? How does it compare with spermatogenesis?
10.29	What occurs during ovulation?
10.30	What is the role of the corpus luteum?

Clinical Perspective

Sexually transmitted diseases (STDs) are infections that are spread through sexual contact. STDs include bacterial diseases, parasite infestation, and viral diseases.

Bacterial Diseases

Nongonococcal urethritis (NGU) is any urethral infection not caused by gonorrhea. NGU accounts for approximately two-thirds of urethral infections. *Chlamydia trachomatis* and *Ureaplasma urealyticum* are the most common organisms that cause NGU. Other bacteria can cause the infection, especially if foreign objects are placed into the urethra. Patients who require frequent catheterizations to remove urine also frequently experience development of urethritis from other types of bacteria. NGU can be acquired through sexual contact.

Chlamydia is a common and curable infection caused by the bacterium *C. trachomatis* (Figure 10.20). The bacteria target the cells of the mucous membranes lining the surfaces of the urethra, vagina, cervix, endometrium, uterine tubes, anus, rectum, eyelid, and less commonly, the throat. In the United States, chlamydia is the most common bacterial

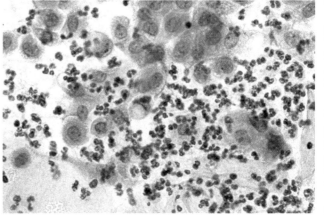

FIGURE 10.20

Chlamydia trachomatis in a cervical smear

FIGURE 10.21

Electron micrograph of *Treponema pallidum* bacteria on cultures of epithelium cells

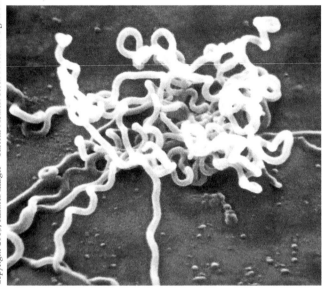

STD, particularly among sexually active adolescents and young adults. Chlamydia is passed primarily when mucous membranes encounter the mucous membrane secretions or semen of an infected person. Approximately 75% of women and 50% of men do not experience symptoms. Symptoms develop within 1 to 3 weeks after exposure to chlamydia, but the symptoms of chlamydia are similar to the symptoms of gonorrhea, and the two infections are often confused.

Chancroid is a highly contagious yet curable STD caused by the bacterium *Haemophilus ducreyi*. Chancroid usually causes genital ulcers, swollen and painful lymph glands, or groin inguinal buboes. Left untreated, chancroid may facilitate the transmission of HIV.

Syphilis is a curable infection caused by the bacterium *Treponema pallidum* (Figure 10.21). The bacteria enter the body through mucous membranes or abraded skin. Once inside the body, syphilis enters the bloodstream and attaches to cells, damaging organs over time. There are four stages through which untreated syphilis progresses, each stage with its own unique signs and symptoms: primary, secondary, latent, and tertiary (or late).

The primary stage of syphilis is usually marked by the appearance of a single sore, known as a *chancre*, within 10 to 90 days after contact with the bacteria at the site of infection. Chancres may be found outside the genitals, including the penis, scrotum, and vagina; inside the vagina or rectum; at or around the anus; or on the lips or in the mouth. The chancre will go away with or without treatment. Without treatment, the person will still have syphilis and can transmit it to others.

The secondary stage of syphilis can develop 17 days to 6 1/2 months after infection. Symptoms can last from 2 to 6 weeks. Symptoms can include a rough, reddish brown rash that appears on the palms of the hands or the soles of the feet that normally does not itch. Rashes on other parts of the body are also common, including the neck, head, and torso. Symptoms of secondary syphilis will clear up with or without treatment, but the disease will still be present if untreated. It will then enter into a latent stage, which has no signs or symptoms.

Latent syphilis is defined as the time where there are no signs or symptoms of the disease. It typically develops from 2 to 30 years after infection. Because there are no signs or symptoms, the only way to test for infection during the latent period is by blood test. A relapse of secondary syphilis can occur once the disease has entered the latent stage. This normally will happen during the first 2 years of latency.

Symptoms of late-stage or tertiary syphilis can occur 2 to 30 years after infection. Complications during this stage can include gummas, small bumps or tumors that can develop on the skin, bones, liver, or any other organ. Many people also suffer from cardiovascular dysfunction or chronic nervous system disorders, such as blindness, insanity, and paralysis. A mother infected with syphilis can pass the disease to her unborn child, either during pregnancy or in childbirth. A newborn infected in this manner has congenital syphilis.

Vaginitis is a name for swelling, itching, burning, or infection in the vagina that can be caused by several different germs. The most common kinds of vaginitis are bacterial vaginitis and yeast vaginitis. Bacterial vaginosis is the most common cause of vaginitis, accounting for 50% of cases. Bacterial vaginosis is caused by a change in the normal vaginal bacteria and overgrowth with organisms such as *Gardnerella vaginalis*. Risk factors include pregnancy, intrauterine device use, and frequent douching. It is associated with sexual activity, possibly a new sexual partner, or multiple sexual partners.

Vaginal yeast infections are caused by the fungus *Candida albicans*. Yeast infection can spread to other parts of the body including skin, mucous membranes, heart valves, esophagus, and other areas. It can cause life-threatening systemic infections, mostly in people with weakened immune defenses (such as pregnant women and people who are HIV positive, have diabetes, or are taking steroids).

Gonorrhea is a curable STD caused by the bacterium *Neisseria gonorrhoeae*. These bacteria attach to epithelial cells and can infect the genital tract, the mouth, and the rectum. In women, the opening to the uterus, the cervix, is the first place of infection. The disease, however, can spread into the uterus and Fallopian tubes, resulting in pelvic inflammatory disease. Gonorrhea is spread during sexual intercourse. Infected women also can pass gonorrhea to their newborn infants during delivery, causing eye infections in their babies. This complication is rare because newborn babies receive eye medicine to prevent this infection. Gonorrhea causes symptoms of painful urination and discharge of pus from the urethra. Symptoms appear within a few days to a week. Recovery can occur without complications, but when complications do occur, they can be serious.

Parasites

Trichomoniasis is a sexually transmitted disease caused by the parasite *Trichomonas vaginalis*. Trichomoniasis is primarily an infection of the urinary and genital tracts. For women, the vagina is the most common site of infection. For men, the urethra is most commonly affected.

Scabies, often called *mites*, is an infestation of the top layer of skin caused by the parasite *Sarcoptes scabiei*. The female parasite burrows under the skin, begins laying eggs within a few hours of infection, and continues to lay two to three eggs daily. It takes approximately 10 days for the eggs to hatch and become adult mites. At this point, the cycle will begin again. Scabies are transmitted through close physical contact.

Pubic lice, or *crabs*, are crab-like parasites, *Pthirus pubis*, that attach themselves to pubic hair and other coarse body hair. Although crabs need blood to survive, they can live up to 24 hours off a human body. Crabs have three very distinct phases: the first phase is the egg, or nit; the second phase is the nymph, or baby form of the adult; and the third phase is the louse, or adult crab. Pubic lice are treated with a shampoo that kills the lice and their eggs.

Viral Diseases

Herpes simplex is a common and usually mild recurrent skin condition caused by either herpes simplex type 1 or herpes simplex type 2 viruses. As with all viruses, there is no cure for herpes. It can cause cold sores or fever blisters on the mouth or face, and in this form is known as *oral herpes*. It can also cause similar symptoms in the genital area, and in this form is known as *genital herpes*. Herpes differs from other common viral infections because it can live in the body over a lifetime. There may be no symptoms or periodic symptoms. The virus can travel the nerve pathways in a part of the body and reside in the nerve roots for long periods before symptoms appear.

In the United States, *human papilloma virus* (HPV), a family of viruses, is the most common STD. Some studies estimate that the majority of the sexually active population is exposed to at least one or more types of HPV, although most do not develop genital warts. Because HPV is so common and prevalent, a person does not need have to have many sexual partners to encounter this virus.

Clinical–In Context Levonorgestrel Releasing IUDs

Intra-uterine devices (IUDs) have been popular contraceptive methods for almost 50 years. Their advantages are that they are inserted once and last 5 to 10 years with no compliance issues. They are also reversible, until tubal ligation. Original IUDs were copper-based and they prevented implantation of the blastocyst in the uterine wall, but had numerous side effects and associated complications. More recently, the introduction of levonorgestrel-releasing (Mirene™) IUDs has led to the wider adoption of IUDs.

Mirene IUDs are T-shaped plastic tubes that release 20-μg/day of levonorgestrel, a progesterone-like steroid, in contrast to the 150-μg/day of steroids released in oral contraceptives. The released steroids are concentrated in the uterus, with very little reaching the blood. After 5 years, the Mirene IUD still releases 14-μg/day levonorgestrel. Levonorgestrel is very effective in preventing pregnancy (99.8% for the first year ad 99.7% after 5 years).

The *molluscum contagiosum virus* (MCV) can cause the formation of small lesions or bumps on the skin. MCV is generally a benign infection and symptoms may self-resolve. MCV was once a disease primarily of children, but it has evolved to become an STD in adults. It is believed to be a member of the poxvirus family. Molluscum contagiosum can be transmitted sexually by skin-to-skin contact, contact with lesions, or both. Transmission through sexual contact is the most common form of transmission for adults. MCV can be transmitted from inanimate objects such as towels and clothing that encounter the lesions. MCV transmission has been associated with swimming pools and sharing baths with an infected person. MCV also can be transmitted by autoinoculation, such as touching a lesion and then touching another part of the body.

AIDS is caused by infection with human immunodeficiency virus HIV, resulting in the destruction of lymphocytes (see chapter 21). HIV is transmitted by sharing needles during administration of illicit drugs, through unprotected anal or vagina sex, and through blood transfusions.

Contraceptive Methods

Many methods are currently used to prevent or terminate a pregnancy. Contraceptive methods prevent fertilization; IUDs (intrauterine devices) prevent zygote implantation into the uterine endometrium; and abortion removes the implanted fetus or zygote. Table 10.2 shows the relative efficacy of each method.

Behavioral Methods

Changes in human behavior can dramatically affect sexual conduct and the chance of pregnancy. The most effective way to prevent pregnancy is abstinence, to refrain from sexual intercourse. Abstinence is not a popular method.

The unreliable coitus interruptus relies on penis removal from the vagina just before ejaculation. The penis must be withdrawn at exactly the right time, before any sperm have been deposited into the vagina, and functional sperm are found in pre-ejaculatory emissions.

Table 10.2 ♦ Failure Rates of Birth Control Methods

Method	Failure Rate* Perfect use[1]	Typical use[2]
None	85%	85%
Abstinence	0%	0%
Vasectomy	0.1%	0.15%
Tubal ligation	0.5%	0.5%
Oral Contraceptives	0.1%	3%
Norplant	0,3%	0.3%
Depo Provera	0.05%	0.05%
IUD-Copper	0.6%	0.8%
IUD-Mirene	0.2%	0.2%
Condom	3%	14%
Diaphragm	6%	20%
Spermicides	6%	26%
Rhythm Method	9%	25%

*Defined as the percentage of women having an unintended pregnancy during the first year of use.
[1] Failure rate when the method is used consistently and correctly.
[2] Includes couples that forget to use the method.

The rhythm method requires abstinence from sexual intercourse for 5 days before ovulation and several days after. This method can be successful if a woman can accurately determine the time of ovulation and whether her partner's sperm last beyond 5 days after ejaculation into the vagina. This method has a low level of reliability.

One of the easiest and most effective ways of preventing pregnancy is to put a barrier in the way, physically stopping the sperm from reaching the egg. Several barrier methods are available. Male and female condoms are probably the best-known barriers, however, the female diaphragm and cap are also available. Male and female condoms are made from very thin rubber (latex) or plastic (polyurethane). Male condoms fit over the erect penis and should always be used before any close genital contact happens. Used properly, they are 98% effective. Female condoms fit inside the vagina, with the open end resting just outside. They are slightly less reliable, at 95%. The female condom consists of a polyurethane sheath attached to a ring of latex that covers the entire vulva, as well as the vagina. Because the sheath is not made of latex, it can be used with a water-based or oil-based lubricant. It has proved effective at preventing the transmission of HIV.

Unlike some other forms of contraception, condoms do not need to be specially fitted and can they can be used without spermicide cream that kills sperm, although some condoms are available with spermicide. They also offer some protection against STDs and HIV. The obvious disadvantage is that they can be damaged if not used carefully.

Diaphragms and caps come in different shapes and sizes (caps are smaller than diaphragms), and are made from silicone or rubber. Because women also come in different shapes, diaphragms have to be fitted to the exact size. Once in place, they fit inside the vagina, covering the cervix. Used in conjunction with a spermicide, diaphragms and caps are about 92 to 98% effective. The diaphragm can be inserted several hours before intercourse. It should be left in place for at least 6 hours after sex to provide complete immobilization of the sperm. The failure rate for the diaphragm is about 5.5%. Disadvantages include an increased risk for bladder infection and cystitis. It also has been linked to toxic shock syndrome, a rare, serious illness caused by a bacterial toxin. A cervical cap is smaller and tighter fitting than a diaphragm, covering only the cervix. It is harder to fit and takes longer to learn how to use, but it can be left in place for up to 72 hours.

An intrauterine device (IUD) is a small piece of metal or plastic that is inserted into the uterus through the cervix. It is not clear how IUDs prevent pregnancy but it is thought that they affect sperm movement, fertilization, or implantation (or a combination of all three). IUDs can be used for as long as 10 years before replacement. IUDs have been shown to increase the risk for pelvic inflammatory disease, ectopic pregnancies (pregnancies outside the uterus), heavy menstruation, irregular menstrual bleeding, and pelvic pain.

Vaginal spermicides are chemicals that kill sperm on contact. Vaginal spermicides are typically composed of a spermicide mixed with a cream, jelly, or foam base. Some spermicides are sold as vaginal suppositories, which are solid at room temperature but melt into liquid at body temperature. Spermicides are only 70% to 80% effective when used alone. They are designed to be used with a barrier type of contraception. It is not clear whether they increase the effectiveness of these methods of contraception, although they do serve as a lubricant, making sex more enjoyable and decreasing the risk that a condom will tear from too much friction. The active ingredient in spermicide preparations is non-oxynol-9, otherwise known as *octoxynol*.

One of the most popular methods of hormonal birth cotrol is the oral contraceptive, or *birth control pill*. Most oral cotraceptives are combination pills containing both estrogen and progestin. Synthetic hormones can also be given by injection. Injected depot-medroxyprogesterone acetate (DMPA) provides protection against pregnancy for 3 months.

Synthetic hormones can also be given through a vaginal ring. A vaginal ring is a flexible, plastic ring that is placed in the upper vagina. The ring releases both estrogen and progestin continuously to prevent pregnancy. It is worn for 21 days, removed for 7 days, and then a new ring is inserted. A contraceptive skin patch is a small adhesive patch that is worn on the skin to prevent pregnancy. It is a weekly method of hormonal birth control.

Synthetic estrogen and progesterone act to inhibit GnRH release by a negative feedback process. Inhibition of GnRH will prevent the LH and FSH surge that is responsible for ovulation. The health risks of using oral contraceptives are much less than the risks of pregnancy and childbearing for almost all women, especially in countries with high maternal mortality rates. Even where maternal mortality is low, pill use is safer than childbearing, except for older women who smoke or have high blood pressure. A major finding of the last decade is the increased risk for heart attack and stroke for older oral contraceptive users with hypertension. For oral contraceptive users who do not smoke and do not have high blood pressure, however, the low doses in today's pills appear to minimize these risks. The major established health risks of oral contraceptives are certain circulatory system diseases, particularly heart attack, stroke, and venous thromboembolism. Other health risks include gallbladder disease in women already susceptible to it and rare, noncancerous liver tumors.

Norplant consists of small silicone capsules filled with levonorgestrel, a synthetic progesterone-like steroid. The silicone capsules were placed under the skin and the levonorgestrel diffused slowly into the blood blocking GnRH release for up to five years. Norplant would also prevent menstruation. Norplant is no longer available in the United States.

Mifepristone (RU-486) is a progesterone antagonist. It is given as a morning-after pill because it can prevent the proliferation of the functional endometrium, and causes the existing endometrium to slough and prevent zygote implantation. It is only about 75% effective in preventing pregnancy.

Vasectomy is a used to make male individuals incapable of fertilization without affecting sexual intercourse. During a vasectomy, the ductus deferens is cut and tied off to prevent the movement of sperm into the epididymis and becoming part of the ejaculate. Since the testes and epididymis provide such a small part of the ejaculate, a vasectomy has little effect on ejaculate volume.

In women, a tubal ligation, cutting and tying off the uterine duct, is an effective surgical technique to make women incapable of becoming pregnant. This procedure blocks sperm from reaching the zygote. The tubal ligation is commonly performed by laparoscopy, in which a special instrument is inserted into the abdomen through the belly button.

The most controversial surgical method to terminate existing pregnancies, instead of preventing their formation, is by abortion. Abortion is performed by scraping the endometrial surface and removing the zygote by suction or by insertion of a special solution into the uterus, which terminates the pregnancy.

Systems Overview: Function, Communication, and Homeostasis

The reproductive system is critical for the continuation of the human species. It is a complex collection of tissues and organs that must work in unison for fertilization to occur and for a fetus to develop normally. Because of its complexity, the reproductive system is affected by, and affects, many other organs and tissues throughout the body.

	Cell-to-Cell Communication	Functional Interaction	Homeostasis
Cardiovascular	During pregnancy, increased demands for maternal blood result in major changes in cardiovascular function; changes are thought to be due to sex steroids; there are also cardiovascular changes during sexual intercourse	Maintenance of sufficient blood pressure	Homeostatic regulation of maternal blood pressure and exchange with fetal tissues
Endocrine	The structure and function of the reproductive organs are under the direct role of the endocrine system.	Maintenance of egg and sperm production	Homeostatic regulation of functional sperm and eggs; regulation of the all the components involved in fertilization and fetal development
Gastrointestinal	During pregnancy, increasing hormone levels affect maternal digestion and energy utilization.	Maintenance of sufficient energy reserves for fetal development during pregnancy	Homeostatic regulation of fetal energy requirements
Integumentary	Blood flow to the skin and tissues associated with the reproductive organs changes during sexual excitement.	Stimulation of the penis and vagina is required for the sexual act	Activation of tissues that bring sperm and egg together for fertilization
Lymphatic/Immune	Sperm production must occur in a compartment separated from the immune system because haploid sperm would be recognized as nonself.	Preventing the attack of the immune system on sperm production	Maintenance of normal sperm production
Muscular	The production of skeletal muscle is under the regulation of sex steroids; the cause of part of the physical differences between men and women.	Maintenance of normal skeletal muscle activity	Homeostatic regulation of skeletal muscle development
Nervous	The nervous system plays a critical role in the production of ova in female individuals and the ability of male individuals to perform the sexual act.	Maintenance of fertility and the steps required in the sexual act	Homeostatic regulation of fertilization
Respiratory	The respiratory system changes dramatically during pregnancy and sex.	Maintenance of sufficient oxygenation of blood during pregnancy	Homeostatic regulation of respiration during pregnancy
Skeletal	Sex steroids affect bone growth and density.	Maintenance of sufficient skeletal strength	Homeostatic regulation of bone density by sex steroids
Urinary	Part of the urinary system in male individuals is required for the movement of sperm out of the body; sex hormones affect the normal functioning of the ureter and urethra.	Maintenance of the tissues required for male individuals to ejaculate	Homeostatic regulation of fertilization

chapter review

Factual
(to define, list, state and name)

1. The role of the corpus luteum is to…
   a. produce hCG.
   b. provide the basis for the placenta.
   c. produce estrogen and progesterone during the follicular phase.
   d. support pregnancy by producing hormones during the last 7 months.
   e. None of the above.

2. Testosterone…
   a. stimulates oogenesis.
   b. induces male secondary sex characteristics in women.
   c. is never found in the blood of women.
   d. is only found in the blood of men.
   e. is found in high concentrations in female blood.

3. Sperm…
   a. do not arise by meiosis.
   b. are produced before puberty.
   c. is one of many human cells with a tail.
   d. are stored within the seminiferous tubule.
   e. have only 23 chromosomes.

4. The blood:testes barrier…
   a. is mediated by seminiferous tubule cells.
   b. is important to increase the testes levels of estrogen.
   c. prevents the immune system from destroying spermatogonia.
   d. is formed by Leydig cells.
   e. is breached by sperm cells as they enter the urethra

5. The primordial follicle contains a(n)…
   a. graafian follicle.
   b. mature ovum.
   c. oogonium.
   d. primary oocyte.
   e. secondary oocyte.

Comprehension
(to explain, summarize, interpret and describe)

6. During the follicular phase of the ovarian cycle, which of the following does not occur?
   a. Granulosa cell growth
   b. Formation of thecal cells
   c. Increased secretion of estrogen
   d. Secretion of progesterone by follicular cells
   e. Secretion of FSH

7. Sperm leaves the body through which path?
   a. Epididymis→ejaculatory duct→ductus deferens→urethra
   b. Seminiferous tubule→ejaculatory duct→efferent ductules→urethra
   c. Seminiferous tubule→rete testes→epididymus→ductus deferens→urethra

d. Epididymus→seminiferous tubule→rete→ductus deferens→ejaculatory duct→urethra
   e. Ejaculatory duct→ductus deferens→rete→urethra

8. The ovulatory surge in the middle of the menstrual cycle is caused by…
   a. a complex relationship between GnRH, FSH, LH, and blood levels of progesterone.
   b. a weakening of the ovarian wall due to the phagocytic action of follicle cells.
   c. the negative-feedback inhibition of LH/FSH release by circulating levels of estrogen.
   d. the positive-feedback stimulation of LH/FSH release by circulating levels of estrogen.
   e. the positive-feedback stimulation of LH/FSH release by circulating levels of progesterone.

9. Completion of the second meiotic division of oogenesis occurs…
   a. before birth.
   b. every month after puberty and before menopause, at ovulation.
   c. every month after puberty, throughout the life of the female.
   d. only if fertilization occurs.

10. During the luteal phase of the ovarian cycle…
    a. granulosa cells grow rapidly.
    b. the ova undergo mitosis.
    c. the corpus luteum secretes progesterone and estrogen.
    d. several follicles are stimulated to divide and grow.
    e. there is decreasing blood levels of progesterone.

Application
(to use, apply, compute, solve, and predict)

11. The production of a functional ova or sperm requires which of the following steps?
    a. Mitosis
    b. Meiosis
    c. Differentiation of a cell into a new form
    d. Movement of the gamete to a different site where fertilization is likely to occur
    e. All of the above.

12. The increased blood levels of progesterone during the menstrual cycle act to…
    a. cause ovulation.
    b. inhibit estrogen production by the ovary.
    c. block release of GnRH in the hypothalamus.
    d. increase the production of estrogen by the ovary.
    e. stimulate the growth of the functional endometrium in the uterus.

13. Which one of the following is not a uterine physiological role?
    a. Mechanical protection to the fetus
    b. The smooth muscle (myometrium) provides contraction to promote parturition
    c. Its rich blood supply provides nutrients for the developing fetus
    d. It is the primary source of estrogen and progesterone during the last 7 months of pregnancy
    e. The endometrium feeds the developing fetus until the development of the placenta

14. Polyspermy is typically prevented by…
    a. a biochemical reaction between the surface of the sperm and follicular cells surrounding the egg.
    b. changes in the sperm flagella when it has moved through the zona pellucida.
    c. the fact that only 10,000 out of 200,000,000 sperm reach the uterine tube alive.
    d. a Na^+ mediated influx and depolarization of the ova membrane.
    e. the exclusion of any additional chromosomes greater than 23 into the ova.

15. All except one of the following are functions of the Sertoli cells. Which is the exception?
    a. Phagocytosis of defective sperm
    b. Produce Androgen Binding Protein
    c. Produce Inhibin
    d. Produce Mullerian Inhibiting Hormone (MIH)
    e. Produce Testosterone

Analysis
(to compare, contrast, and distinguish)

16. Which of the following statements is false?
    a. Both men and women produce testosterone.
    b. Testosterone induces male secondary sex characteristics.
    c. Testosterone induces and maintains differentiation of female accessory organs.
    d. Testosterone concentration is much higher inside the seminiferous tubule than in blood.
    e. Men are more aggressive than women, in part, because of their higher testosterone levels.

17. Which of the following statements is true?
    a. Primordial oocytes are arrested during meiosis I.
    b. After meiosis, an oocyte and polar body are formed.
    c. Women may be born with all of the ova that they will ever have.
    d. Men continuously produce sperm throughout their lives (if they continue to produce testosterone).
    e. All of the above.

18. During this portion of the development of functional gametes in women, there is an increase in the blood levels of progesterone.
    a. During stem cell selection
    b. Follicular phase
    c. Luteal phase
    d. Ovulatory phase
    e. Puberty

19. All except one of the following is caused by the positive feedback effect upon anterior pituitary production of LH (as in the previous question). Which is the exception?
    a. Completion of meiosis I
    b. Formation of the corpus luteum
    c. Mitosis of oogonia
    d. Ovulation
    e. Production of a secondary oocyte

20. Anterior pituitary production of Follicle Stimulating Hormone (FSH) is decreased by...
    a. inhibin from interstitial (Leydig) cells.
    b. inhibin from Sertoli cells.
    c. Luteinizing Hormone (LH) from granulosa cells.
    d. testosterone from Sertoli cells.

Synthesis
(to develop, create, propose, design and invent)

21. During the ovulatory phase of the menstrual cycle, a woman's body temperature spikes one degree higher than normal. Why might this be the case?

22. After a tubal ligation, would a woman's GnRH, LH, FSH, and estrogen levels change? Would she still ovulate?

23. If a prepubescent boy took estrogen treatments, would he develop as a woman during puberty (formation of breast, deposition of fat on the hips, change in voice, etc.)?

24. Is an oral birth control pill possible for men? If so, what step in spermatogenesis should this pill affect?

Evaluation
(to judge, justify, defend, criticize and evaluate)

25. A new male contraceptive drug downregulates LH receptors in the male testes. This drug would block production of spermatocytes by which mechanism?
    a. The testes levels of FSH would plummet.
    b. The levels of GnRH would be decreased.
    c. Interstitial cells would not produce testosterone.
    d. Inhibin release would be dramatically increased.
    e. Sustenacular cells would not secrete androgen binding protein, reducing testosterone within seminiferous tubules.

26. A woman carrying a fertilized egg in her body can be assured that she will not go through an additional menstrual cycle because...
    a. trophoblast cells secrete hCG, which directly inhibits the release of GnRH from the hypothalamus.
    b. once the fertilized egg reaches the blastocyst stage it secretes a powerful inhibitor of ovulation.
    c. trophoblast cells secrete hCG, which maintains blood levels of estrogen and progesterone at high levels.
    d. the corpus luteum secretes inhibin into the surrounding ovaries, which directly blocks further follicle development.
    e. zygotes secrete inhibin into the uterine fluid, which in turn, causes the release of hCG into the blood, which blocks further ovulation.

Answers to end-of-chapter questions

1. a	6. b	11. e	16. c	25. c
2. b	7. c	12. e	17. e	26. c
3. e	8. d	13. d	18. c	
4. c	9. d	14. d	19. c	
5. d	10. c	15. e	20. b	

21. Though this has not been conclusively proven, it is thought that the increased body temperature increases the likelihood of a successful fertilization of the egg by the sperm.

22. A tubal ligation would not disrupt the normal hormonal regulation of oogenesis, it merely prevents the oocyte from moving into the uterus, blocking fertilization. A woman with a tubal ligation would still ovulate, and still go through menstruation.

23. He already has a penis and the required internal male sex organs. Estrogen would not allow these organs to develop, and he would be sterile. The estrogen would change body fat deposition and would prevent the boy's larynx from changing and giving him a deeper voice.

24. Oral birth control pills would need to decrease spermatogenesis without decreasing the normal sex drive. The drug would have to specifically target spermatogenesis within the seminiferous tubules without affecting testosterone actions within the body.

Thought Questions

1. Spermatogenesis requires slightly decreased temperatures because at elevated temperatures, the number of non-functional sperm cells increases as does the likelihood of cancer.

2. It allows for a different environment around the developing sperm cells—elevated testosterone levels—and their normal production.

3. Sperm cells do not develop until after birth and after the immune system has differentiated between self and non-self antigens.

4. Skeletal muscle is under conscious control and its use around the ampulla would require sensory information about when ovulation has occurred to stimulate contractions to drive the oocyte toward the uterus.

5. The cells could down regulate the expression of the testosterone receptor.

6. Absolutely, which is why couples trying to conceive are advised to have the husband wear loose underwear that keeps the testes cool.

7. Smooth muscle can maintain contractions for long periods of time without fatigue (see chapter 10), making the birthing process much more energy efficient.

8. Because the oocytes were present when the immune system differentiated between self and non-self antigens.

9. The male and female reproductive systems develop from similar embryonic tissues and carry out similar roles—as during sexual excitation, so it is not surprising that they are regulated by similar neural pathways.

Chapter 11
Fetal Development

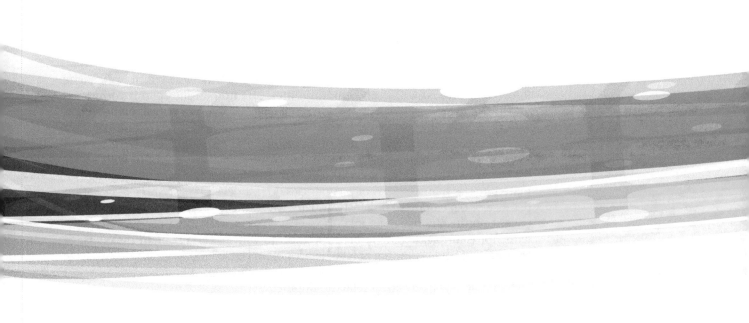

OBJECTIVES

Students will be able to:
- Outline the process of conception through pregnancy.
- Align pregnancy stages with fetal development.
- Identify the major life stages and events, and how the body changes with age.

From *Human Anatomy and Physiology, 2nd Edition* by Joseph F. Crivello. Copyright © 2013 by Kendall Hunt Publishing Company. Reprinted by permission.

KEY TERMS

allele	codominant	genotype	mesoderm	polyspermy
amniocentesis	diploid	germinal phase	morula	progeria
amniotic fluid	dominant	haploid	mutation	recessive
aneuploidy	ectoderm	heredity	neonate	senescence
Apgar score	embryonic phase	heterozygous	parturition	somites
autosomal chromosomes	endoderm	homozygous	pedigree	teratogen
blastocyst	fetal phase	karyotype	perivitelline	trophoblast
blastomere	fetus	lactation	phenotype	vitelline
chorion	gene	locus	placenta	zygote
chorionic villi	genetics	mesenchyme	pleiotropy	

Introduction

Human life begins with the fertilization of an oocyte by a sperm. The fertilized oocyte undergoes a rapid series of cell divisions and becomes a zygote. The zygote develops and becomes a fetus.

Fetal development is a remarkable process that is under the control of genetic information stored within the chromosomes. The human life span is considered the time between birth and death, but fetal development is a critical part of a person's existence. The processes and steps that occur during fetal development have an enormous impact on the rest of a person's life. Although fetal development occurs normally in most pregnancies, at least 3% of infants are born with severe birth defects that require medical treatment within the first year of life. It is estimated that as many as 50% of pregnancies do not develop past 8 weeks' gestation because of developmental defects, usually without the pregnant woman's knowledge. Later in life, many people discover previously unknown physiologic problems, for example, a genetic predisposition to cancer, certain brain disorders, or asthma. Our understanding of fetal development has led to identification of a large number of teratogens (G. *teras*, monster; *gen* producing), chemicals, and physical factors that negatively affect fetal development.

Improvements in nutrition and medical treatment have lengthened the life span of a child born today, from 47 years in 1900 to almost 80 years in 2010. In addition, the percentage of the United States population older than 65 years has increased from 5% in 1900 to 16% in 2010. Older people are healthier and more active than they ever have been.

This chapter begins with a discussion of fetal development. The chapter then discusses parturition, the newborn, lactation, the first year of life, life stages, aging, death, and genetics.

Embryonic Development

Embryonic development begins immediately after fertilization of the oocyte by sperm. It is followed after 8 weeks by fetal development and ends after 36 weeks with parturition (birthing). There are three major phases of development: (1) the germinal phase is the first 2 weeks of development, during which the primitive germinal layers are formed; (2) the embryonic phase is from the second to the eighth week, during which the major organs are formed; and (3) the fetal phase is the last 30 weeks of gestation, during which the organ systems grow and become mature, and the fetus is capable of living outside the womb.

A gynecologist or obstetrician usually dates the clinical age of a developing fetus using the mother's last menses. Most embryologists date development from the day of ovulation, which is known as the postovulatory age. Postovulatory age is used in this book to describe the steps in embryonic and fetal development. Because ovulation occurs 14 days after the last menses, and fertilization occurs near the time of ovulation, it is assumed that the postovulatory age is 14 days less than the clinical age.

Stages of Germinal Development

Stage 1, Fertilization, Day 0

Fertilization occurs on day 0 of postovulatory age (Figure 11.1). The oocyte and associated follicular cells are less than 0.1 mm in diameter (much too small to be seen with the human eye). The typical male ejaculate has 100 to 300 million sperm, and only a few thousand sperm actually reach the ampulla of the uterine tube.

A sperm can survive for up to 48 hours after ejaculation, and it takes a sperm about 10 hours to move through the vaginal canal, the cervix, and then to the ampulla of the uterine tube, where fertilization should optimally occur. Although 300 million sperm may enter the upper part of the vagina, only 1% (3 million) enters the uterus. The few sperm that reach the follicle in the uterine tube must penetrate the corona radiata (Figure 11.2). The movement of sperm toward the egg is thought to be through chemotaxis, and a likely chemoattractant agent is the peptide formyl-Met-Leu-Phe (fMLP). Sperm have a receptor on their surface for fMLP.

Sperm must bind to and penetrate the corona pellucida, zone pellucida, and ova cell membrane to complete fertilization. The penetration of the zona pellucida, a tough membrane surrounding the oocyte, takes about 20 minutes. The process begins when a sperm head binds to specific proteins on the zona pellucida. The sperm surface enzyme galactosyltransferase binds to N-acetylglucosamine residues within the zona pellucida. This binding initiates the acrosomal reaction, with enzymes and proteins released from the sperm acrosomal membrane. The acrosomal reaction is mediated by the opening of Na^+ channels in sperm cell membrane. The resulting depolarization of the acrosomal membrane opens voltage-sensitive Ca^{2+} channels. Ca^{2+} ions flow into the sperm head and cause fusion of acrosomal vesicles with the cell membrane, releasing acrosomal vesicle contents into the extracellular space. One of the released enzymes, acrosin, is a protease that digests the zona pellucida. The first sperm that binds to receptors on the ova plasma membrane initiates a cortical reaction that prevents polyspermy (see later

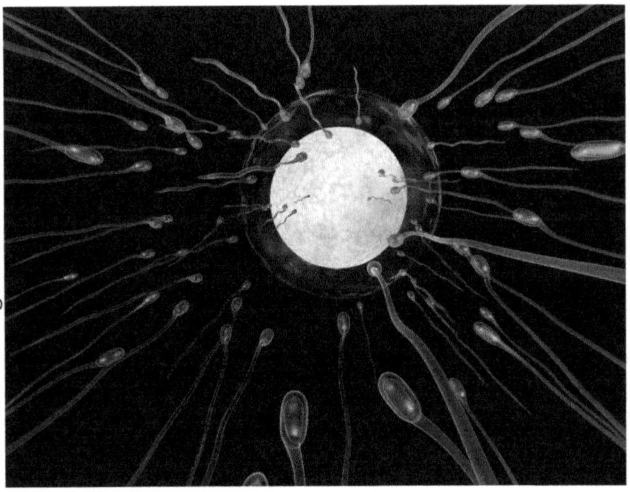

FIGURE 11.1

Sperm approaching an egg for fertilization

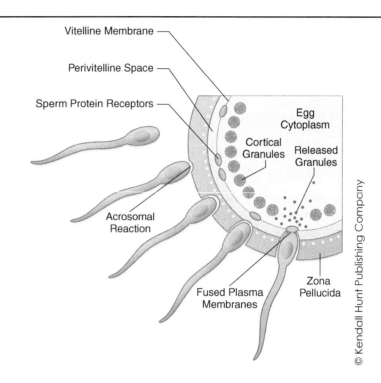

FIGURE 11.2
Sperm binding to sugars in the zona pellucid results in the acrosomal reaction and the release of enzymes into the zona pellucida, allowing sperm to move toward the vitelline membrane. Binding to receptors on the vitelline membrane triggers the release of cortical granule contents within the egg and changes to the vitelline membrane

discussion). Acrosomal vesicular digestive enzymes digest zona pellucida proteins, whereas movement of the flagellum drives the sperm forward. Ca^{2+} also activates a Na^+/H^+ ion exchanger, which pumps H^+ out of the sperm, increasing intracellular pH. This pH change causes the polymerization of actin subunits into microfilament cables that thrust acrosomal processes toward the egg plasma membrane. A protein (bindin) released from the acrosomal vesicle coats the acrosomal membrane. The bindin protein is a species-specific marker protein that prevents the sperm of other mammals from fertilizing a human egg.

After many sperm make their way through the zona pellucida (and perish in the process), finally a sperm makes contact with the **vitelline** (L. *vitellus*, egg yolk) or ovum plasma membrane, which has sperm-binding receptors. The vitelline membrane is separated from the oocyte cell membrane by a narrow gap of fluid, the **perivitelline** space (Figure 11.2). Sperm-binding proteins bind to specific receptors on the vitelline membrane. This binding interaction allows the sperm head to move through the vitelline membrane and make contact with the oocyte cell membrane. The sperm and egg plasma membranes then fuse, and the sperm nucleus enters the egg cytoplasm.

Polyspermy

Although many sperm attach to the zona pellucida surrounding the egg, it is important that only one sperm fuses with the egg plasma membrane and delivers its nucleus into the egg. Two mechanisms are used to ensure that only one sperm fertilizes a given egg: the fast block and the slow block to **polyspermy**. Polyspermy refers to the fertilization of an oocyte by more than one sperm.

> Why is it important that only one sperm cell fuses with the oocyte cell membrane?

The fast block to polyspermy occurs within a tenth of a second of the fusion of the sperm and oocyte cell membranes, and involves opening of oocyte cell membrane Na^+ channels (Figure 11.3). An inward Na^+ current depolarizes the oocyte cell membrane, eliciting an inward Ca^{2+} current that causes exocytosis of cortical vesicles into the space between the oocyte and vitelline membranes. Enzymes released from the cortical vesicles remove sperm receptors from the

Fetal Development

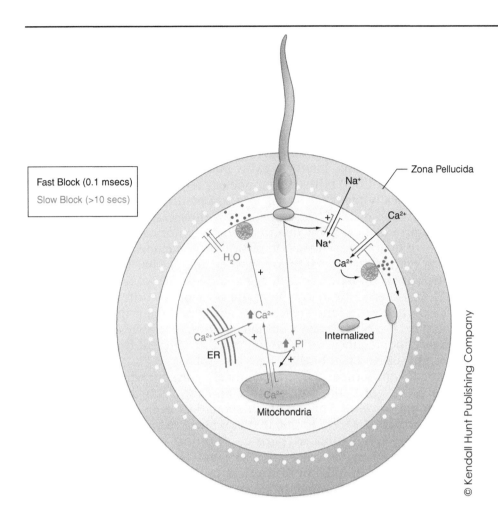

FIGURE 11.3

The fast block to polyspermy involves opening of sodium channels and the rapid depolarization of the vitelline membrane. A subsequent calcium influx causes cortical granules to be released, downregulating the expression of sperm receptors on the vitelline membrane. The slow block occurs about 10 seconds after sperm binding and involves the formation of inositol triphosphate (IP_3) and release of calcium from internal stores. The increased chytosolic calcium causes more cortical granules to be released into the perivitelline space, drawing water out of the cytoplasm, pulling the vitelline membrane away from the zona pellucida and making polyspermy more difficult. ER, endoplasmic reaction

vitelline membrane. This action prevents additional sperm from fusing to the vitelline membrane. The oocyte cell membrane is restored to its normal resting membrane potential of –70 mV potential within minutes of fusion as the Na^+ channels close. If depolarization is prevented, polyspermy occurs.

The slow block to polyspermy begins within 10 seconds of fusion of the sperm and oocyte cell membranes. Sperm contact with the oocyte cell membrane induces the formation of inositol triphosphate (see chapter 17), a second messenger that causes the release of Ca^{2+} from intracellular stores in the oocyte endoplasmic reticulum. Ca^{2+} is first released at the site of sperm fusion, and during the next minute, a wave of free Ca^{2+} passes through the egg. This Ca^{2+} fuses cortical vesicles with the egg cell membrane, releasing their contents into the perivitelline space. Cortical vesicle solutes increase the osmolarity of the perivitelline space and cause the movement of water out of the oocyte into the perivitelline space. This pushes the vitelline membrane away from the oocyte cell membrane, inactivating binding receptors on the vitelline membrane. As a result, additional sperm are released from the vitelline membrane and no more can bind.

Fusion of the oocyte and sperm cell membranes allows the entire contents of the sperm to be drawn into the egg cytosol. Even though the sperm's mitochondria enter the egg, they are usually destroyed and do not contribute their genes to the embryo. Thus, human mitochondrial DNA is usually inherited from mothers only.

Soon the nucleus of the fused sperm enlarges into the male pronucleus. At the same time, the oocyte (now a secondary oocyte) completes meiosis II, forming a second polar body and the female pronucleus. The male and female pronuclei move toward each other through the actions of the cytoskeleton.

> What is the significance of only maternal inheritance of mitochondria?

When they make contact, their nuclear envelopes disintegrate. A spindle is formed (after replication of the sperm's centriole), and a full diploid set of chromosomes assembles on it. The fertilized egg known as a **zygote** (G. *zygotes*, yolked) is ready for its first mitosis. The fusion of the oocyte and sperm nuclei marks the creation of the zygote and the end of fertilization.

Multiple Births

Multiple births occur in certain pregnancies (Figure 11.4). The most common multiple births are twins. Identical twins (*monozygotic*) are formed at a frequency of four per 1,000 births, when one egg fertilized by one sperm splits into two genetically identical halves in a process that is still poorly understood. Identical twins are of the same sex; only three cases have been recorded of boy and girl identical twins due to a genetic abnormality. Identical twins share 100% of their genes and are, in reality, natural clones.

Fraternal twins (*dizygotic*) are formed when the mother releases two eggs, and a different sperm fertilizes each egg at a frequency of 8 per 1,000 births (Figure 11.5). Fraternal twins can be of the same or opposite sex. They share up to 50% of their genes, and are no more alike or different than any two siblings would be. It is also possible, but uncommon, for fraternal twins to be of different races or parentage, or to be conceived at different times, resulting in a large weight difference at birth.

Conjoined twins form exactly like identical twins, but when the single egg splits into two zygotes, the process stops and the twins develop physically attached to one another. This occurs in about 1 of every 40,000 births, but only once in every 100,000 to 200,000 live births. More than 60% of conjoined twins are either stillborn (born dead) or lost in utero. Girls are conjoined more often than boys, at a ratio of 3:1, and female conjoined births are more likely to occur in India or Africa than in China and the United States, suggesting a possible environmental factor.

Polar body twinning occurs during oocyte development, even before fertilization. Normally, a polar cell forms during meiosis and usually degenerates or dies because of a lack of cytoplasmic components. In some cases, especially in older women, formation of the polar body is abnormal and it receives a larger percentage of cytoplasmic contents. Under these circumstances, the polar body does not degenerate and can act as a second oocyte. The polar body and the egg share identical genes from the mother but may be fertilized by two

FIGURE 11.4

Triplets occurred at a frequency of 1.4/1,000 live births in 2007, while twins occurred at a frequency of 32.2/1,000 live births

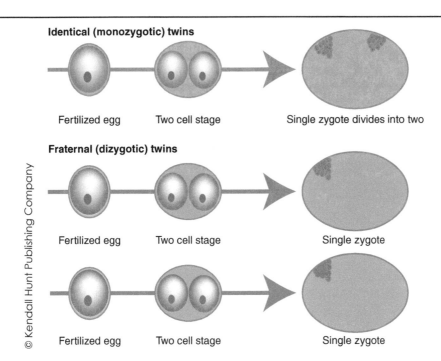

FIGURE 11.5
A monozygotic twin occurs when a fertilized egg splits in two. A dizygotic twin occurs when two eggs are fertilized at the same time

separate sperm. This will result in twins who are greater than 50% similar but not 100%, like identical twins. They receive 50% of the identical genes from their mother and the other genes, different ones, from their father. This may explain why these twins look as alike as some identical twins.

Many women being treated for infertility are given drugs that induce ovulation. In many cases, these drugs induce *superfecundation*, the ability to ovulate more than one egg. In this case, multiple eggs can be fertilized by sperm and always produces fraternal twins that are genetically half siblings, despite being twins.

It is also possible for women to ovulate multiple eggs but at different times over two menstrual cycles. In this case, *superfetation*, oocytes can be ovulated 24 days apart. The resulting babies have different conception dates, so the babies may be very different in size.

Identical triplets are thought to occur in almost the same manner as identical twins, but with the developing zygote splitting into three identical zygotes. In the rare cases of identical quadruplets, the splitting zygote would split into halves to form four identical babies. There is one recorded case of identical quintuplets, the Dionne quintuplets of Canada, but there is no evidence of larger numbers of identical twins.

In the United States, there were more than 137,000 multiple births in 2006 (32 per 1,000 live births), 6,100 triplets, 355 quadruplets, 67 births of quintuplets, and no sextuplets or septuplets.

Stage 2, Zygote Formation, Days 1 to 4

After approximately 90 minutes after fertilization, the zygote undergoes mitosis creating two cells, about half the size of the original ova. These cells continue to divide by mitosis creating a **blastomere** (G. *blastos*, germ; *meras*, part) (Figure 11.6). Even before the first round of mitosis, the fused cell has begun to unevenly divide cytoplasmic contents between the two daughter cells. This is the first step in the differentiation of cells into specific developmental paths. Mitotic division occurs approximately every 20 hours. Each cell becomes smaller and smaller with each subsequent division. When cell division generates 16 cells, after the fifth mitotic division, the zygote becomes a **morula** (L., mulberry shaped). The morula leaves the uterine tube and enters the uterine cavity 3 to 4 days after fertilization. The cells of the dividing embryonic mass are known as *pluripotent*, which means they have the capability to form a wide range of tissues.

Stage 3, Blastocyst Formation, Days 4 to 5

Approximately four days after fertilization the morula enters the uterine cavity. Cell division continues, and a cavity known as a **blastocele** (G. *blasto*, germ; *koilos*, hollow) forms in the center of the morula (Figure 11.7). At this point, there are about 32 cells within the developing embryo. Cells flatten and compact on the inside of the cavity, whereas the zona pellucida remains the same size. With the appearance of the cavity in the center, the entire structure is now called a **blastocyst** (germ; G. *kystis*, bladder). The presence of the blastocyst indicates that two cell types are forming: the embryoblast, the inner cell mass on the inside of the blastocele, and the **trophoblast**, the cells on the outside of the blastocele.

Stage 4, Implantation, Human Chorionic Gonadotropin Secretion, Days 5 to 6

During stage 4, the blastocyst is about 0.2 mm in diameter. The blastocyst hatches (pushes out) from the zona pellucida around the sixth day after fertilization, as the blastocyst enters the uterus. In the uterus, trophoblast cells secrete enzymes that erode the epithelial uterine lining and create a blastocyst implantation site. As the blastocyst invades the uterine wall, two populations of trophoblast cells develop and form the embryonic portion of the **placenta** (L., flat cake) (Figure 11.8).

FIGURE 11.6

The blastomere forms when the egg begins rapid division after fertilization

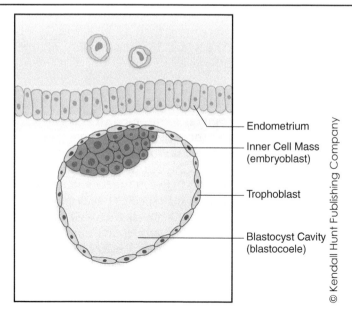

FIGURE 11.7

Four days after fertilization, a cavity—the blastocyst—forms together with new cell types—trophoblasts

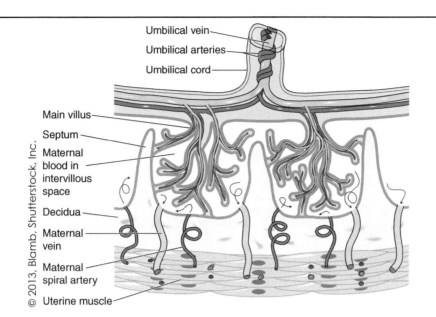

FIGURE 11.8
Placental formation occurs as the trophoblasts begin to digest the uterine wall and form blood-filled spaces known as *lacunae*

The placenta will eventually exchange nutrients and waste products between the embryo and the mother. The first type of trophoblast cell is a proliferating population of individual trophoblast cells known as *cytotrophoblasts*, and the other is a non-dividing multinucleated *syncytium* (cell mass) known as the *syncytiotrophoblast*. The roles of cytotrophoblast and syncytiotrophoblast cells are not the same. Cytotrophoblast cells are associated with embryonic tissues, whereas syncytiotrophoblast cells digest the uterine wall. Syncytiotrophoblast cells do not elicit a maternal immune response because they are nonantigenic.

As the functional endometrium is digested, maternal blood vessels are destroyed and they form *lacunae* (L. *lacus*, lake), or pools of maternal blood. Maternal blood continues to flow through the lacunae and not clot. The syncytiotrophoblast cells are immersed in maternal blood within the lacunae. Cytotrophoblast cells continue to divide and eventually surround syncytiotrophoblast cells and lacunae with cordlike structures. **Chorionic villi**, or *lacunae-like fingers*, develop from these cords. Cytotrophoblast and syncytiotrophoblast cells that surround maternal tissues are now known as the **chorion** (G. *chorion*, membrane). Fetal blood vessels will eventually follow the chorionic villi as the placenta develops.

During this stage, trophoblast cells begin to secrete human chorionic gonadotropin (hCG), which stimulates the corpus luteum to continue to produce estrogen and progesterone (see chapter 27). Estrogen and progesterone stimulate functional endometrial growth and vascularization, and if estrogen and progesterone levels decrease, the pregnancy is terminated.

> How is it possible for the syncytiotrophoblast to be nonantigenic?

Stage 5, Implantation Complete, Placental Circulation Begins, Days 7 to 12

By stage 5, trophoblast cells continue to engulf and destroy cells of the uterine lining creating lacunae, stimulating both new capillary growth, placenta growth, and the umbilical cord containing the umbilical vein and artery (Figure 11.9). The inner cell mass divides rapidly, forming a two-layered disk, with the top layer of cells becoming the embryo and amniotic cavity, and the lower cells becoming the yolk sac.

Ectopic (L., out-of-place) pregnancies can occur at this time and sometimes continue for up to 16 weeks of pregnancy before being noticed. Most ectopic pregnancies occur within the uterine tubes, or less commonly within the peritoneal cavity. Diagnosed quickly, ectopic pregnancies can be treated pharmacologically without surgery, reducing danger to the mother and preserving the site of the ectopic pregnancy.

FIGURE 11.9

Further development of the placenta leads to the formation of the umbilical arteries, allowing for a more efficient exchange of nutrients between maternal and fetal blood

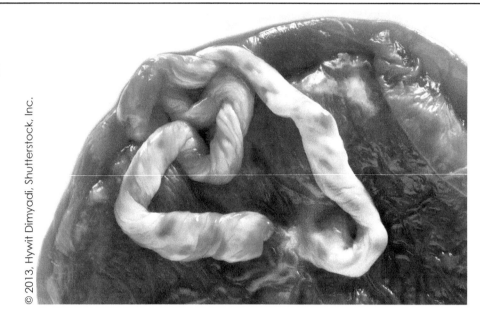

Clinical in Context: In Vitro Fertilization and Embryo Transfer

Infertility affects 6.1 million American women and their partners, about 10% of the reproductive age population. In nearly 90% of these infertility cases, the uterine tubes are incapable of transporting a zygote to the uterus or allowing the sperm to reach the oocyte. In vitro fertilization and embryo transfer (IVF-ET) is a fertility procedure that first succeeded in the late 1970s in England (Figure 11.10). Since then, the technology has been refined and developed, with more than 20,000 babies born worldwide using IVF-ET.

Because of improvements in ultrasound imaging, surgery is no longer necessary to recover of eggs from the ovary. In a procedure called *transvaginal oocyte retrieval*, physicians use an ultrasound-guided needle to collect eggs from the ovary. This method does not require hospitalization or general anesthesia. Before collection of eggs, women are treated in a variety of ways to increase the chances of recovering several healthy and mature eggs: Leuprolide (Lupron) is an injectable drug that blocks leutenizing hormone (LH) and follicle- stimulating hormone (FSH) secretion from the pituitary gland, optimizing the number of oocytes retrieved. Human menopausal gonadotropin (Pergonal) or FSH stimulates ovarian activity; hCG induces ovulation; and progesterone and serophene are used to promote egg development.

Blood tests and ovarian ultrasounds determine the optimal time to retrieve the eggs from the ovary, just before ovulation when the oocytes are almost ready for fertilization. At the proper time, in an outpatient procedure under local anesthesia, the female's eggs

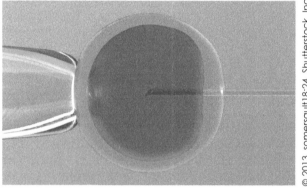

FIGURE 11.10

In vitro fertilization involves the harvesting of eggs from the ovary and allowing them to be fertilized in a Petri dish by sperm

are visualized by ultrasound and retrieved from the ovary by placing a needle through the vaginal wall. The eggs are then mixed on a Petri dish with donor sperm and allowed to develop into a eight-cell mass before implantation in the mother. Using a catheter, the embryo is passed through the vagina and into the uterus.

According to the latest statistics, the success rate for IVF is 29.4% deliveries per egg retrieval. This success rate is similar to the 20% chance that a healthy, reproductively normal couple has of achieving a pregnancy that results in a live born baby in any given month. Studies suggest that IVF-ET is a safe technology. A recent study found that the children, measured from birth to age 5, were as healthy as children conceived naturally.

Stages 6a and 6b, Gastrulation, Chorionic Villi Formation, Day 13

By stage 6, the blastocyst is 0.2 mm. Chorionic villi in the forming placenta now anchor the site to the uterus (Figure 11.8). The embryonic formation of blood and blood vessels begins in this stage. The blood system appears first in the area of the placenta surrounding the embryo, whereas the yolk sac begins to produce hematopoietic, or non-nucleated, blood cells. By the end of stage 6a, the embryo is attached to the developing placenta by a connecting stalk that will later become part of the umbilical cord. Stage 6b begins when a narrow line of cells appears on the surface of the embryonic disk. This primitive streak is the future axis of the embryo and it marks the beginning of *gastrulation* (G. *gastrula*, belly), a process that creates all three layers of the embryo: **ectoderm, mesoderm,** and **endoderm** (Figure. 11.11).

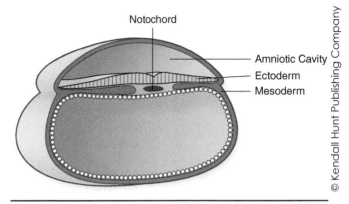

FIGURE 11.11

Gastrulation occurs by day 14 with the formation of the ectoderm, mesoderm, and endoderm layers along with the notochord

Embryonic Development

Neurulation and Notochord Formation, Days 14 to 16

By days 14 to 16, the embryo is 0.4 mm and gastrulation continues with the formation of the ectoderm and mesoderm, which develop from the primitive streak, changing the two-layered disk into a three-layered disk with

Clinical-in Context: Stem Cell research

Stem cells (see chapter 3) have the capability of replacing tissues and organs, or entire individuals. Research on stem cells is advancing knowledge about how humans develop from a single cell and how healthy cells replace damaged cells in adult organisms. This promising area of science is also leading scientists to investigate the possibility of cell-based therapies to treat disease, which is often referred to as regenerative or reparative medicine. Stem cells are one of the most fascinating areas of biology today, but like many expanding fi elds of scientifi c inquiry, research on stem cells raises scientifi c and ethical questions as rapidly as it generates new discoveries. Stem cells have two important characteristics that distinguish them from other types of cells. First, they are unspecialized cells that renew themselves for long periods through cell division. Second, under certain physiologic or experimental conditions, stem cells can be induced to become cells with special functions, such as heart muscle or pancreas insulin-producing cells.

Two types of stem cells are being investigated: embryonic stem cells and adult stem cells. These cell types have different functions and characteristics. Embryonic stem cells are isolated from embryos created for in vitro fertilization procedures. These cells are donated for research with the informed consent of the donor. Embryonic stem cells may become the basis for treating diseases such as Parkinson disease, diabetes, and heart disease.

The issue of stem cell research is a controversial, politically charged issue, with opponents arguing that the blastocyst (source of embryonic stem cells) is a living human being killed to provide embryonic stem cells. Their argument is that life begins at the earliest stages of development, and law should protect the blastocyst in the same way as a newborn baby. Proponents of stem cell research argue that the potential of this research to relieve the suffering of people who have debilitating or fatal diseases justifi es the use of stem cells. They also argue that a blastocyst is not a human being. This debate will continue.

Adult stem cells exist but are more diffi cult to isolate, and in many cases, their origin is unknown. The central nervous system contains stem cells, in small numbers, that are stimulated to replace astrocytes and oligodendrocytes, and perhaps neurons, after injury and disease. Adult stem cells have been identifi ed in bone marrow, blood, blood vessels, skeletal muscle, liver, and skin epithelium. Bone marrow stem cells (hematopoietic stem cells) can create a wide range of blood cells. Stem cells in the other tissues are stimulated to replace damaged or destroyed tissue. Isolation and characterization of adult stem cells may become a means by which diseases and disorders are treated.

the endoderm. The cells in the central part of the mesoderm release a chemical that causes a dramatic change in the size of the cells in the top layer (ectoderm) of the flat, disk-shaped embryo. The ectoderm grows rapidly over the next few days, forming a thickened area. The three layers will eventually create all body tissues. Endoderm cells will form the lining of lungs, tongue, tonsils, urethra and associated glands, bladder, and digestive tract. Mesoderm cells will form muscles, bones, lymphatic tissue, spleen, blood cells, heart, lungs, and reproductive and excretory systems. Ectoderm cells will form skin, nails, hair, lens of eye, lining of the internal and external ear, nose, sinuses, mouth, anus, tooth enamel, pituitary gland, mammary glands, and all parts of the nervous system.

Formation of the Notochord, Days 17 to 19

By days 17 to 19, the embryo is 1 to 1.5 mm and is shaped like a pear, with the head region broader than the tail end. The ectoderm has thickened to form the neural plate (Figure 11.12). The edges of this plate rise, like two ocean waves coming together, to form a concave area known as the *neural groove*. This groove is the precursor of the embryo's nervous system and is one of the first organs to develop. The underlying notochord stimulates the folding of the neural plate at the neural groove. The neural tube becomes the brain and spinal cord, and the cells of the neural tube are known as *neuroectoderm*. As the neural folds come together and fuse, a population of cells breaks away from the neuroectoderm all along the crests of the fold. Neural crest cells eventually create the peripheral nervous system and adrenal medulla by moving along the sides of the developing neural tube. Skin melanocytes are also derived from neural crest cells that migrate below the ectoderm. Neural crest cells that migrate to the head contribute to the skull, tooth dentin, small skeletal muscles, and general connective tissue. Because neural crest cells behave in the head the same way mesoderm cells behave in the trunk, they are both commonly referred to as **mesenchyme**. Embryonic blood cells are already developed and they begin to form channels along epithelial cells that form consecutively with the blood cells.

> Why is the CNS one of the first organs to develop?

FIGURE 11.12

Formation of the neural tube. (a) The neural plate and notochord are visible in a day 17 embryo. (b) The neural folds form by day 18. (c) By day 19, the neural fold has led to the neural tube. (d) By 20 days, neural crest cells are migrating and the neural tube is fully formed

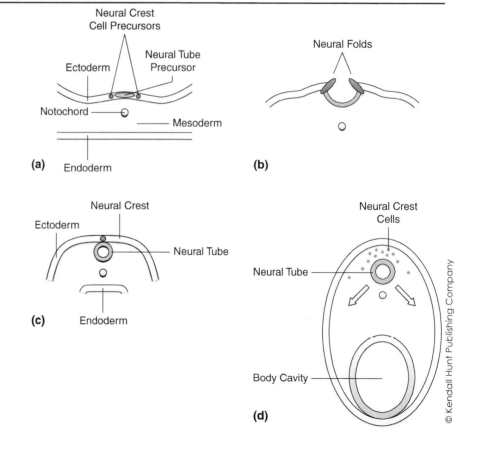

Appearance of Somites, Days 19 to 21

By days 19-21, the embryo resembles the sole of a shoe. The head end is wider than the tail end, and the middle is slightly narrowed in a superior view. During this stage, **somite** (G. *soma*, body) condensations composed of mesoderm appear on either side of the neural groove (Figure 11.13). The first pair of somites appears at the tail and progresses to the middle. One to three pairs of somites are present by stage 9. The first few somites never become clearly divided but develop into indistinct segmented structures known as *somitomeres*. The somites and somitomeres eventually create parts of the skull, the vertebral column, skeletal muscle, and many of the head muscles.

Every ridge, bump, and recess now indicates cellular differentiation. A head fold rises on either side of the primitive streak, which now runs between 1/4 to 1/3 of the length of the embryo. Secondary blood vessels now appear in the chorion and placenta. Hematopoietic cells appear on the yolk sac simultaneously with endothelial cells that will form blood vessels for the newly emerging blood cells. Endocardial muscle cells begin to fuse and form into the early embryo's two heart tubes.

Fusion of the Neural Fold, Days 21 to 23

By days 21-23, the embryo is 1.5 to 3.0 mm in length. This stage is characterized by rapid growth and change as the embryo becomes longer and the yolk sac expands. On each side of the neural tube, between 4 and 12 pairs of somites can exist by the end of stage 10. The cells that become the eyes appear as thickened circles just to one side of the neural folds, and ear cells are present. Neural folds rise and fuse at several points along the length of the neural tube concomitant with the budding somites that appear to "zipper" the neural tube closed. Neural crest cells will eventually contribute to the skull and face of the embryo.

The two endocardial tubes formed in stage 9 fuse in stage 10 to form one single tube derived from the roof of the neural tube, which becomes S-shaped and makes the primitive heart asymmetric. As the S shape forms, cardiac muscle contraction begins.

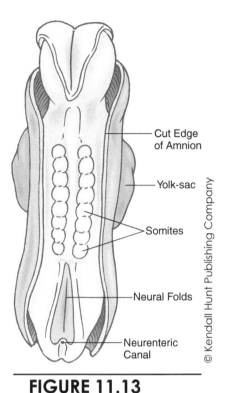

FIGURE 11.13

By 22 days, there are numerous somites along the neural tube

Days 23 to 25

By days 23 to 25, the embryo is 2.5 to 3 mm in length. Thirteen to 20 pairs of somites are present in stage 11, and the embryo is shaped in a modified S curve. The embryo has a bulblike tail and a stalk that connects to the developing placenta. A primitive S-shaped tubal heart beats, and peristalsis begins propelling fluids throughout the body. However, this is not true circulation because blood vessel development is still incomplete.

At this stage, the neural tube determines the embryo form. Although the primary blood vessels along the central nervous system are connecting in this stage, the central nervous system is the most developed system.

Days 25 to 27

By days 25 to 27, the embryo now curves into a C shape. The arches that form the face and neck become evident under the enlarging forebrain. By the time the neural tube is closed, both the eye and the ear have begun to form. At this stage, the brain and spinal cord together are the largest and most compact embryo tissue. The blood system continues to develop. Blood cells follow the surface of yolk sac where they originate, move along the central nervous system, and to the chorionic villi. Valves and septa may appear in the heart.

Days 27 to 29

At this stage, the embryo has four limb buds (lens disks and optic vesicles) and the first thin surface layer of skin appear covering the embryo (Figure 11.14). There are between 30 and 40 embryo somite pairs now.

FIGURE 11.14

Stages in embryonic development. Top line—29-day-old embryo, 5 mm in length with arm and leg buds; middle line—36-day-old embryo, 11 mm in length; bottom line—44-day-old embryo with clear fingers and toes, about 14 mm in length

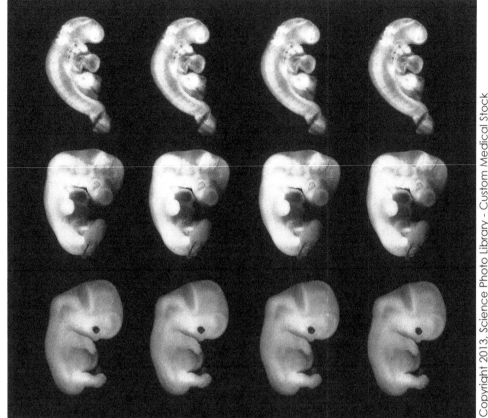

The brain is differentiated into the three main parts: the forebrain, midbrain, and hindbrain. The forebrain consists of lobes that translate input from the senses and will be responsible for memory formation, thinking, reasoning, and problem solving. The midbrain will serve as a relay station, coordinating messages to their final destination. The hindbrain will be responsible for regulating the heart, breathing, and muscle movements. The thyroid continues to develop, and the lymphatic system, which filters out bacteria, starts to form.

The *otic placode* (local thickening of the ectoderm) invaginates and forms the *otic vesicle*, which will develop into the structures needed for hearing and maintenance of equilibrium. A retinal disk presses outward and touches the surface ectoderm. In response, the ectoderm proliferates, forming the lens disk. Specific parts of the eye, such as the retina, the future pigment of the retina, and the optic stalk are identifiable. The primitive mouth with a tongue can now be recognized.

Heart chambers are filled with plasma and blood cells, making the heart seem distended and prominent. The heart and liver combined are equal in volume to the head by this stage. Blood circulation is well established, though true valves are not yet present. The villous network is in place to accommodate the exchange of blood between the mother and the embryo. Aortic arches may appear during this stage. The lung buds continue to form.

The gallbladder, stomach, intestines, and pancreas continue to form, and the metanephric bud, the precursor of the kidneys, appears in the chest cavity. The stomach is spindle-shaped, and the pancreas may be detected near the intestinal tube. By this stage, the developing liver receives blood from the placenta via the umbilical cord. The amnion encloses the connecting stalk, helping to fuse it with the longer and more slender umbilical vesicle, the remnant of the yolk sac.

Upper limb buds are visible as ridges and the lower limb buds begin to develop. These limb buds will create the arms and legs. The apical ectodermal ridge, a specialized ectodermal thickening, develops on the lateral margin of each limb bud and stimulates outgrowth. As the buds elongate, limb tissues are laid down in a proximal-to-distal sequence, with the hands appearing last.

Folding is complete, and the embryo is now three-dimensional and is completely enclosed by the amniotic sac. The somites are now involved in building bones and muscles. A thin surface layer of skin covers the embryo.

Days 29 to 33

Through the fourth and sixth weeks of development, the brain and head grow rapidly. The mandibular and hyoid arches are noticeable. Ridges mark the three sections of the brain (midbrain, forebrain, and hindbrain). The spinal cord at this stage contains three zones: the ventricular, the mantle, and the marginal. The ventricular zone will form neurons, glial cells, and ependymal cells; the mantle will form neuron clusters; and the marginal zone will contain processes of neurons. The adenohypophyseal pouch, which will develop into the anterior pituitary, is visible.

> Why does the kidney appear so late in fetal development?

The lens vesicle opens to the surface and is nestled within the optic cup. The otic vesicle increases its size by approximately one fourth and is more defined. During this stage, the nasal plate can be detected by thickened ectoderm.

The esophagus forms from a groove of tissue that separates from the trachea, which is also visible. Semilunar valves begin to form in the heart. Four major subdivisions of the heart, the left and right ventricles, the conus cords, and the truncus arteriosus, are clearly defined. Two sprouts, a ventral one from the aortic sac and a dorsal one from the aorta, form the pulmonary arch. Right and left lung sacs lie on either side of the esophagus.

The ureteric buds, precursors of the urethra and ureter, appear at this stage of development. The metanephros, which will eventually form the permanent kidney, is developing. The upper limbs are elongated into cylindrically shaped buds, tapering at the tip to eventually form the hand. The distribution of nerves and the innervation of the upper limbs begin.

Days 33 to 38

At this stage, the brain has increased in size by one third since the last stage, and it is still larger than the trunk. Four pairs of pharyngeal arches are visible now, though the fourth one is still quite small. The stomodeum, a depression in the ectoderm that will develop into the mouth and oral cavity, appears between the prominent forebrain and the fused mandibular prominence. Swellings of the external ear begin to appear on both sides of the head, formed by the mandibular arch.

The lens pit has closed and retinal pigment appears in the external layer of the optic cup. Lens fibers form the lens body. Two symmetrical and separate nasal pits appear as depressions in the nasal disk. In the thorax region, the esophagus lengthens.

Blood flow through the atrioventricular canal is now divided into left and right streams, which continue through the outflow tract and aortic sac. The left ventricle is larger than the right and has a thicker wall.

In the lung, there is development of lobar buds in the bronchial tree. In the abdomen and pelvic regions, the intestine lengthens, the ureteric bud lengthens, and its tip expands, beginning the formation of the final and permanent set of kidneys.

In the limbs, distinct regions of the hand, forearm, arm, and shoulder may be discerned in the upper limb bud. The lower limb bud begins to round at the top and the tip of its tapering end will eventually form the foot. The innervation of the lower limb also begins in this stage. In the spine, the relative width of the trunk increases from the growth of the spinal ganglia, the muscular plate, and the corresponding mesenchymal tissues.

Days 38 to 41

By the ninth week of fetal development, the embryo is 11 mm in length and the brain is well marked by its cerebral hemispheres. The hindbrain, which is responsible for heart regulation, breathing, and muscle movements, begins to develop. The future lower jaw, the first part of the face to be established, is now visible, and the future upper jaw is present but not demarcated. The skull and face continue to form from mesenchymal cells originating in the primitive streak, the neural crest, and the prechordal plate. In the eye, external retina pigment is visible, and the lens pit has grown into a D shape. The nasal pits are still separate, but they rotate to face ventrally as the head widens.

In the thorax, the cardiac tube separates into aortic and pulmonary channels, and the ventricular pouches deepen and enlarge, forming a common wall with their myocardial shells. The mammary gland tissue begins to mature. In the abdomen and pelvic region, the mesentery is now clearly defined, and the ureter continues to lengthen. In the limbs, the hand continues to develop and the central carpal region and finger buds begin to form. In the lower limbs, the thigh, lower leg, and foot can be distinguished.

Days 41 to 43

By days 41-43, the embryo is 13 mm in length, and the jaw and facial muscles are developing. The nasofrontal groove becomes distinct and an olfactory bulb forms in the brain. The formation of the olfactory bulb indicates that the fetus has a sense of smell. Auricular (ear) hillocks become recognizable. The dental laminae or *teeth buds* begin to form. The pituitary begins to form, as do the trachea, the larynx, and the bronchi. In the thorax, the heart begins to separate into four chambers, and the diaphragm forms. In the abdomen, the intestines begin to develop within the umbilical cord, and will later migrate into the abdomen when the embryo's body is large enough to accommodate them. In the pelvic region, primitive germ cells arrive in the genital area and start to develop into either female or male genitals. In the lower limbs, toes begin to develop and in the upper limbs, the fingers are more distinct. The spine also becomes straighter.

Days 44 to 46

By days 44 to 46, the embryo is 14 mm in length and the ossification of the skeleton has begun. In the head and neck, nerve plexuses begin to develop in the scalp region, the eyes are clearly pigmented, and eyelids begin to develop and fold. In the thorax, within the heart, the trunk of the pulmonary artery separates from the trunk of the aorta. Nipples appear on the chest, and the body appears more like a cube.

In the abdomen, the kidneys begin to produce urine for the first time. In the pelvic region, the genital tubercle, urogenital membrane, and anal membrane appear. In the limbs, the critical period of arm development ends and the arms are at their proper location, roughly proportional to the embryo. However, the hands are not finished, but develop further in the next 2 days. The wrist is clearly visible and the hands already have ridges or notches indicating the future separation of the fingers and the thumbs. In the spine and skeleton, ossification begins and bones become clearly visible.

Days 47 to 48

By days 47 to 48, the embryo is 18 mm in length, brain waves can be detected, and the head is more erect. Inner ear semicircular canals start to form, and soon the fetus will be able to sense balance and body position.

In the thorax, the septum primum fuses with septum intermedium in the heart, continuing the process of the formation the heart's four chambers. In the pelvic region, the gonads continue to form and embryo sex is recognizable in the form of testes or ovaries. In the lower limbs, the knee and ankle locations are indicated by indentations. Legs are now at their proper location, proportional to the embryo. The critical period for the lower limbs is about to end. Toes are almost completely notched and toenails begin to appear. Joints grow more distinct. The trunk elongates and straightens, and bone cartilage begins to form a more solid structure. Muscles develop and get stronger.

Days 49 to 51

By day 51, spontaneous involuntary movements begin, indicating both the development of skeletal muscles and the formation of neuromuscular regulatory circuits. By now, the brain has innervated muscles throughout the embryo. Nerve plexuses become more common and the scalp plexus forms. The nasal openings and the tip of the nose are fully formed. The upper limbs become longer and continue to bend at the elbows and extend forward. Skin on the foot folds down between the future toes, each distinguishable from the other. In the pelvic region, the urogenital membranes differentiate in male and female embryos, and the testes or ovaries are clearly distinguishable.

Days 52 to 53

By day 53, the embryo is 22 mm in length and the intestines begin to recede into the body cavity. Intestines begin migration within the umbilical cord toward the embryo. The liver now causes a ventral prominence of the abdomen. The eyes are well developed but are still located on the side of the embryonic head. As head development continues, they will migrate forward. External ears are still low on the embryo's head, but will move up as the head enlarges. Tongue development is complete. The fingers lengthen whereas distinct grooves form between the fingers, and the hands approach each other across the abdomen. Feet also approach each other but are still fan-shaped, and the toe digits are still webbed.

Days 53 to 55

By day 55, the critical period of heart development ends. The heart will continue to develop, but not at such a quick pace. In the head, fissures become apparent, and eyelids and the external ear more developed and the upper lip fully formed. By now, the brain can consciously move muscles. In female embryos, the clitoris begins to form; in male embryos, the penis begins to form. In the limbs, primary ossification centers appear in the long bones, directing the replacement of cartilage by bone. This process usually begins in the upper limbs. Fingers overlap those of opposite hand, and the digits of the fingers fully separate. The feet lengthen and become clearly defined. The embryo still appears to have a stubby tail, but it is beginning to become smaller.

Days 56 to 57

In this last stage of embryonic development, all of the essential external and internal structures are present. The head is erect and rounded, and the external ear is completely developed. The eyes are closed, but the retina of the eye is fully pigmented. Taste buds begin to form on the surface of the tongue, and palate bones begin to fuse. In the abdomen, the intestines continue to migrate from the umbilical cord into the body cavity. In the pelvic region, the external genitals are still difficult to recognize. The upper and lower limbs are well formed, the fingers continue to grow longer, the toes are no longer webbed, and all digits are separate and distinct. A layer of flattened cells, the precursor of the surface layer of the skin, replaces the thin ectoderm of the embryo, and the embryonic tail has disappeared.

In Review...

28.1 What events occur during fertilization?

28.2 What occurs during the acrosomal reaction?

28.3 How is polyspermy prevented?

28.4 What steps are involved in the fusion of an oocyte and a sperm?

28.5 What is a zygote? A morula? A blastocyst?

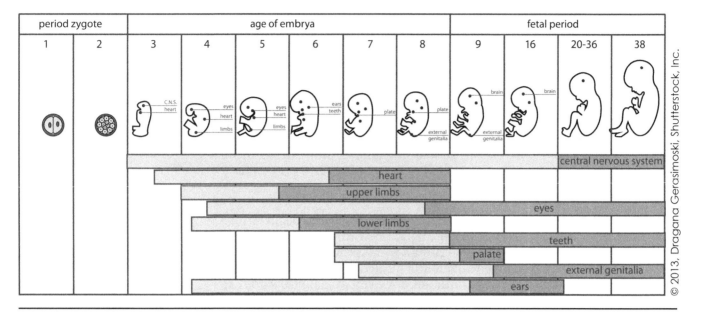

FIGURE 11.15
Stages of fetal development

28.6 How does uterine implantation occur?

28.7 What occurs during each stage of embryonic development?

Fetal Development

By week 10, the growing embryo, now a **fetus** (L. offspring), floats in about 50 mL of amniotic fluid (Figure 11.15). The brain structure is now complete and its mass increases rapidly. The sockets for all 20 teeth are formed in gums, and the face has a normal human appearance. Within the mouth, separate folds begin to form the palate, and early facial hair follicles begin to develop.

Within the thorax, vocal cords form in the larynx and the fetus could make sounds. Within the abdomen, the intestines have completely migrated from the umbilical cord, and digestive tract muscles are functional and begin to contract. Within the intestinal tract, villi begin to form and the liver begins to secrete bile, which is stored in the gallbladder. The development of the thyroid, pancreas, and gallbladder is complete and the pancreas starts to produce insulin.

Within the pelvis, the genitalia begin to show female characteristics, labium minus, urogenital groove, labium majora, or male characteristics, glans penis, urethral groove, scrotum, but they are not fully formed. In the hands, the fingernails begin to grow from nail beds. The fetus begins to develop reflexes and the skin becomes sensitive to contact.

Amniotic Fluid

Amniotic **fluid** is the fluid in which the fetus floats. It is produced partially by the amniotic membrane and partially by fetal urine production. It is thought to replace itself every 3 hours during pregnancy. Amniotic fluid allows the fetus to:

- Grow uniformly
- Allow bones and muscles to develop correctly by allowing their movement
- Helps the fetus to maintain a constant temperature
- Prevents the fetus from sticking to the amniotic membrane
- Allows for fetal inter-uterine movement
- Provides protection against mechanical trauma whereas the mother moves
- Is breathed in and out of the lungs, allowing the lungs and the respiratory muscles to develop and grow.
- And finally, amniotic fluid contains both maternal and fetal cells, which makes it useful for fetal testing.

If there is an abnormal over production of amniotic fluid, this can cause an increase in uterine size and premature labor. Abnormally increased amniotic fluid production can occur in multiple births (and may be one of the reasons why many multiple births are premature), in diabetic mothers (because of osmoregulatory problems), and in other cases. The fetus normally swallows amniotic fluid, which passes through the fetal stomach and small intestines. An esophageal blockage that prevents fetal swallowing will cause an increase in amniotic fluid. Too little amniotic fluid can cause adhesion of the fetus to the amniotic membrane, difficulty in fetal movement, and the incorrect development of many tissues and organs.

Amniocentesis and Fetal Monitoring

Amniocentesis is a common prenatal test in which a small sample of the amniotic fluid surrounding the fetus is removed and examined (Figure 11.16). It was first used in 1882 to remove excess amniotic fluid, and has long been performed in late pregnancy to assess anemia in babies with Rh disease and to find out whether the fetal lungs are mature enough for the baby to be delivered. Today, amniocentesis often is used in the second trimester of pregnancy (usually 16 to 20 weeks after a woman's last menstrual period) to diagnose certain birth defects. Amniocentesis is the most common prenatal test used to diagnose chromosomal and genetic birth defects. It does have some risks associated with the procedure, with most minor, but miscarriages occur in 0.06% of patients.

Chorionic villus sampling in which chorionic villi cells are collected can diagnose most, but not all, of the same birth defects as amniocentesis. Chorionic villus sampling can be done earlier in pregnancy (usually between 10 and 12 weeks) than amniocentesis, but it poses a slightly greater risk for miscarriage and other complications, about 1%.

The maternal α-fetoprotein test is used to detect the presence of neural tube defects. α-Fetoprotein is produced by the fetus and travels through the placenta to the maternal blood. α-Fetoprotein levels normally increase at 12 to 15 weeks of development and then decrease to a very low level in fetal and maternal blood. A high level after 16 weeks is usually due to the presence of spina bifida or anencephaly. The test is 95% accurate and is recommended for all pregnant women. A newer test probes the maternal blood for α-fetoprotein and three other proteins, and is used to screen for Down syndrome, neural tube defects, and helps to predict the delivery date and the presence of twins.

Recent work has led to the discovery of fetal cells and DNA in maternal blood, opening up the possibility of much safer fetal testing in the future (see Clinical–In Context: Detection of Fetal Abnormalities in Maternal Blood)

Fetal Monitoring

Almost one third of all United States childbirths are marked by a period of uncertainty about the well-being of the fetus because of an abnormal fetal heart rate pattern, possibly caused by a lack of O_2. This uncertainty can be extremely stressful for parents as well as physicians. A number of approaches have been developed to provide indirect or direct measurement of fetal health during parturition. Fetal distress is the term commonly used

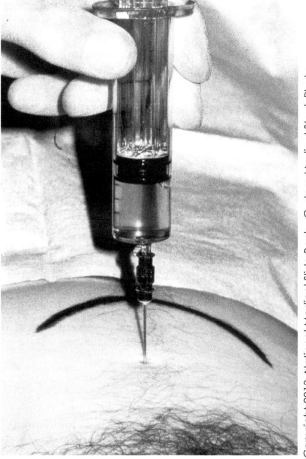

FIGURE 11.16

While using ultrasound as a guide a physician punctures the abdominal wall of a pregnant woman to withdraw amniotic fluid to analyze for various chromosomal abnormalities

Clinical–In Context: Detection of Fetal Abnormalities in Maternal Blood

Extensive research over the past 20 years has discovered that maternal blood contains both fetal cells and fetal DNA. Isolation of fetal cells from maternal blood is technically difficult by FISH (fluorescent in-situ hybridization) using fetal specific cell-surface markers. A more promising approach has to been to isolate cell free fetal DNA from maternal blood to test for genetic defects. This approach has been used to detect fetal trisomy. Both of these approaches if further developed will radically change how fetal genetic defects are detected during pregnancies and greatly reduce the risks associated with current methods of fetal monitoring. A further area of research will be to determine why maternal blood has fetal cells or cell free fetal DNA, and the significance to the fetus and mother.

> What components are needed to develop a reflex?

> How does the fetal cardiac output compare with an adult cardiac output?

to describe fetal hypoxia. It is a clinical diagnosis made by indirect methods. Hypoxia can result in fetal damage or death if not reversed or if the fetus is not delivered immediately.

There are potentially limitless causes for fetal distress, but several key mechanisms are usually involved. Contractions temporarily reduce placental blood flow and can compress the umbilical cord. A long labor can cause fetal distress. All babies are at risk for becoming distressed. Fetal blood sampling is a useful tool for the diagnosis of fetal distress. The fetal scalp is first visualized with an amnioscope inserted vaginally. The fetal scalp is cleaned and a small cut is made in the scalp from which blood is collected. The blood is immediately analyzed for pH. If the pH is less than 7.2, there is significant hypoxia and the baby should be delivered by the fastest means possible.

A *cardiotocograph* is a record of the fetal heart rate measured either from a transducer on the mother's abdomen or from a probe on the fetal scalp. In addition to the fetal heart rate, another transducer measures the uterine contractions over the fundus. Babies in fetal distress typically have reduced heart rates, especially during a uterine contraction. A fetoscope is a special type of stethoscope used for listening to a baby's heart rate in utero. A fetoscope can usually be used after about 18 weeks.

A Doppler is a handheld ultrasound device that transmits the sounds of the baby's heart rate either through a speaker or into earpieces. This device can generally pick up heart tones after 12 weeks' gestation. Electronic fetal monitoring is an ultrasound device used during labor and birth, or during certain testing (nonstress test, contraction stress test, etc.) to record the baby's heart rate and sometimes the mother's contractions. It can be used intermittently or continuously.

Internal fetal monitoring records the baby's heart tones via an electrode on the baby's head and records contractions via a pressure catheter. This method is also used during labor and birth; however it is used intermittently.

Week 12 to 13

The fetus begins to move within the amniotic fluid, but the mother cannot sense these movements. The head is about half of the crown-to-rump length and rests on the well-defined neck instead of the shoulders. Mouth muscles fill out cheeks, tooth buds continue to develop, and salivary glands begin to function. The scalp hair pattern is now discernible. The heartbeat can now be detected with external instruments and the fetal heart pumps about 17 mL of blood per minute and increases to 210 mL per minute by the time of delivery. The lungs have developed and the fetus inhales and exhales amniotic fluid. The movement of amniotic fluid in and out of the lungs is essential for the normal development of lung alveoli. The fetus can now swallow fluid and is able to suck on its fingers.

The abdomen has a fully functional spleen that begins removal of old red blood cells and produces antibodies. The sex organs are now clearly visible. The arms have almost reached final proportion and length, but the legs are still quite short relative to the length of the fetus's body. The hands, particularly the thumbs, have become more functional.

The fetus is now much more flexible and has advanced movements of head, mouth and lips, arms, wrists, hands, legs, foot, and toes. This advanced movement is due to the continued development of the muscular and nervous systems. Sweat glands have begun to appear and body hair begins to grow.

Week 14 to 15

By the 14th week of development, the fetus is more flexible and has the ability to move its head, mouth, lips, arms, wrists, hands, legs, feet, and toes. The head and neck are straighter and almost erect as muscles strengthen and additional bone forms in the back. The eyes have moved closer to their final position in the face and so have the ears. The philtrum, the vertical groove on the surface of the upper lip, is formed. By this time, body torso grows rapidly, increasing in proportion to the head. The limbs are now well developed and more defined. Toenails begin to grow from their nail beds.

Week 16 to 17

Fetal growth continues its rapid pace, but no new structures form after this point. The eyes are at their final destination on the face. Reflexes, such as blinking, begin to develop.[1] Ears also move to their final position and stand out from the head. Meconium, the product of cell loss, digestive secretion, and swallowed amniotic fluid, begins to accumulate in the bowels. Fingertips and toes develop the unique swirls and creases of fingerprints and toe prints. Nerves are myelinated and the circulation is completely functional. The umbilical cord system continues to grow and thicken as blood travels with considerable force through the body to nurture the fetus. The placenta is now almost equal in size to the fetus.

Week 18 to 19

This is a dramatic period of fetal growth. The fetus now has periods of sleep and waking. Temporary hair, lanugo, appears on the head, and eyebrows begin to form. Fetal ovaries now contain primary oocytes and the uterus is fully formed. Brown fat now fills the neck, chest, and crotch areas. Vernix (dead skin, lanugo cells, and oil from glands) is formed and visible on the skin. The placenta is now fully formed and grows in diameter but not in thickness.

Clinical–In Context: Maternal Epigenetics

Epigenetics refers to the heritable process that modifies gene expression without changes in DNA sequence. This is commonly caused by methylation of cytosine to form CpG islands, by modification of histones or production of microRNAs. The formation of CpG islands is transmitted from mother to offspring and can reduce gene transcription of selected genes. This normally occurs during chromosome X inactivation in female fetuses and during normal cell development and gene silencing.

More recently it has been realized that a mothers environmental exposures during pregnancy can induce epigenetic changes in fetal DNA, through dietary changes, drug exposures, stress, or environmental factors. One of the major impacts is the nutritional status of mothers while pregnant. Male sperm DNA is also methylated but 100% of the methylation is removed by the fertilized ova, while ova methylation is removed to a much lower level. This explains in part, why identical twins can be subtly different from each other if the intrauterine environments and epigenetic changes are different for each fetus. This promises to be an important area of research in the future.

Week 20 to 21

By this period, the fetus may suck on its thumb. This is also a period of extremely rapid brain growth, which lasts until 5 years after birth. Eyebrows and scalp hair become more visible and the fetus blinks more often. Lanugo covers the body completely, although it is concentrated around the head, neck, and face. Male testes begin descending from the pelvis into the scrotum. The legs are almost at their final length and are proportional relative to fetus's body size. Arms and legs move with more force, as muscles strengthen and the skeleton hardens, and hand strength improves.

Week 22 to 23

As the fetus continues to grow, it has less space to move around in the uterus. The inner ear bones ossify, making sound conduction possible. The fetus is now capable of recognizing maternal sounds such as breathing, heartbeat, voice, and digestion. The respiratory system is still developing, but it is not yet able to transfer O_2 to the bloodstream and release CO_2 by exhaling. Blood vessels, bones, and organs are now visible underneath a thin layer of wrinkled, translucent, pink skin.

Week 24 to 25

The fetal brain has begun to activate auditory and visual systems, allowing for the input of sensory information. The eyes now respond to light, and the ears respond to sounds originating outside the uterus. Permanent teeth buds now appear in the gums and nostrils begin to open. Blood vessels begin to develop in the lungs, and alveoli begin to produce surfactant. Fingernails and toenails continue to grow. The vertebrae continue to develop and can support fetal body weight.

Week 26 to 27

Brain wave patterns now resemble those of a full-term baby at birth. The forebrain enlarges to cover all other developed brain structures, whereas still maintaining its hemisphere divisions. Eyes are partially open and eyelashes are fully formed. Sucking and swallowing reflexes improve. The lungs are now capable of breathing air but are filled with amniotic fluid. The fetus now has 2% to 3% body fat. The male testes are now completely descended.

Week 28 to 29

The fetal brain now has a wrinkled surface as sulci and gyri begin to form. The brain now controls rhythmic breathing and body temperature. Lanugo hair has almost completely disappeared, except on the back and shoulders, and head hair begins to grow. The production of red blood cells is now entirely centered in bone marrow. Skin becomes smooth as fat deposits accumulate underneath. Body fat content increases and forms in places designed to insulate and to be used as an energy source.

Week 30 to 31

Rapid brain growth continues and head size increases as the growing brain pushes the skull outward, creating more surface convolutions. This quick growth increases the number of interconnections between individual nerve cells. The iris is colored and the pupil reflexes respond to light. Head hair grows thicker, and the toenails are fully formed. Because of the lack of space in the uterus, the legs are drawn up in what is known as the fetal position.

Week 32 to 33

The eyes are now open during waking periods and closed during sleep. Eye color is usually blue, regardless of the permanent color because pigmentation is not fully developed. Final formation of eye pigmentation requires exposure to light and usually happens a few weeks after birth. The fetus now begins to develop its own immune system. Fingernails reach over fingertips and the fetus can scratch itself. Fat begins to build up underneath skin, making the fetus appear lighter in color.

Week 34 to 35

The placenta is now one-sixth of total fetal weight. Within the mouth, the gums appear ridged and may look like teeth. The head may now be pointed downward in the position for parturition. The gastrointestinal system is very immature and will stay that way until 3 or 4 years after birth. The fetus now has about 15% body fat content, which is primarily used as insulation. The limbs begin to dimple at the elbows and knees, and creases form around the wrists and neck. The skin appears light pink because of blood vessels close to its surface.

Week 36 to 37

The fetal body now is round and plump because of increases in fat storage. The fetal body temperature is warmer than the maternal body temperature. The fetus now turns toward sound sources in what is known as the *orienting response*. Intestines accumulate a considerable amount of meconium that is usually eliminated shortly after birth. If birth is delayed, the fecal material will appear in the amniotic fluid.

Space limitations now restrict fetal movement. Limbs are bent and drawn close to body. Bones are flexible but ossification progresses. At birth, the tibia, the long bone of the leg, is usually completely ossified into bone.

Week 38 to 39

The skull is not fully solid because the five fontanels are still filled with connective tissue. The birthing process may mold and elongate the fetal head, a safety precaution to reduce the skull's diameter for an easier birth without damaging the fetal brain. After delivery, the baby's head returns to a rounded shape. Eyes have no tear ducts yet; they appear a few weeks after birth.

The chest is now more prominent and the lungs increase their production of surfactant. The fetal abdomen is large and round mainly because of liver production of red blood cells. The last of the vernix usually disappears, but may remain until birth. Skin becomes thicker and paler (white or bluish pink), and each day the fetus gains 1/2 ounce (14 g) of fat.

Week 40

The baby is now considered full term and has 15% body fat, 80% underneath the skin, and 20% around the organs. At the time of birth, the baby has 300 bones. Some bones will fuse together later, which is why an adult has only 206 bones. At term, a baby has more than 70 autonomic reflex behaviors that are necessary for survival.

> Why would the fetal body temperature be greater than the maternal body temperature?

Clinical–In Context: Congenital Birth Defects

Mistakes in development can result in a large number of congenital (present at birth) defects. These defects can be very serious, but in some cases can be repaired before birth.

Cardiovascular System Defects

Atrial septal defects are a group of common congenital abnormalities that occur more often in female newborns. An atrial septal defect is a hole between the two atria. These defects occur in a number of different forms: *patent foramen ovale* is an anatomical interatrial connection with potential for a left-to-right atrial shunt; *sinus venosus defect* occurs when the sinus venosus is not reabsorbed. Treatment for most atrial septal defects involves surgical repair but at a significant risk. Atrial septal defects usually cause hypertrophy of the right side of the heart. Increased pressure within the pulmonary blood vessels and a decreased systemic blood flow result in pulmonary edema, cyanosis, or heart failure.

Tetralogy of Fallot is named after Etienne-Louis Arthur Fallot, who in 1888 described this condition as "la maladie blue" (F., blue malady). The syndrome consists of a number of a cardiac defects possibly stemming from abnormal neural crest migration: ventricular septal defect, pulmonary stenosis, and right ventricular hypertrophy. They all result in diminished cardiac output and newborns appear blue (cyanotic).

Patent ductus arteriosus is the persistence of a normal fetal structure between the left pulmonary artery and the descending aorta. Persistence of ductus arteriosus beyond 10 days of life is abnormal. Surgical treatment acts to tie off the connection and usually has little risk. The operation is always recommended even in the absence of heart problems. It can often be deferred until early childhood.

A *ventricular septal defect* describes one or more holes in the muscular wall that separates the right and left ventricles of the heart and is the most common congenital heart defect. The defect allows the movement, or shunting, of blood from the left to the right ventricle. The volume of shunted blood depends on the size of the hole. Small defects may close spontaneously, but larger defects usually result in infant congestive heart failure.

Transposition of great vessels occurs when the two major arteries leaving the heart are connected to the wrong ventricles. The result is that oxygenated blood is pumped back into the lungs, whereas systemic circulation receives unoxygenated blood. This transposition occurs when the aorta arises from right ventricle and pulmonary artery from the left ventricle, and is often associated with other cardiac abnormalities.

The *hypoplastic left heart syndrome* is characterized by the poor formation of the left side of the heart, and the left ventricle cannot generate enough pressure to drive systemic circulation. The left ventricle and aorta are abnormally small (hypoplastic). This is among the most severe forms of a heart defect. Most babies are very ill in the early days of life and need urgent surgery to survive.

Dextrocardia is an abnormality in which the primitive heart tubes folds to the left in a mirror image of a normal bulboventricular loop. This usually occurs when all the organ systems are reversed, a condition called *situs inversus*. This results in the heart and major blood vessels being reversed from their normal position.

Syndactyly occurs when there is fusion of fingers or toes. There may be single or multiple fusions that affect the skin only, or the skin and soft tissues, or the bone.

Scoliosis occurs when there is asymmetrical growth of the vertebrae, resulting in spinal column deviation. Scoliosis can cause a lateral or forward flexion of the spine and, in some cases, the vertebral column is rotated on its long axis.

Congenital limb reduction defects typically occur when a pregnant mother ingests a teratogen. This congenital defect results in babies being born without arms or legs. Thalidomide was given to pregnant mothers in the 1950s and 1960s to control nausea and caused limb-reducing defects and other deformities. Limb reduction defects are caused by interruption of limb blood supply or innervation during development.

Muscular dystrophies are congenital muscular defects caused by mutations in genes essential for normal muscle development. The most common form of muscular dystrophy is a result of a mutation in dystrophin, a protein that lies under the muscle fiber membrane and maintains the cell's integrity. Fetal skeletal muscles are not used extensively, nor do they have significant loads placed on them. As a result, muscle wasting does not occur until after birth when the baby begins to use skeletal muscles. This is a progressive disease usually detected between 3 and 5 years of age. The dystrophin gene is very large (one of the biggest), is located on the X chromosome, and has several places where mutations have been shown to occur.

Spina bifida is a neural tube disorder involving incomplete development of the brain, spinal cord, and/or their protective coverings caused by the failure of the fetus's spine to close properly during the first month of pregnancy. Infants born with spina bifida sometimes have an open lesion on their spine where significant damage to the nerves and spinal cord has occurred. Although the spinal opening can be surgically repaired shortly after birth, the nerve damage is permanent, resulting in varying degrees of lower limb paralysis. Even when there is no lesion present, there may be improperly formed or missing vertebrae and accompanying nerve damage. In addition to physical and mobility difficulties, most affected individuals have some form of learning disability.

Down syndrome is caused by trisomy (possessing three copies) of chromosome 21. It occurs in approximately 1 out of every 650 births. Down syndrome was first characterized in 1866, when a physician named John Langdon Down described a set of children with common features who were distinct from those of other children with mental retardation. Children with Down syndrome exhibit a range of congenital defects: a wide range of mental retardation, heart defects, epilepsy, hypothyroidism, and celiac disease.

Agenesis of the corpus callosum is a rare congenital abnormality in which there is a partial or complete absence (agenesis) of the corpus callosum. The corpus callosum is the area of the brain that connects the two cerebral hemispheres. In most patients, this defect is diagnosed within the first 2 years of life. It may occur as a severe syndrome in infancy or childhood, as a milder condition in young adults, or as an asymptomatic incidental finding. The first symptoms of this defect are usually seizures, which may be followed by feeding problems and delays in holding the head erect, sitting, standing, and walking. There may be impairments in mental and physical development, hand-eye coordination, and visual and auditory memory.

Consuming alcohol during pregnancy is the cause of *fetal alcohol syndrome*, a leading preventable cause of birth defects and mental retardation. Alcohol is able to cross the placenta from maternal circulation into fetal circulation. Embryo exposure to ethanol simulates premature differentiation of prechondrogenic mesenchyme of the facial primordia. This result explains some of the facial abnormalities associated with fetal alcohol syndrome, but the mechanism is still unknown. The associated neurologic effects appear to be due to cell death in the outer layer of the developing neural tube at an early developmental stage. These developmental abnormalities are maternal in origin and are not congenital.

Neural tube defects occur because of a defect in the neurulation process. Since the anterior and posterior portions of the neural tube close last, they are the most vulnerable to defects. Neural tube defects can be classified as open or closed types, based on embryologic considerations and the presence or absence of exposed neural tissue. Open defects frequently involve the entire CNS, for example, hydrocephalus, and are due to failure of primary neurulation. Damage can occur because neural tissue is exposed and cerebrospinal fluid can leak out. Closed defects are localized and confined to the spine, and result from a defect in secondary neurulation. Research since 1990 has suggested a relationship between maternal diet and the birth of an affected infant. Recent evidence has confirmed that folic acid, a water-soluble vitamin found in many fruits, particularly oranges, berries, and bananas, leafy green vegetables, cereals, and legumes may prevent the majority of neural tube defects.

Gastrointestinal System Defects

Intestinal malrotation is a congenital defect that occurs at the 10 week of gestation. During this stage of development, the intestines normally migrate back into the abdominal cavity after a brief period when they are temporarily located at the base of the umbilical cord. As the intestine returns to the abdomen, it makes two rotations and becomes fixed into its normal position, with the small bowel centrally located in the abdomen and the colon (large intestine) draping around the top and sides of the small intestine. When rotation is incomplete and intestinal fixation does not occur, this creates an anomaly known as *intestinal malrotation*. With intestinal malrotation, the large intestine resides to the right of the abdomen, whereas the small intestine resides to the left of the abdomen. The cecum, the beginning of large intestine, and the appendix, which are normally attached to the right lower abdominal wall, are unattached and located in the upper abdomen. In patients with malrotation, the blood supply to the intestine is channeled through a very narrow supportive mesentery. Because the intestine is not properly fixated, the bowel may twist on its own blood supply (volvulus) and can result in the loss of most of the intestine. In some cases, it may also result in death. Chronic problems include abdominal pain, malabsorption and malnutrition, and subsequent growth disturbances.

A *situs inversus viscera* is a congenital disorder in which the visceral organs are transposed through the sagittal plane so that the heart, for example, is on the right side of the body.

Urogenital System Defects

Renal agenesis is a congenital defect in which the baby is born without kidneys. Typically, the child dies within a few days of birth.

Polycystic kidney disease is a genetic disorder characterized by the growth of numerous, fluid-filled cysts in the kidneys. These cysts can slowly replace much of the mass of the kidneys, reducing kidney function and leading to kidney failure.

Urinary tract outflow obstructions occur when a congenital defect prevents the normal formation of the urinary tract. Genital tract outflow is important for the expulsion of normal secretions from the cervix and vagina. Outflow also is critical for menstrual efflux. Outflow obstruction may occur at different levels with resultant variation of clinical presentation. Embryologically, the lower two thirds of the vagina develop from the urogenital sinus. The upper vagina, cervix, uterus, Fallopian tubes, and ovaries form from the Müllerian duct system. Failure of vertical fusion or canalization of the two systems in utero may result in urinary tract outflow obstructions.

The *horseshoe kidney* is the most common type of renal fusion defect. It consists of two distinct functioning kidneys on each side of the midline, connected at the lower poles by an isthmus of functioning renal tissue or fibrous tissue that crosses the midline of the body. By itself, the horseshoe kidney does not produce symptoms. However, by virtue of its embryogenesis and anatomy, it is predisposed to a greater incidence of disease when compared with the normal kidney.

Cryptorchidism is the most common genital problem encountered in pediatrics. Cryptorchidism means hidden or obscure testis, and generally refers to an undescended or maldescended testis. Many aspects of cryptorchidism are not well defined and remain controversial. Untreated cryptorchidism clearly has deleterious effects on the testis over time.

Male baby hypogonadism, poorly functioning gonads, occurs at an incidence of approximately 1 in 1,000 male births, and affects male babies who have an extra X chromosome (XXY male syndrome). The supplementary X chromosome is of maternal origin in 60% of cases and of paternal origin in the remainder. The presence of the additional X chromosome causes accelerated atrophy of germ cells before puberty, resulting in sterility with small, firm testes. Many patients are tall with relatively long legs. Behavioral disorders and delayed speech development are common. Testosterone therapy may be used to improve the development of secondary sexual characteristics. Boys tend to be shy, clumsy, and have low self-esteem. XXY male individuals have a normal life span.

Respiratory distress syndrome, an acute lung disease present at birth, usually affects premature babies. Layers of hyaline membranes within the lung prevent O2 transport into the blood. Without treatment, the infant will die within a few days after birth, but if O2 can be provided and the infant receives modern treatment in a neonatal intensive care unit, complete recovery with no long-term effects can be expected.

Repeating triplets of nucleotides in certain genes are known to cause more than a dozen different trinucleotide repeat diseases. In each case, a specific sequence of three DNA bases, for example, CAG (cytosine-adenosine-guanosine), that normally are repeated several times within a gene become rapidly expanded with each succeeding generation, causing severe diseases that start at an early age.

One such trinucleotide repeats disease is *Huntington disease*. People normally have an average of 18 CAG repeats within each allele of the *HD* gene that encodes the huntington protein. The CAG sequences codes for the amino acid glutamine, and in the presence of 35 to 40 copies of the CAG repeat, the additional glutamines cause distortion of the huntington protein. The normal function of the huntington protein is not known, but the abnormal form is toxic to neurons. *HD* is a dominant disease, and only one altered allele is required for symptoms to develop.

Amniotic Sac

On implantation, the blastocyst begins to develop into the embryo and the amniotic sac, a membrane filled with amniotic fluid. The amnion lining the amniotic cavity fills with amniotic fluid. The amniotic sac and fluid provide a protective, shock-absorbing cushion for the developing embryo (see earlier). The fetus moves freely until the late stages of pregnancy, when it becomes too large. In 95% of all pregnancies, the fetus ends up with its head pointing toward the cervix. In the remaining 5%, the fetus is in a breech position, which can cause complications during parturition.

Placenta

The placenta is a key organ and pregnancy is not possible without a normal functioning placenta (Figure 11.8). In humans, all layers of the placenta are of fetal origin. At 6 to 8 days after fertilization, trophoblast cells initiate implantation, and, together with endometrial tissue, develop into the placenta. At the time of implantation, trophoblast cells pass deep into the endometrium and fuse to form a syncytium, the syncytiotrophoblast. The attachment to the uterus triggers dramatic changes in the underlying uterine stroma, also known as the *decidua*. As the trophoblast cells invade the uterus, maternal cells are absorbed and lacunae appear. The lacunae become filled with maternal blood from eroded capillaries and a slow circulation is evident by 12 days. Placental growth and the development of fetal circulation continue through pregnancy as the fetus develops. The placenta is the site of gas and nutrient exchange between maternal and fetal blood supplies. In addition, hormone production by the placenta and fetus plays a role in the maintenance of pregnancy and initiation of labor.

During pregnancy, the human placenta produces several unique protein hormones, as well as estrogen and progesterone. As a result, ovarian production of estrogen and progesterone is limited to early pregnancy, and the placenta takes over steroid production after the first trimester. Increased blood estrogen and progesterone levels suppress LH and FSH levels through a negative feedback loop, blocking the development of additional follicles. Increased levels of progesterone inhibit uterine contractions that would expel the fetus prematurely. To prevent premature labor, circulating progesterone levels increase dramatically during pregnancy. The source of progesterone in the first trimester is the corpus luteum; in the second and third trimesters, it is the placenta. Even though the corpus luteum is no longer needed to produce progesterone, it is maintained throughout pregnancy, suggesting that it produces unknown essential factors. hCG maintains the structure and secretory activity of the corpus luteum.

The placenta is also the site of the production of human chorionic somatomammaotropin (hCS), also known *placental lactogen*, a protein hormone that has a similar structure to growth hormone. The concentration of hCS increases gradually throughout pregnancy and peaks during the last month. hCS has both weak growth-promoting and lactogenic activities, and may act on the mammary gland to induce the enzymes needed for milk synthesis and produce important maternal metabolic changes. During pregnancy, maternal glucose utilization decreases and cells become less sensitive to the hypoglycemic action of insulin. These metabolic changes may shunt larger glucose quantities to the fetus, and because growth hormone (GH) can produce similar effects, increasing levels of hCS may cause them. Because pregnant women are relatively insensitive to insulin, diabetes mellitus often first becomes evident during pregnancy. Consequently, urinary glucose levels are routinely determined during pregnancy to monitor the possible development of diabetes.

Yolk Sac

In humans, the yolk sac provides a source of the blood islands, endothelial cells lining blood vessels, and adult blood vessels.

Allantois

The *allantois* is a membranous sac growing out of the ventral surface of the hindgut. It combines with the chorion to form the mammalian placenta. It is also associated with the development of the urinary bladder and parts of the umbilical cord.

Gut and Body Cavities

The gut and body cavities start to form during day 18 of embryonic development. On day 18, the embryo forms a tubelike structure along the yolk sac upper region. The two ends of this tube, the foregut and hindgut, are the beginning of the digestive tract. As the digestive tract develops, it remains attached to the yolk sac, but begins to seal on the end. Eventually, the primitive digestive tract will have to move into the peritoneal cavity during a later stage of development.

The mouth forms from the oropharyngeal membrane, a point of contact between the foregut and ectoderm. The urethra and anus develop from the cloacal membrane forming a tube, the digestive tract, that is open at both ends.

The anterior pituitary, thyroid gland, liver, pancreas, and urinary bladder develop from evaginations (L., outpocketing) formed by the early digestive tract. The pharynx develops from expanded pockets of the brachial arches. The brachial arches form as tissue along the lateral sides of the head and foregut. The auditory tubes, tonsils, parathyroid glands, and thymus also develop from brachial arch tissue pockets.

Body cavities form from embryonic disk mesoderm that forms a hollow celom (L., hollow). Along the celom, four cavities form (two on other side) that become continuous. These cavities create the peritoneal, pleural, cranial, and thoracic body cavities. The cavities start as continuous but separate to form distinct body cavities.

The Face

The face develops from a number of embryonic tissues derived from ectoderm and mesoderm. The nasal and oral cavities are derived from ectoderm around the stomodeum, skull bones are derived from the mesenchyme and neural crest cells, the mouth is derived from small buds of tissue known as *facial primordia* (mostly neural crest cells), and the upper jaw from several tissue buds (median frontonasal, lateral frontonasal, and maxillary). The lower jaw develops from mandibular tissue buds. Neural crest cells create all facial connective tissue (cartilage, bone, and ligaments). The forehead and part of the nose arise from the frontonasal tissue bud, the sides of the nose from the lateral frontonasal tissue buds, whereas the cheeks form from the maxillary tissue buds. Nasal placodes (thickened tissue) form about the fourth week of development and will create the olfactory epithelium.

By the sixth week of development, many of these structures come together as the brain increases in size to form the primary palate. The fusion of these structures is an important part of face development. If these structures fail to fuse, a facial cleft can form and extend from the eye to the mouth. Cleft palates arise if the secondary palate does fuse later in embryonic development. These steps create a recognizable human face by 50 days of embryonic growth.

Skin

The ectoderm creates skin epidermis, whereas the mesoderm or neural crest cells create the dermal layer. The epidermis creates nail, hair, and glands (see chapter 3). Neural crest cells also create melanocytes and skin sensory cells.

Skeleton

Mesoderm and neural crest cells create the skeleton. Bones form through intramembranous or endochondral bone formation (see chapter 6). Most bones form from somites (skull, vertebral column, and ribs) or somitomere-derived mesoderm. Neural crest cells create facial bones, and mesoderm creates the appendicular skeleton.

Muscle

Skeletal muscle fibers are derived from myoblasts, early embryonic cells. Myoblast cells migrate to sites of future muscle formation, where they divide and fuse to form myotubes. Myotubes eventually enlarge to become muscle fibers. Muscle formation follows the formation of the primitive nervous system. This is important because shortly after their formation, motoneurons migrate to the growing muscle fibers and innervate them. The total number of muscle fibers is established before birth and does not change significantly during life. Muscle fiber length does change as the body increases in size. Muscle diameter changes with muscle use (see chapter 9).

Nervous System

The nervous system is one of the first embryonic structures to develop. It arises from neural tube and neural crest cells (Figure 11.12). The primordial brain develops from the superior end of the neural tube. After closure of the neural tube, the superior region expands to form the primordial brain. The brain ventricles and spinal cord canal develop from the central cavity of the neural tube. Somatic motor neurons also develop from the neural tube, as do preganglionic neurons of the autonomic nervous system. Neural crest cells create sensory neurons and postganglionic neurons of the autonomic nervous system.

Clinical–In Context: Cleft Lips and Palate

A cleft refers to a separation in a body structure. Clefts that occur in the oral-facial region often involve the lip, the roof of the mouth (hard palate), or the soft tissue in the back of the mouth (soft palate). Two major types of oral-facial clefts are cleft lip/palate and isolated cleft palate. Babies with cleft lip/palate have a cleft lip that usually is accompanied by cleft palate. In isolated cleft palate, the cleft palate occurs by itself, without cleft lip or other malformations. These two forms of oral-facial clefts are considered separate birth defects.

There are about 400 syndromes in which babies have some form of oral-facial cleft along with a wide variety of other birth defects. These separations normally are present in early fetal development. The lip usually closes by 5 to 6 weeks after conception and the palate by 10 weeks. A lip or palate fails to close in approximately one in every 1,000 babies born because of a failure of apoptosis. A cleft lip/palate occurs more often among Asians and among certain groups of American Indians. It occurs less frequently among African-Americans. Males are affected more frequently than females.

The causes of cleft lip/palate are not well understood. Studies suggest that a number of genes, as well as environmental factors, such as drugs (including several different antiseizure drugs), infections, maternal illnesses, maternal smoking, and alcohol use and, possibly, deficiency the vitamin folic acid may be involved.

Cleft lip/palate may occur alone or with other abnormalities that may be hidden or obvious. Some cases involve genetic syndromes that may pose specific problems for the baby and may have a high risk of affecting others in the family.

Respiratory System

Lung development begins on approximately day 28 of embryonic development. Respiratory epithelium develops from the endoderm, whereas connective tissue, cartilage, and muscle develop from the splanchnic mesoderm and neural crest cells. Foregut evaginations create the lungs and respiratory tissues. The two lung buds form from the evaginations. The lung buds elongate and branch through 15 to 20 generations (through the sixth month) to form bronchi, bronchioles, and alveoli. After birth, there is an additional four to six generations of branching.

By 17 weeks, all respiratory tissues have developed except for those involved in gas exchange. Thus, a 17-week-old fetus will not survive outside the uterus at this stage of development. Surfactant is not produced until 20 weeks of fetal development. By 25 weeks, there has been development of respiratory tissues, so a fetus at this age could survive. In the last 15 weeks of development, there is an increase in the number of functional alveoli and in the ability to exchange gas with the environment because of the rapid growth of capillary beds.

Urinary System

Kidney development goes through several steps that clearly demonstrate our evolutionary relationship to other creatures (Figure 11.17). After 21 days of the development, a *pronephros* (earliest kidney) forms from cervical region mesoderm. These mesoderm cells are located between somites and lateral portions of the embryo. The pronephros has a similar structure to kidney tissue found in primitive vertebrates and consists of a duct and simple tubules that open to the celom. The pronephros is nonfunctional and plays no role in removal of waste products from fetal blood, which is lost to the mother through the placenta.

The *mesonephros* (middle kidney) forms at the end of the fourth week, caudal to the degenerating pronephros. The mesonephros is similar to the kidney tissue found in fish and amphibians, and is more advanced in structure than the pronephros. The mesonephros has a primary duct with glomeruli and short tubules that connect to this duct. The mesonephric vesicle, a collection of mesoderm, elongates to form a tube. The medial end is invaginated with capillaries, creating a Bowman's capsule, whereas the lateral end of the tubule drains into the mesonephric duct. The caudal end of the mesonephric duct connects with the urogenital sinus. This junction will be important later in the developing male genital structures. The mesonephric duct is also known as the *Wolffian duct*. The mesonephros is a functional kidney and eventually degenerates.

The *metanephros* (final kidney) arises in the fifth week of development and becomes functional at the ninth week of development. The metanephros arises from a tissue bud form on the mesonephric duct. This tissue bud triggers kidney formation from the surrounding mesoderm (the ureteric bud). The ureteric bud splits to form major calyces, minor calyces, and collecting ducts. A junction of the digestive, urinary, and genital systems forms from the cloaca (L., sewer), the enlarged caudal end of the hindgut. The rectum and digestive tract arises from the urorectal septum that divides the cloaca into two parts. The ureter forms from the junction of the metonephric duct with the cloaca. As the kidney develops in the elongating fetus, it "ascends" from its original location adjacent to the developing bladder to its mature location in the retroperitoneum just caudal to the diaphragm. As the kidney moves relative to the bladder, it takes new arterial supply from the aorta and new venous drainage into the vena cava.

Cardiovascular System

The human heart develops in the mesenchyme of precordal plate region, an area known as the *cardiogenic region*. The heart begins at three weeks of development as a pair of endothelial tubes that fuse into a single, midline heart tube. The heart begins to contract between days 22 and 23 of fetal development. By the fourth week of embryonic development, the heart continues to elongate and starts to form an S shape, with the apex becoming the ventricles. The sinus venosus (tube that allows blood entry to the heart), atrium, ventricle, and bulbus cordis (tube that drains blood from the heart) develop from dilations that form along the primitive heart tube. By the fifth week, the heart tube begins to divide into an atria and ventricle. The atria and ventricle then begin to form right and left divisions by the development of septa. The ventricle is separated by an interventricular septum and the atria by an interatrial septum composed of the septum primum

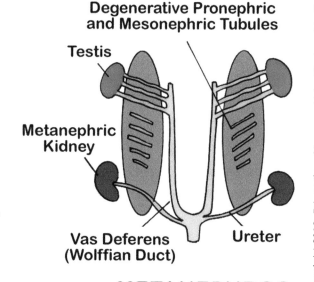

FIGURE 11.17
Human fetal kidney formation begins with the formation of a pronephros (21 days), followed at 28 days by the formation of a mesonephros, and then at 35 days for the formation of a metanephros (final form)

and septum secundum. There is a hole between the two atria, the foramen ovale, which remains open until after the baby is born. The foramen ovale allows blood to flow from the right to the left atrium in the embryo and fetus, because there is no reason to send blood to the lungs because they are filled with amniotic fluid.

The atria and ventricles expand rapidly during development. The atria absorb the sinus venosus, and the ventricle absorbs the bulbus cordis. The sinoatrial node develops from the sinus venosus.

Blood vessels start to form in the third week in the extraembryonic mesoderm. Blood vessel formation (angiogenesis) starts in blood islands, clusters of cells on the surface of the yolk sac. These blood islands extend and fuse with each other to form a network of cells. There are two populations of cells within this network: peripheral cells that will form endothelial cells, and core cells that will eventually form red blood cells. All vessels, arteries, and veins initially appear the same. Blood cell formation occurs in week 5 throughout the mesenchyme. Initially, blood cells are produced by the liver, but then by the spleen, bone marrow, and the lymph nodes.

During development, there are three different types of vascular systems: vitelline, embryonic, and placental. Each system is composed of arteries and veins. Vitelline blood vessels cover the entire surface of the yolk sac and connect to the embryo through the yolk stalk. Vitelline arteries contribute to the baby's gastrointestinal tract arteries, whereas vitelline veins empty into the sinus venosus and eventually contribute to the adult portal system. Embryonic blood vessels will eventually form nearly all of the cardiovascular system blood vessels. During fetal development, embryonic arteries will form the aortic sac, leading to the aortic arches, then the dorsal arteries, and finally, the umbilical artery. Three pairs of embryonic veins empty into the fetal heart sinus venosus.

> Is it dangerous for maternal and fetal blood to directly mix?

Placental blood vessels form initially in the connecting stalk and then in the umbilical cord. Paired placental arteries carry deoxygenated blood from the dorsal aorta and waste products to the placental villi. Paired embryonic veins carry oxygenated blood to the embryo from the sinus venosus.

Reproductive System

Gonads (male and female) arise from ridges along mesonephros ventral borders. Oocytes and spermatozoa develop from primordial germ cells on the surface of the yolk sac. Primordial germ cells enter the gonadal ridge after migrating into the embryo. Müllerian ducts, or *paramesonephric ducts*, develop lateral to mesonephric ducts and create uterine tubes, the uterus, and part of the vagina.

> Can an XY male fetus develop as a fully functional female?

Male babies have an X and a Y chromosome. The SRY locus on the Y chromosome produces a protein that stimulates the developing reproductive tissue to produce testosterone. In the absence of the Y chromosome, or if the SRY locus protein is not produced, the male reproductive tissues will not form. Increasing testosterone production causes the mesonephric duct system to enlarge and form the ductus deferens, the seminal vesicles, and the prostate gland. The developing testes then secrete Müllerian-inhibiting hormone, which causes the Müllerian ducts to degenerate.[1]

In men and women, ovaries and testes initially develop high up in the abdomen and then descend toward the pelvis. The testes descend farther than the ovaries. The inguinal canals are stimulated to form when the descending testes make contact with the lower abdominal wall. The inguinal canals provide a pathway for the testes to move into the scrotum after seven months of development. The testes move through the inguinal canal and reach the scrotum at 36 weeks' gestation.

Male and female external genitalia begin in the same structures but then diverge. The genital tubercle, a tissue enlargement, creates urogenital folds in the embryonic groin area. Labioscrotal swellings develop lateral to urogenital folds on each side of a urogenital opening. The urethra develops as a groove on the genital tubercle ventral surface. The genital tubercle can create penis in males, and the clitoris in females. The sex of the fetus is determined by the presence of absence of testosterone (and DHT). In the presence of a Y chromosome, testosterone is made and converted to DHT. The penis is formed, when DHT induces the genital tubercle and urogenital folds to close over the urogenital opening, and growth of the urethral groove. The scrotum arises from the labioscrotal swelling. In DHT cannot be produced, because of a lack of a Y chromosome or some developmental error, a penis does not form and a clitoris develops instead. The urogenital folds and genital tubercle do not close over the urogenital opening and the urethral groove disappears. The urethra now is positioned in front to the vaginal opening but behind to the clitoris. The labia minora arise from the urogenital and the labia majora arise from the labioscrotal folds.

In Review...

28.8 What are the major stages in fetal development? What occurs during each stage?

28.9 During what stage has a fetus enough control over its muscles to move within the uterus?

28.10 What steps occur during the fetal development of the gastrointestinal tract?

28.11 How does the nervous system develop?

Maternal Changes during Pregnancy

There are extensive physiologic, biochemical, and anatomical changes that occur in a mother during pregnancy. These changes are designed to increase the likelihood that a mother will carry a healthy baby to parturition without compromising her own health (Figure 11.18).

Digestive System

During pregnancy, nutritional requirements, including those for vitamins and minerals, are increased and several maternal alterations occur to meet this demand. In the early stages of pregnancy, the first trimester, it is not uncommon for women to actually lose weight because of morning sickness. Morning sickness refers to an increased feeling of nausea in the morning and greater incidence of vomiting. Morning sickness is thought to be a protective

FIGURE 11.18 Maternal changes throughout pregnancy

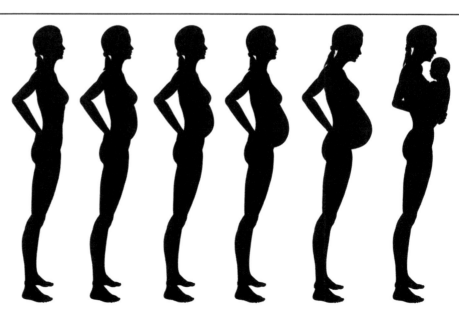

mechanism against the ingestion of teratogens during a critical period in fetal development. Many foods can contain potentially harmful chemicals to the fetus. Increasing levels of estrogen and progesterone are thought to act as an irritant to the digestive tract. During the second and third trimesters, a mother's appetite and food intake increase. The increase in appetite and food intake is mediated, in part, by increased levels of hCG.

Stomach and small intestine motility may be reduced during pregnancy due to increased progesterone levels, which, in turn, decrease the production of motilin, a peptide hormone known to stimulate gut smooth muscle. The transit time of food throughout the gastrointestinal tract decreases to the point where too much water is reabsorbed, leading to constipation.

Gastric production of HCl becomes more variable during pregnancy, especially during the first trimester. Production of the hormone gastrin increases significantly, resulting in increased stomach volume and decreased stomach pH. Gastric production of mucus may be increased. Esophageal peristalsis is deceased, accompanied by gastric reflux because of the slower emptying time and dilatation or relaxation of the cardiac sphincter. Gastric reflux is more prevalent in later pregnancy because of stomach elevation by the enlarged uterus. All of these alterations make the use of anesthesia more hazardous because of the increased possibility of regurgitation and aspiration.

During pregnancy, the large and small bowels move upward and laterally, and the appendix is displaced superiorly in the right flank area. These organs return to their normal positions after parturition. Gallbladder function is also altered during pregnancy because of the weakening (hypotonia) of the smooth muscle wall. Emptying time is slowed and is often incomplete. Bile can become thick, and bile stasis may lead to gallstone formation. There are no apparent morphologic changes in the liver during normal pregnancy, but there are functional alterations. Commonly, there is a change in blood albumin levels.

Oral Cavity

Increased salivation in pregnant women is related to the difficulty in swallowing associated with morning sickness. If the oral cavity pH decreases as a result, tooth decay may occur. Tooth decay during pregnancy is not due to lack of calcium because dental calcium is stable and is not mobilized during pregnancy, as is bone calcium.

The gums may become hypertrophic (swollen), hyperemic (increased blood flow leading to frequent bleeding), and friable (sensitive and easily damaged) because of increased systemic estrogen. Vitamin C deficiency also can cause tenderness and gum bleeding. The gums usually return to normal soon after the baby is born.

Kidneys and Urinary Tract

During pregnancy, each kidney increases in length by 1 to 1.5 cm, with a corresponding increase in weight. The renal pelvis and ureters become dilated and the ureters elongate, widen, and become curved. These changes can lead to a decreased evacuation of the bladder, increasing the risk of bladder infections. The glomerular filtration rate (GFR) increases during pregnancy by about 50%. The renal plasma flow rate increases by as much as 25 to 50%. Even thought the GFR increases dramatically during pregnancy, the volume of the urine passed each day is not increased. Thus, the urinary system appears to be even more efficient during pregnancy. With the increase in GFR, there is an increase in endogenous clearance of creatinine. The concentration of creatinine in serum is reduced in proportion to the increase in GFR, and concentration of blood urea nitrogen is similarly reduced.

Glucosuria, or glucose in the urine, during pregnancy is not necessarily abnormal, and may be explained by the increase in GFR with impairment of tubular reabsorption capacity for filtered glucose. Increased levels of urinary glucose may also contribute to increased susceptibility of pregnant women to urinary tract infection.

During pregnancy, renin, produced in the kidneys, increases early in the first trimester, and continue to rise until parturition. Renin increases the blood levels of angiotensin II, a potent blood vessel vasoconstrictor. Normally, pregnant women are resistant to the vasoconstrictive effect of elevated angiotensin II levels, but pregnant women who suffer from pre-eclampsia are not.

As the uterus enlarges, the urinary bladder is displaced upward and flattened anteriorly, posteriorly, or in diameter. Pressure from the uterus leads to an increase in micturition, whereas an increase in bladder vascularization and a decrease in muscle tone raise its capacity up to 1,500 ml.

Perhaps the most striking maternal physiologic change during pregnancy is the increase in the blood volume. The magnitude of the increases varies according to the size of the woman, the number of pregnancies she has had, the number of infants she has delivered, and whether there is one or multiple fetuses. Increases in blood volume progress until term; the average increase in volume at term is 45 to 50%. Increased blood volume is needed for extra blood flow to the uterus, the extra metabolic needs of fetus, and increased perfusion of others organs, especially the kidneys. Extra volume also compensates for maternal blood loss during parturition. The average blood loss with vaginal delivery is 500 to 600 mL, and with cesarean section, 1000 mL.

There is a corresponding increase in red blood cell production, approximately by a one third, but increases in plasma volume lead to a decline in hematocrit levels until the end of the second trimester, when the increase in the red blood cells is synchronized with the plasma volume increase. The hematocrit then stabilizes or may increase slightly near term.

With the increase in erythrocyte production, there is an increased dietary need for Fe^{3+}. If supplemental iron is not added to the diet, Fe^{3+} deficiency anemia will result. Maternal requirements can reach 5 to 6 mg Fe^{3+}/day in the latter half of pregnancy. If Fe^{3+} is not readily available, the fetus uses maternal Fe^{3+} stores. Thus, production of fetal hemoglobin is usually adequate even if the mother is severely Fe^{3+} deficient. Anemia in the newborn is rarely a problem; instead, maternal Fe^{3+} deficiency more commonly may cause premature labor and late spontaneous abortion.

The total blood white cell count increases during pregnancy from a pre-pregnancy level of 4,300 to 4,500/mL to 5,000 to 12,000/mL in the last trimester. Lymphocyte and monocyte numbers stay the same throughout pregnancy; but there are large increases in polymorphonuclear leucocytes.

During pregnancy, levels of several essential coagulation factors increase: there are marked increases in fibrinogen and factor VIII. Factors VII, IX, X, and XII are also increased, but to a lesser extent. Fibrinolytic activity, the removal of clots, is depressed during pregnancy and labor, although the precise mechanism is unknown. The placenta may be partially responsible for this alteration. Plasminogen levels increase concomitantly with fibrinogen levels, causing an equilibration of clotting and lysing activity. Changes in clotting dramatically affect the treatment of hemorrhaging during pregnancy or parturition.

As the uterus enlarges and the diaphragm becomes elevated, the heart is displaced upward and somewhat to the left with rotation on its long axis, so that the apex beat is moved laterally. Cardiac capacity increases to 70 to 80 mL and may be caused by increased volume or hypertrophy of cardiac muscle. The size of the heart appears to increase by about 12%.

Cardiac output usually increases 40% during pregnancy, reaching its maximum at 20 to 24 weeks' gestation and continuing at this level until parturition. The increase in cardiac output can be as much as 1.5 L/min over the nonpregnant level. Cardiac output in pregnant women is very sensitive to changes in body position, because the uterus can impinge on the inferior vena cava, decreasing blood return to the heart. Mean arterial blood pressure usually declines slightly during pregnancy, partially because of the obstruction posed by the uterus. The decreased mean arterial blood causes edema in the lower extremities.

Respiratory System

Pregnancy produces anatomic and physiologic changes that affect respiratory performance. Early in pregnancy, capillary dilatation occurs throughout the respiratory tract, leading to engorgement of the nasopharynx, larynx, trachea, and bronchi. This causes the voice to change and makes breathing though the nose difficult. As the uterus enlarges, the diaphragm is elevated and the rib cage is displaced upward and widens, increasing the lower thoracic diameter and circumference. Abdominal muscles have less tone and are less active during pregnancy, limiting forced inspiration and expiration.

Alterations in lung volumes and capacities during pregnancy include an increase in dead volume as a result of airway dilation; an increase in tidal volume; a decrease in total lung capacity because of diaphragm elevation; a decrease in functional residual capacity, residual volume, and respiratory reserve volume; and an increase in alveolar ventilation. There is also an increase in the respiratory rate leading to hyperventilation, and a decrease in alveolar CO_2. This decrease lowers maternal blood PCO_2; however, maternal PO_2 is maintained within normal limits. Maternal hyperventilation is considered a protective measure that protects the fetus from the exposure to excessive levels of CO_2.

Physical Changes

As the fetus and placenta grow and place increasing demands on the mother, large physical changes occur. The most obvious physical changes are weight gain and altered body shape. Weight gain is due not only to increases in uterus size but also to increases in breast tissue, blood, and water volume. Deposition of fat and protein is increased and there is an increase in intracellular fluid. The average weight gain during pregnancy is 12.5 kg.

During a normal pregnancy, approximately 1 kg of weight gain is attributable to an increase in protein content. Half of this increase is in the fetus and the placenta; the rest occurs in the uterus, breast glandular tissue, plasma protein, and hemoglobin.

Total body fat increases during pregnancy, but the amount varies with total weight gain. During the second half of pregnancy, plasma lipid levels increase, but triglycerides, cholesterol, and lipoprotein levels decrease soon after delivery. The ratio of low-density lipoproteins to high-density lipoproteins increases during pregnancy.

Exercise and Pregnancy

Pregnant women do not have to stop their exercise routine if they are cautious. Some will tire more easily and may suffer from morning sickness that can interfere with exercise. As the pregnancy progresses, shifts in body weight and posture require additional expenditures of energy for certain movements. Sudden stops, changes in direction, and rapid movements can be difficult to carry out. The pubic symphysis is less stable because of the release of relaxin, causing mothers to move with widely spread legs and a shuffling motion.

Clinical—In Context: Fetal Surgery

Surgery on the fetus in utero (still in the uterus) was first performed in the United States in 1979 to drain excess fluid associated with hydrocephalus. The hydrocephalus surgeries proved to be ineffective in correcting the underlying neurologic defects associated with hydrocephalus and have been discontinued. Fetal anatomic malformations have traditionally been managed after birth. However, advances in prenatal diagnosis, including prenatal ultrasound, have led to a new understanding of the progression and physiologic outcomes of certain congenital anomalies. In addition to this understanding, advances in anesthesia, drugs to prevent uterine contractions, and hysterotomy (surgical incisions of the uterine wall) have made maternal-fetal surgery more common. Fetal surgery as an option for early intervention for fetal malformations may produce life-saving results. Whereas controversial for nonlethal conditions, fetal surgery has been utilized to treat spina bifida, diaphragmatic (into the thorax) hernia, urinary tract obstructions, lung malformation, sacrococcygeal teratoma (coccygeal tumors), and complications of twin pregnancies. These surgeries are dangerous and done only when the benefits outweigh the risks and if it is easier to correct the congenital defect in utero than after birth.

Spina bifida used to be treated immediately after birth, but in utero repair leads to far less nerve damage and an improved outcome at birth. Congenital diaphragmic hernia is a diaphragm defect that results in abdominal protrusion into the chest, causing displacement of the lungs and the heart in the thoracic cavity. Repair of the diaphragm in utero allows for normal lung growth and a better outcome. Urinary tract obstruction can result in failure of the normal development of the lower urinary tract, leading to renal impairment, renal failure, and death. In utero surgeries to correct this problem have dramatically increased long-term survivorship.

Fetal surgery jeopardizes the pregnancy and puts both the mother and fetus at risk. Specialized surgical techniques used to access and operate on the fetus require a multidisciplinary approach. Once the uterus is opened, either through a traditional cesarean surgical incision or through single small incisions, the fetus is removed and the surgery is completed. The fetus is then returned to the uterus and the uterus is closed.

Even though blood flows to exercising muscles, there is no apparent reduction of placental blood flow. Exercise has no effect on lactation, provided women drink plenty of fluids and wear a supporting bra. Overall, moderate exercise does not pose a risk to a pregnant mother or her fetus, and increases a mother's sense of well-being and recovery from the birthing process.

In Review...

28.12 What are the metabolic changes that occur in a mother during pregnancy?

28.13 What are the cardiovascular changes during pregnancy? Respiratory changes? Gastrointestinal changes?

28.14 How does exercise affect pregnant women?

Parturition and Postnatal Life

Parturition, or *labor*, is the steps involved in childbirth (Figure 11.19). Near the end of the pregnancy, the uterus starts to exhibit Braxton-Hicks (J. Braxton Hicks, British gynecologist, 1823-1897) contractions. Braxton-Hicks contractions are a sign of the increasing irritability of the uterus smooth muscle that become stronger and more frequent until parturition has begun. Braxton-Hicks contractions are not strong enough for the expulsion of the fetus from the uterus. During parturition, powerful contractions begin at the superior portions of the uterus and travel toward the cervix. These contractions press the fetal head against the cervix. As a result, the cervix gradually dilates (thins). Before birth, contractions cause the amniotic sac to rupture and amniotic fluid is lost through the vagina (also known as when a woman *breaks her water*).

Initiation of Parturition

The factors that trigger parturition are not clearly understood, but hormones from the fetus, placenta, and mother appear to play key roles (Figure 11.20). One key signal is the increase in uterine size throughout pregnancy, in response to increasing estrogen levels. Normally, when smooth muscle is stretched it contracts and stiffens. Hormonal factors prevent this from occurring during the second and third trimesters whereas the uterine myometrium expands and stretches. Throughout most of the pregnancy, weak Braxton-Hicks contractions occur, but they become stronger and more frequent just before the onset of parturition. High levels of progesterone act to inhibit uterine contractions, but elevated estrogen levels have the opposite effect. During pregnancy, the progesterone/estrogen ratio prevents uterine contraction during fetal development. Parturition may be initiated by a change in this ratio by a decreased production of progesterone in late pregnancy. Parturition may also be initiated by increasing levels of prostaglandin PGF_{2a}, a potent stimulator of uterine contractions. Another possible factor may be increased fetal adrenal glucocorticoid production.

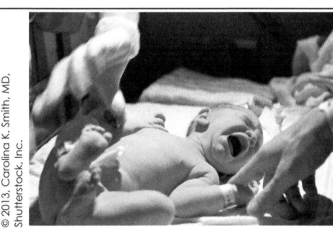

FIGURE 11.19
A newborn baby

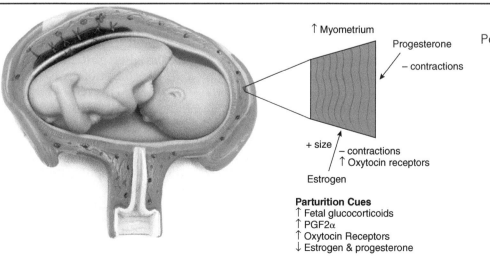

FIGURE 11.20
Parturition

Even though a single trigger for the initiation of parturition has yet to be identified, there is an increased uterine sensitivity because of increased oxytocin receptor number at the end of the pregnancy. Increased uterine sensitivity to oxytocin does not initiate parturition, but it clearly regulates and promotes parturition. A neuroendocrine reflex involving oxytocin maintains uterine contractions during parturition. Strong uterine contractions begin in the fundus and spread and weaken as they move toward the cervix. Uterine contractions force the fetus toward the cervix and fetal pressure stretches the cervix. Cervical stretch produces a nervous signal that travels up the spinal cord to the anterior hypothalamus and triggers oxytocin release. Oxytocin then acts on the uterine myometrium to stimulate further uterine contractions. This is a positive feedback loop because each step in the reflex is stimulatory. Uterine contractions will continue, and increase, until the pressure to the cervix is relieved by delivery.

Throughout most of the pregnancy, the cervix acts as a plug, keeping the fetus in and infectious agents out of the uterus. At the end of pregnancy, the cervix begins to soften as its collagen fibers dissociate and decrease in number. This softening of the cervix may be caused by relaxin, a polypeptide hormone produced by the corpus luteum and by the placenta (see chapter xx).

Parturition

Parturition has three stages with the first lasting from the onset of labor and finishing when the cervix is completely dilated, the second stage is finished when the baby is born, and the third stage ends with the expulsion of the placenta.

Stage 1: This stage is initiated by uterine changes brought about by fetal growth and hormonal changes. This stage has no diskrete beginning; some woman will have painful contractions that last for a few hours, stop, and begin again later. The length of the first stage can be as short as a few hours or last several days. Stage 1 is usually shorter with the second and third pregnancy. The powerful contractions generated at the top of the uterus act to drive the fetus downward against the cervix. The uterine contractions and fetal pressure on the cervix cause it to dilate and provide a larger opening for the head of the fetus to pass through the birth canal. Most women start active stage 1 parturition with a cervix that is dilated to 3 cm. In most women, the cervix dilates at the rate of 1 cm/hr. The cervix is completely dilated at 10 cm. Cervical dilation can only be determined by an internal vaginal exam.

Stage 2: Powerful uterine contractions in stage 2 drive the fetal head through the cervix and into the vaginal canal. This stage, also known as *transition stage*, can be short (minutes) or as long as an hour. As soon as the baby's head is visible from the external vagina (crowning), physicians may or may not perform an episiotomy through the peritoneum to widen the vaginal opening. The episiotomy is easier to repair than if the perineum was torn by fetal pressure but there is accumulating clinical evidence that women without episiotomies do not suffer more than women with episiotomies after birth. During this stage, uterine and abdominal contractions are powerful enough to temporarily compress the placenta and halt fetal blood flow. This temporary fetal hypoxia is usually not long enough to cause harm. This stage ends when the entire fetus has moved through the vaginal opening.

Stage 3: After stage 2, the baby is still attached through the umbilical cord to the placenta. The umbilical cord is tied off and severed and the placenta is eventually sloughed off the uterine wall by the remaining uterine contractions. Placental loss can include severe maternal bleeding as uterine blood vessels are damaged. Maternal bleeding is limited by uterine contractions compressing uterine arteries. This stage ends with the expulsion of the placenta and the perineum is rapidly stitched to seal the episiotomy.

Post Parturition

Maternal changes that occur after stage 3 are driven by the loss of placental estrogen and progesterone. Without these hormones, the uterus shrinks in size, but not to its prepregnancy size. It is common for women to have bloody vaginal discharges for a week after delivery as dying uterine cells and vessels are sloughed off.

Postnatal Life

The **neonate**, or *newborn baby*, goes through rapid, and dramatic, changes at the time of birth (Figure 11.21). These major changes are initiated by the neonatal chemical, mechanical, and thermal stimuli, and because of the separation from the maternal environment (maternal circulation and an aqueous environment).

Initiation of Breathing

The removal from an aqueous environment to a gaseous atmosphere and the removal of amniotic fluid from the fetal lung cause lung inflation (because of its elastic recoil and presence of surfactant) (see chapter 21). Lung expansion is a key trigger that causes important changes within the neonate's circulatory system. Pulmonary ventilation is stimulated by sensory, mechanical, chemical, and thermal stimuli. At birth, approximately 80 to 110 mL of amniotic fluid remains in the respiratory passages of a normal full-term fetus. The removal of this fluid is a key mechanical event that initiates respiration. The movement of the fetus through the vagina acts to compress the fetal chest, squeezing out approximately one third of this fluid from the lungs. Chest recoil is thought to produce a negative intrathoracic pressure that sucks air in to replace the fluid that was forced out. When the newborn exhales, a positive intrathoracic pressure is created. The remaining fluid is absorbed into interstitial space, fetal capillaries, and the lymphatic system. The movement of the remaining amniotic fluid to the interstitial space is rapid, but the movement into lymph and blood vessels may take several hours.

The fetus begins to experience brief periods of asphyxia during movement through the vagina, causing increased fetal PCO_2, decreased pH, and PO_2. These chemical changes stimulate aortic arch and carotid chemoreceptors, initiating reflexes that stimulate the medulla's respiratory center. This induces the neonate to inhale as soon as it has moved through the vagina. The decrease in environmental temperature after birth is also considered to initiate breathing.

Cardiopulmonary Changes

Expansion of the lungs with air reduces the resistance to blood flow through the pulmonary arteries and veins. In addition, the movement of O_2 into the alveoli triggers the relaxation of the pulmonary arteries and a decrease in the pulmonary vascular resistance. At the same time, the lowered surface tension

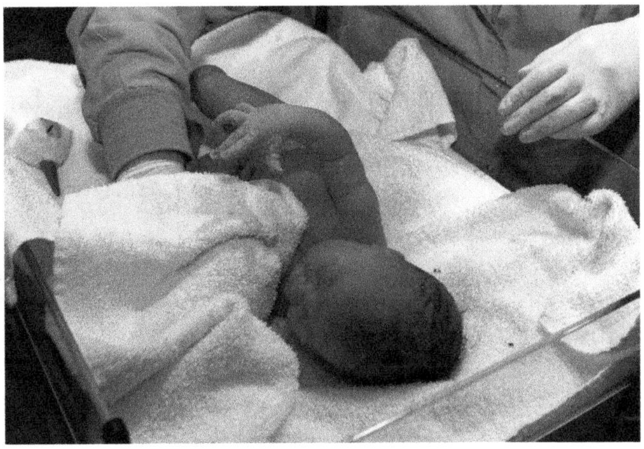

FIGURE 11.21

A newborn baby being cleaned prior to be tested for skin color, heart rate, reflexes, muscle tone, and respiratory effort or the **APGAR** (appearance, pulse, grimace, activity, respiration) score

> What is the significance of an open ductus arteriosus during the first two days of birth?

decreases interstitial pressure. As pulmonary vascular resistance decreases, the vascular flow in the lung increases to maximal levels within 24 hours.

Because of the increased pulmonary blood flow, blood flow increases through the right atrium to the right ventricle, and less blood flows through the right atrium and the foramen ovale into the left atrium. In addition, the left atrium is expanded by an increased pulmonary venous return. The increased left atrial pressure and decreased right atrial pressure, resulting from decreased pulmonary resistance, force blood against the septum primum, causing the foramen ovale to close. The closing of the foramen ovale separates the neonate heart into two separate pumps. The foramen ovale eventually becomes the fossa ovalis.

The ductus arteriosus, the connection between the pulmonary truck and aorta that allows pulmonary blood to be shunted to fetal systemic circulation, closes within two days after birth. This closure occurs because of local changes in blood pressure and P_{O2} that cause arterial constriction. After closure, the ligamentum arteriosum is formed as the ductus arteriosus is replaced by connective tissue. A transient murmur is often heard in newborns before the ductus arteriosus closes completely.

During fetal development, fetal blood passes through the placenta and returns through the umbilical vein. Fetal blood then passes through the liver via the ductus venosus, which then joins the inferior vena cava. When the umbilical cord is tied off and severed, umbilical artery and vein blood flow ceases, and they eventually degenerate. The liver ligamentum teres arises from the remnants of the umbilical vein, and the ligamentum venosus arises from the remnants of ductus venosus.

Approximately 70% to 90% of neonate hemoglobin is the fetal variety (Hb_F; see chapter xx). Hb_F has a higher affinity for O_2 than maternal Hb and the neonate's Hb O_2 saturation is greater than in an adult, but the amount of O_2 available to tissues is less. Because of this high concentration of O_2 in the blood, newborn hypoxia is difficult to recognize.

The normal neonate respiration rate is 30 to 50 breaths/min. Newborn breathing is largely diaphragmatic, shallow, and irregular, with the abdomen moving synchronously with the chest. Newborns normally experience short periods of apnea (lasting 5 to 15 seconds) that are rarely associated cyanosis or heart rate changes. Newborns are obligatory nose breathers immediately after birth, and for 2 hours after birth, respiration rates of 60 to 70 breaths/min are normal.

At birth, the neonate heart rate goes up to 175 to 180 beats per minute (bpm) and averages 125 to 130 bpm for the first week of life. The newborn's heart rate declines to approximately 100 bpm but increases quickly when awake.

Thermoregulatory Changes

A newborn does have physiologic mechanisms that increase heat production. The removal of the newborn from the warm, protected uterus to the outside world causes an immediate drop in body temperature that is compensated for by increased basal metabolic rate, muscular activity, and nonshivering thermogenesis. Newborn babies have brown adipose, which is absent in adults, surrounding key tissues and is the source of nonshivering thermogenesis. Newborns do not shiver to keep warm.

The main source of energy in the first four to six hours after birth is glucose, and there is an increased need for glucose for energy to maintain breathing and body temperature. The normal blood glucose level of a newborn is 50 to 100 mg/dL.

Digestive Changes

After birth, the neonate it is abruptly separated from maternal nutrients. This separation and the requirements of thermoregulation, cause a baby to lose 5% to 10% of its body weight in the first few days of life. The neonate digestive tract is very immature in comparison to an adult digestive tract, and the neonate can digest only a limited number of food types.

At birth, swallowed alkaline amniotic fluid raises the stomach pH near 7.0. Within the first 8 hours of life, stomach acid secretion increases rapidly and stomach pH decreases to very low levels. It takes about a week for stomach acidic production to stabilize at normal levels and stomach pH gradually rises over the first month of life.

The newborn's liver lacks the necessary enzymes to excrete bilirubin, until 14 days postpartum, causing temporary jaundice (a buildup of bilirubin), which is common in premature babies. The immature neonate digestive tract has the ability to digest lactose, the primary breast milk sugar, from birth. Neonate pancreatic secretions support lactose digestion but it takes several months before the digestive system and secretions can support the ingestion of solid foods. Newborns can also become allergic to new foods and caution should be exercised when exposing babies to new foods.

Salivary gland and pancreatic amylase secretion remains low until after the first year of life. Lactase activity, required for lactose digestion, remains high during the first year of life but then decreases during infancy, but at levels that are still higher than those in adults. Many adults become lactose-intolerant because of a lack of lactase activity (see chapter xx).

> Why does placental blood first go to the fetal liver?

Integumentary Changes

Until a baby's liver begins to fully function, bilirubin can build up in the bloodstream, resulting in jaundice. This condition usually occurs after the first 24 hours of life, and bilirubin levels peak after 3 to 5 days. In the majority of instances, this jaundice is normal, harmless, and temporary. Bilirubin levels also can rise when a mother begins to produce mature breast milk. This type of jaundice, known as breast-milk jaundice, peaks at 2 to 3 weeks of age.

Babies are also born with many different common birthmarks. Stork bites are pale pink or red spots found on eyelids, nose, and lower occipital bone and nape of the neck. These birthmarks are more noticeable during periods of crying; they have no clinical significance and usually fade. Mongolian spots are macular areas of bluish black or gray-blue pigmentation that are common in babies of dark skinned parents. These birthmarks gradually fade but may be mistaken for bruises. Portwine stain is caused by a capillary angioma directly below the epidermis. This birthmark is a nonelevated, sharply demarcated, red-to-purple collection of dense capillaries. It commonly appears on the face but does not grow, fade, or blanch. A strawberry mark is caused by a capillary hemangioma in the dermal and subdermal layers of the skin. It is a raised, clearly delineated, dark red, rough-surfaced birthmark, commonly found in the head region. It grows and becomes fixed in size by eight months.

Immunological Changes

During fetal development, the fetus is protected from external threats by the mother's immune system and the separation of the maternal and fetal blood supplies by the placenta. Maternal transfer of IgG antibodies to the fetus in utero, passive active immunity, gives the fetus some level of protection against bacterial toxins. IgG antibodies are transferred primarily during the third trimester, and premature babies are at greater risk to develop infections in part because these antibodies were not transferred. Newborn babies have high levels of maternal IgG antibodies until 5 to 6 months old. The fetus produces its own IgM antibodies. Maternal IgM and IgA antibodies do not cross the placenta. Colostrum, the first breast milk, has very high concentrations of IgA antibodies. Newborn begin to produce secretory IgA at about four weeks after birth.

Apgar Scores

Newborn babies are evaluated to assess their physiologic condition. This assessment is known as the **Apgar score**. The Apgar score (Virginia Apgar, U.S. anesthesiologist, 1909-1974) refers **a**ppearance, **p**ulse, **g**rimace, **a**ctivity, and respiratory effort of the neonate (Figure 11.21). Each characteristic is rated on a scale of 0 to 2, with 2 denoting normal function, 1 denoting reduced function, and 0 denoting seriously impaired, or no, function. The total Apgar score is the sum of all five individual scores and ranges from 0 to 10. A normal Apgar score within 5 minutes of birth is 8 to 10.

Lactation

Lactation refers to new mother breast milk production. Breast milk is normally formed in a mother after childbirth and, if given the proper tactile and sensory stimulation, last for 2 or 3 years.

Lactation does not occur until there has been substantial development of mammary tissues and ducts. During pregnancy, mammary tissue undergoes rapid hypertrophy under the control of high estrogen and progesterone concentrations and other important hormones. During mammary gland hypertrophy, ducts that carry milk to the nipple also grow and repeatedly branch to form an extensive network. The increase in mammary tissue, and adipose tissue, normally cause an increase in breast size. The high levels of estrogen and progesterone can also cause the areola to darken. Estrogen primarily causes an increase in breast size, whereas progesterone causes the development and enlargement of breast secretory alveoli, but they do not secrete milk during pregnancy. Mammary tissue development also requires maternal insulin, growth hormone, glucocorticoids, thyroid hormones, and prolactin. Placental hormones, human placental lactogen (hPL) and somatotropin (hPST) pass through the placenta and stimulate mammary tissue growth and development.

Estrogen, progesterone, and a suite of other maternal and fetal hormones are required to develop the mammary tissues required to support lactation. Lactation itself is under the regulation of anterior pituitary prolactin (see chapter xx). During pregnancy, prolactin release could be stimulated by high estrogen levels, but milk production is blocked by high progesterone levels (through a downregulation of mammary tissue prolactin receptors). After parturition both progesterone and estrogen (and prolactin) levels decrease, and *normal* prolactin levels stimulate milk production. The overlap between the CNS and endocrine system is clearly demonstrated when a baby begins to breast-feed. The tactile (suckling) and sensory (mother seeing the baby nurse) information is relayed to the hypothalamus, which blocks the release of PIF (prolactin-inhibitor factor or dopamine). As a result, there is the release of a putative prolactin-releasing hormone (an unknown peptide) and blood prolactin levels rise and the breasts are stimulated to produce milk.

Immediately after parturition, mammary tissue produces **colostrum**, or *first milk*, a white, milky substance that has little fat and lower lactose (milk sugar) levels than later breast milk. The later milk, with its higher fat and lactose levels, is a much more nutritious milk. Babies have an underdeveloped digestive tract and cannot digest more complex foods than breast milk, or an artificial substitute. Breast milk contains more than readily digestible lactose; it also contains maternal IgA, which helps to protect the baby against infections. Ingested IgA antibodies are able to enter fetal circulation because of the incomplete development of the neonate's digestive tract.

Repeated stimulation of prolactin release makes breast-feeding possible for 2 to 3 years. If nursing stops, prolactin levels drop rapidly and milk production stops. Milk production is not instantaneous; it takes time to produce breast milk. The prolactin released during a morning feeding session will produce milk for the afternoon feeding. At the time of nursing, mechanical, olfactory, and visual stimuli cause the release of oxytocin from the posterior pituitary. Oxytocin stimulates cells surrounding breast alveoli to contract, releasing milk from the breasts in a process known as milk letdown.

In Review...

28.15 What events occur during parturition? During labor?

28.16 What factors initiate labor?

28.17 What is the role of the cervix in parturition?

28.18 What events characterize the first few minutes of life?

28.19 What factors initiate respiration?

28.20 What changes occur in the newborn baby's cardiovascular system?

28.21 What digestive changes occur after birth?

28.22 What is the Apgar score? What does it signify?

28.23 What hormones are responsible for the production of breast milk?

28.24 What hormones stimulate breast development during pregnancy?

28.25 Why is breast milk important for the development of a newborn?

Growth and Development in the First Year of Life, Life Stages, and Aging

The first year of life is a time of many changes, and the order and timing of these changes vary from child to child. In the first month of life, arm and leg movements are reflexes and the head flops if not supported. The baby can only see objects about 10 inches away from its face. A one-month baby is only awake 1 hour out of every 10 and cries if under or over stimulated. The 1-month old baby does respond to voices, especially the mothers (remember that the fetus was capable of hearing the mother's voice in utero).

By the second month, the baby has begun to develop more sophisticated skeletal muscles control and can lift its head by 45 degrees. It may be also able to briefly hold a rattle. CNS development allows a baby to begin to associate sensory information with physical needs, for example, if hungry or upset, it might only be comforted by a mother and no one else. The baby begins to develop language skills and makes primitive noises.

By the third month, a baby can roll on its back and hold up its head.[1] It may even be able to briefly stand if supported. The baby has begun to discover its own body parts (like hands and feet) and begins to make extended vowel sounds. A mother can begin to notice that a baby makes different cries for different needs. The baby is more aware of its social surroundings and can now smile at faces.

By the fifth month, the baby has developed motor skills and can reach for objects, stand up with help, and being to wiggle on the floor. CNS development now allows babies to show interest in colors and display expressions and decision-making skills. A baby at this age will try to mimic sounds and will turn toward speakers.

By the ninth month, a baby has developed the skills to crawl and stand when supported. A baby is now capable of walking with support. CNS development lets a baby now begin to understand simple phrases, and the baby begins to explore the physical relationship between objects by building and disassembling blocks and toys. A nine-month baby should be able to say simple words and understand the concept of no and a mother's anger. The baby should also be able to point to things that it wants and like to play games.

The normal development of these skills is dependent, in a large part, on the rapid development of the CNS during the first year of life. It was previously thought that the total number of adult neurons was present in the brain at the time of birth, but this is not the case, new neurons are added during the first few years of life and some scientists feel that new neurons are added throughout life. Brain development and growth is not thought to be due to the growth of large numbers of new neurons, but growth and maturation of neuroglial cells, formation of myelin sheaths, and formation of neuron-to-neuron connections that form circuits. Any development deficit (the inability to carry out a normal age-related task) may be an indication of a CNS development problem or a problem with the development of the skeletal muscles.

Life Stages

The prenatal and neonatal life periods represent a small part of the entire human life span. The life stages from fertilization to death are:

1. Germinal period: from fertilization to 14 days in utero.
2. Embryonic development: from 14 days to 8 weeks in utero.
3. Fetal development: from 8 weeks in utero to parturition.
4. Neonatal development: from parturition to 1 month after birth.
5. Infancy: from one month to 1 or 2 years (usually when the baby begins to walk).
6. Childhood: from infant until puberty.
7. Adolescence: from puberty until 20 years.
8. Adulthood: from age 20 until death.

It is beyond the scope of this text to discuss all of the physical, emotional, biochemical, and structural changes that occur during childhood, adolescence, and adulthood. An adult is formed by experience; its environment, genetic makeup, and the changes that occur in the womb. Many personality traits are formed as an infant and retained throughout life. The enormous changes that occur during puberty (see chapter 27) also affect an individual's behavior and emotions. During adulthood, the human body undergoes a number changes that characterize aging.

Aging

Humans may or may not have a preprogrammed life span. The life span of people living in the United States has increased over the last century and is now over 79 years for men and women (Table 28.1). Women have a slightly longer predicted life span than men. The increase in lifespan over the past 20

TABLE 28.1 ♦ Historical Changes in United States Longevity Rates

Year	Male (years)	Female (years)
1900	48	50
1920	54	55
1940	62	66
1960	67	73
1980	70	77
2000	74	79
2010	75	80
2020	76*	81*

*Estimate From Centers for Disease Control.

years was as great as the increase in the life span the previous 80 years (from 1900 to 1980 compared with 1980 to 2000). This increase was a result of better sanitation, better diets, and better medical care.

Many theories have been proposed for why humans age. Several of the most well known are: (1) the accumulation of deleterious mutations theory, (2) and the antagonistic pleiotropy theory, and (3) the recently developed reliability theory. According to the deleterious mutation theory, aging is an inevitable result of the declining force of natural selection with age. For example, a mutant gene (see later section in this chapter) that kills young children will be strongly selected against (not be passed to the next generation) whereas a lethal mutation that affects only people over the age of 80 will experience no selection pressure because people with this mutation will have already passed it on to their offspring. Over successive generations, late-acting deleterious mutations will accumulate, leading to an increase in mortality rates late in life.

The antagonistic pleiotropy theory states that late-acting deleterious genes are favored by selection and will be actively accumulated if they have any beneficial effects early in life. This means that a gene will be retained if it provides some benefit during the childbearing years even if it causes death when a person is over 60 years of age. These two theories of aging are not mutually exclusive, and both evolutionary mechanisms might operate simultaneously. The main difference is that in the mutation accumulation theory, genes with negative effects during old age accumulate passively from one generation to the next, whereas in the antagonistic pleiotropy theory, these alleles are actively kept in the gene pool by selection.

The reliability theory was developed as an engineering approach. In this theory, aging is caused by an increased failure rate of body tissues or organs. The failure rate of human tissues and organs is U-shaped. Failure rates are very high during fetal and infant development, low during adolescence, and increasing during adulthood. As humans approach the age of 100, the risk for death no longer increases exponentially but plateaus. The theory suggests that there is no fixed upper limit to human longevity and assumes that the human body is a collection of redundant parts, that themselves do not age. The theory also assumes that humans start with many defective tissues and organs and that correction of these problems could have a major impact on human longevity.

Whatever the reason for aging, there are clear physiologic changes that occur within cells. The obvious age-related physiologic changes are all due to cellular changes (remember that structure dictates function on a cellular level). One of the most important is the development of cell **senescence**, or the inability of cells to divide and replace themselves. Throughout life, epithelium, liver cells, and olfactory neurons (to name a few) proliferate and replace dead or damaged cells, but many other cells show signs of senescence. It had been thought that neuronal cells cease to form after birth, but this is not the case. Neuronal cells are continually replaced within the olfactory epithelium and may even grow within the brain, but this theory remains controversial. The inability to replace damaged cells means that tissues and organs begin to lose their ability to carry out physiologic tasks.

Another common age-related change occurs within mitochondrial DNA. The transcription of mitochondrial genes decreases with age in a tissue-specific pattern. This leads to particular diseases, disorders, and premature aging in a small group of people in which this loss occurs faster.

Connective tissues decrease their flexibility with age because of a decrease in hyaluronate and an increase in cross-linked collagen (caused by products of cellular respiration), which leads to wrinkles. Fetal collagen has very few cross-links but as cells age, many connective tissue proteins become highly cross-linked, making the tissues more rigid and less elastic. The tissues with the highest concentration of collagen and related proteins are those most severely affected by cross-linking. The eye lens is one of the first tissues to show evidence of cross-linking. Near-vision becomes increasing difficult with age and older people require reading glasses. Loss of elasticity greater affects joints, blood vessels, kidneys, lungs, and the heart, greatly reducing their functional ability.

Mature muscle cells do not proliferate after terminal differentiation and formation of muscle fibers. As a result, a baby is born with all of its muscle fibers and their number declines with age when cells become damaged or die. The peak of skeletal muscle strength is between the ages of 20 and 30; strength declines steadily thereafter. Furthermore, skeletal muscle proteins undergo changes with age that makes them less functional. A good exercise program can slow or even reverse this process.

The age-related decline in muscular function also contributes to a decline in cardiac function. Connective tissue changes cause the heart to lose elastic recoil coupled with decreases in muscular contractibility. As a result, cardiac output declines and less O_2 and nutrients reach cells throughout the body, contributing to their decline (a positive feedback loop).

In addition to a decrease in cardiac output, changes to arteries further reduce the flow of blood to tissues. Atherosclerosis is the deposit of a soft, gruel-like material in large- and medium-sized artery lesions. These deposits become fibrotic and calcified, resulting in arteriosclerosis, or a hardening of the arteries. Arteriosclerosis interferes with normal blood flow and can result in a thrombus, a clear plaque, or clot formed within a vessel. If a piece of the clot breaks off and becomes an embolism, it can lodge in smaller arteries and cause myocardial infarction or strokes. Atherosclerosis occurs to some extent in all middle-aged and older people and even may occur in younger people. People with high blood pressure are at an increased risk for the formation of atherosclerosis. There is also a clear genetic predisposition for atherosclerosis in certain families.

All of the large visceral organs (liver, pancreas, stomach, and colon) undergo age-related degenerative changes. The ingestion of harmful substances may accelerate these changes, like excessive consumption of alcohol, smoking cigarettes, or exposure to noxious compounds in the environment.

Cells can also be damaged by exposure to radiation and toxic chemicals in the environment. Some substances, such as vitamins C and E, have beneficial effects in repairing or preventing cell damage. Vitamin C also stimulates collagen production and may slow the loss of tissue elasticity associated with aging collagen. Because of poor diets, many people over the age of 50 do not get enough vitamins and minerals. Feeling "bad" is not necessarily a part of aging but may be a result of a person not consuming the right amounts of vitamins and minerals in their diets or failure to get enough exercise.

There are decreases in the effectiveness of the immune system with aging. The aging immune system loses its ability to respond to foreign antigens, whereas at the same time becoming more sensitive to self-antigens. This results in an increase in autoimmune disorders, like arthritis, chronic glomerular nephritis, and hyperthyroidism. The change in immune function is demonstrated by the differences in the most common infections for people over the age of 65 compared to those in the general population. In the general population, the most common infections are sexually transmitted diseases, but in older adults, the most common infections are *Salmonella* poisoning, Lyme disease, and tuberculosis.

The best evidence for a genetically predetermined lifespan is found in patients suffering from **progeria** (G. *pro*, before; *geras*, old age). These patients suffer from an accelerated aging process that occurs shortly after the first year of life. By age 7, progeria patients may look like a very old person.

Death is not a result of old age. Other problems, like heart failure, renal failure, or stroke, are usually listed as the cause of death. The clinical definition of death is the permanent cessation of life functions and the cessation of integrated tissue and organ functions. The most widely clinically accepted indication of human death is brain death, which is defined as irreparable brain damage manifested clinically by the absence of response to stimulation, the absence of spontaneous heartbeat and respiration, and a isoelectric (flat) electroencephalogram for at least 30 minutes in the absence of known CNS poisoning or hypothermia.

In Review...

28.26   What are the normal developmental events that occur during the first year of life?

28.27   What changes occur during aging?

28.28   What are the two models of aging?

28.29   What are the major causes of death?

Human Genetics

Genetics is the scientific study of how particular traits are transmitted from parents to offspring. **Heredity** is a scientific examination of process of handing down certain traits from parents to their children; like skin color, height, features, etc. Heredity is based on the handing of genetic information in the form of genes. A person's physical traits are determined by their genetic makeup but are influenced by the environment. A person who suffered from malnutrition, as a child during the critical growing years, will be shorter than their potential height. The susceptibility of people to a disease like cancer, his or her physical abilities, and lifespan are all affected in a complex way by heredity and the environment. Because many of the infectious diseases are now controllable by modern medicine, diseases that have a genetic basis are quickly becoming much more of a health concern. However, in many cases, it is very difficult to separate the genetic and environmental factors responsible for a disease or disorder.

Chromosomes

Within each cell nucleus (except for mature red blood cells) is a collection of deoxyribonucleic acid (DNA) and proteins known as chromosomes. DNA is a polymer composed of four bases: ATP, CTP, GTP, and TTP. Chromosomes are responsible for the control of all cellular activities through the cell's production of proteins. The great majority of cells, known as somatic cells, contain 23 pairs of chromosomes, or 46 total chromosomes. Because they contain two copies of each chromosome, they are known as **diploid** cells. Gametes contain only copy of each chromosome and are known as **haploid** cells. Somatic cells are all of the cells of the body except for oocytes and spermatozoa. Chromosomes and their associated proteins (like histones) become visible as densely stained bodies during cell division (see chapter 2).

It is possible to identify specific human chromosomes through a **karyotype**. A karyotype is created by staining chromosomes with dye and photographing them through a microscope. The photograph is then cut up and rearranged so that the chromosomes are lined up into corresponding pairs (Figure 11.22). The 46 chromosomes are divided into 22 pairs of **autosomal chromosomes**, all of the chromosomes except the sex chromosomes, and the sex chromosomes, which determine individual sex. The autosomal chromosomes are numbered in pairs from 1 to 22, based on size, with the sex chromosomes denoted as the X or Y chromosome (based on their shapes). A normal female has 22 autosomal chromosomes and two copies of the X chromosome and is denoted as *XX*. A normal male has 22 autosomal chromosomes and one X and one Y chromosome, and is denoted as *XY*. Chromosomes differ in size: chromosome 1 is the largest and chromosome 21 is the smallest. The X chromosome is three times

FIGURE 11.22

A karyotype of human chromosomes with trisomy of the 21st chromosome

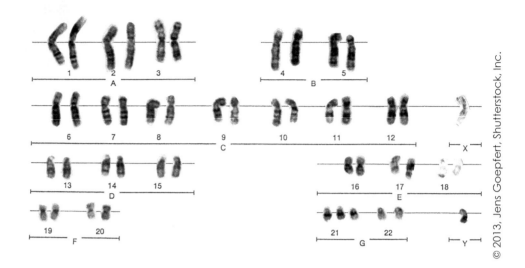

as large as the Y chromosome. The recently completed human genome project has determined the sequence of each base within all 23 chromosomes (http://www.ncbi.nlm.nih.gov/genome/guide/human/). One potential impact of the human genome project is a better understanding of the genetic basis of disease.

Chromosomes contain instructions for the synthesis of all of the body's proteins organized in genes. The gene is the functional unit of hereditary and is found within a specific area of a chromosome known as a **locus**. Homologous chromosomes (both members of a pair of chromosomes) contain similar, but not always identical, genes at a locus. An **allele** is the gene form inherited from each parent and found on the same locus on homologous chromosomes. **Homozygous** alleles encode for the same trait (the alleles are identical), whereas **heterozygous** alleles encode for different traits (the alleles are not identical). It is estimated that there are, ~30,000 genes within all 23 chromosomes. Chromosome 1 (the largest chromosome) has 2,475 known genes and chromosome 21 (the second smallest chromosome) has 315 genes. The exact number of genes is not known because of the difficulty in identifying a stretch of DNA as a gene.

All genes can be transcribed into mRNA, which travels to the ribosomes and is translated into proteins (see chapter 2). Regulatory genes produce proteins that regulate the transcription of other genes. Structural genes produce enzymes, hormones, collagen, actin, myosin, and other proteins. Even though each somatic cell contains identical copies of each gene, not all proteins are produced in all cells, and the same proteins are produced at different times within different cells. This tissue-specific production of proteins is responsible for the differences among nerve, muscle, connective, and epithelial cells.

An important law of genetics (Mendel's Law of independent assortment) is that allele pairs separate independently during the formation of gametes. This means that traits are transmitted to offspring independently of one another. This independent assortment is a random process and the distribution of alleles within gametes is unique (because no two gametes get exactly the same assortment of alleles). A genetic examination of heredity reveals that some alleles seem to be linked to others during independent assortment. During meiosis, when tetrads are formed (see chapter 2), homologous chromosomes can share large sections by crossing-over. Crossing-over occurs when a region of one non-sister chromatid is exchanged for another. Crossing-over provides for genetic recombination and a means for repairing damage to chromosomes.

Linked alleles are inherited together rather than as individual genes because of chromosomal segregation during meiosis. The closer that two genes are on the same chromosome, the more likely it is that they will segregate together during meiosis. This linkage trait has been used successfully to identify genes responsible for a number of diseases.

During meiosis, segregation errors can occur because as the chromosomes separate during meiosis, the two members of a homologous pair may adhere strongly and not segregate properly. Unequal crossing-over can result in **aneuploidy**, when a gamete has one more or less chromosome than normal. In most cases, aneuploidy is a lethal error and an aneuploidic oocyte may not be able to be fertilized, or if fertilized may not be able to develop normally. Aneuploidy is thought to be a major reason why as many as 50% of embryos do not survive past 10 weeks of development. Not all aneuploidies are immediately lethal; chromosome 15 deletions are associated with Prader-Willi syndrome and fragile X syndrome is characterized by the lower tip of the X chromosome being partially disrupted.

> Why would it be lethal if a fertilized oocyte had 45 or 47 chromosomes instead of 46?

Linkage Analysis

Modern medicine is interested in identifying the genes responsible for certain diseases in an effort to develop new treatments, like a treatment for cystic fibrosis. The identification of genes responsible for disease can be made through a linkage analysis approach. This approach depends on a number of factors: the disease and its prevalence, the age of onset, the number of genes involved, the nature of their contributions, the effects of gender, and the effects of the environment.

Linkage analysis requires information about families or populations in which a specific disease occurs at significant levels. Scientists look at the linkage between known chromosomal markers, specific known locations on chromosomes that act like street signs, and the disease in individuals. For example, a women suffering from a disease marries a man and they have four children. Two of the children also have the disease when they become adults and they pass it on to some, but not all, of their children. Through this type of information, it is possible to develop a pedigree tree of the transmission of the gene responsible for the particular trait. Scientists then look for chromosomal markers that are linked with the inheritance of the defective gene. Using this signpost, it is possible to determine on which chromosome the defective gene is found, and to close in on that gene. If good chromosomal markers are available for a particular chromosome, it is possible to find the gene, as happened in the case of breast and ovarian cancers. The genes *BRAC1* on chromosome 17 and *BRAC2* on chromosome 13 have been shown to play a key role in the development of breast cancer through their discovery by linkage analysis (Figure 11.23).

Dominant and Recessive Genes

In his classic experiments, Mendel examined the inheritance of traits between pea plants. These experiments determined that not all traits are inherited in the same way; some traits seemed to dominate over other traits, whereas others could be masked. The inheritance of human traits was first motivated by a desire to identify the genes responsible for defective tracts, like albinism. Another defective trait creates phenylketonuria (PKU) in children. Children suffering from PKU lack the necessary enzyme to metabolize phenylalanine. Phenylalanine accumulates in the brain and these children suffer from developmental abnormalities. The gene was localized to a locus on chromosome 12. One of the main promises of genetic medicine in the 21st century will be the ability to correct these defective alleles in utero.

FIGURE 11.23

One family's inheritance of the *BRCA* gene mutation

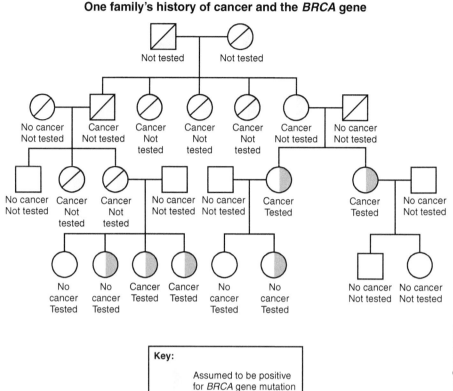

For many genetic traits, the two alleles are not normal and defective, but alleles that code for slightly different proteins. In this case, the effects of one allele may mask the effect of the other. When this occurs, one allele is said to be **dominant** and the other **recessive**.

By convention, dominant traits are designated by uppercase letters and recessive traits are designated with lowercase letters. For example, a heterozygous trait would be designated as *Aa*, and a homozygous trait would be designated as *aa* (homozygous recessive) or *AA* (homozygous dominant). The actual set of alleles a person has at a specific locus is known as the person's **genotype**. The expression of those alleles and the resulting changes they cause in a person is known as the **phenotype**.

It is possible to predict inherited traits in the offspring of two parents if their genotypes are known. For example, Tay Sachs disease is caused by a genetic defect in the hexosaminidase gene. If a child has defective copies of both alleles for the hexosaminidase gene, glycosphingolipids accumulate in **CNS** cellular lysosomes. The glycosphingolipid accumulation causes developmental problems leading to blindness, seizures, paralysis, and death. If a heterozygous normal father (*Hh*) mates with a heterozygous normal mother (*Hh*), the probability of their children having Tay Sachs disease can be predicted in a Punnett square (Figure 11.24). The probability of one of their children having Tay Sachs disease is 25%. If the mother is homozygous normal (*HH*) for the trait, the probability that her children will have Tay Sachs disease is 0%. Both heterozygous normal parents are carriers with a normal phenotype because they have a dominant allele.

Sex-Linked Traits

If a trait is localized to the X or Y chromosome, then those traits are sex-linked traits. Almost all sex-linked traits are found on the X chromosome (X-linked) and only one, for maleness, is found on the Y chromosome (Y-linked). A classic example of an X chromosome sex-linked trait is green or red color blindness. The allele for the red or green eye pigment is found only the X chromosome. A normal homozygous woman ($X^B X^B$) has two copies of the normal pigment gene and is not color blind. A heterozygous woman ($X^B X^b$) has one normal and a defective allele but is not color blind. Only homozygous recessive women ($X^b X^b$) are colorblind. The male offspring of a normal homozygous woman ($X^B X^B$) and a normal man ($X^B Y$) are not colorblind ($X^B Y$), nor are the female offspring ($X^B X^B$). The cross between a carrier mother ($X^B X^b$) and a normal man ($X^B Y$) will result in normal female offspring but 50% of the males will be colorblind ($X^b Y$).

Another X chromosome sex-linked disorder is Duchenne muscular dystrophy. Duchenne muscular dystrophy is caused by a defect in the muscle protein dystrophin, leading to muscle degeneration and death. The gene for dystrophin is found on the X chromosome. The carriers for Duchenne muscular dystrophy are only female, because males with a defective copy of the dystrophin gene on the X chromosome usually do not live long enough to have offspring. Hemophilia is another example of an X chromosome sex-linked trait.

A Y-linked recessive trait will only be passed on to sons and never to daughters because only male children will receive the Y chromosome from their fathers. This means that a mother can never express the Y-linked recessive phenotype, nor can she pass it on to her children. In a cross between a normal female and a male expressing a Y-linked trait, 100% of the male children are at risk for inheriting the recessive trait. If this couple does not have male children, the trait will be lost.

Exceptions to Simple Inheritance

Since the landmark experiments of Mendel, our understanding of human genetics has increased enormously. We now understand that genetics is more complex than the simple inheritance pattern of dominant and recessive alleles. The simple rules of Mendelian inheritance do not always apply. Below are examples when the simple rules do not apply.

Sex Limited Traits

Not all traits are equally expressed in both sexes. Women and men can inherit the same alleles from their parents that produce facial hair, but the presence of sex steroids affects the expression of the trait so that men hair heavy facial beards and women have sparse and very fine facial hair. This is also an example of incomplete penetrance (see later).

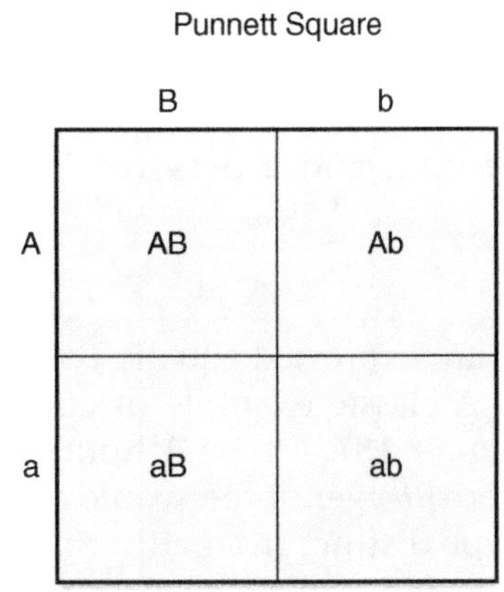

Punnett Square

	B	b
A	AB	Ab
a	aB	ab

FIGURE 11.24

A Punnett square analysis of the inheritance of a somatic allele

Sex Controlled Traits

In contrast to sex limited traits, in sex controlled traits, men and women expressed the same alleles but in a different manner. This is because sex steroids (testosterone and estrogen) affect the expression of traits in a different way. Sex controlled traits are responsible for the altered fat deposition in men and women and affects the severity of diseases. Gout is a disorder characterized by the accumulation of uric acid crystals within joints. Men are 8 to 10 times more likely to suffer severe symptoms than women.

Sex Imprinting

Sex imprinting, or *genome imprinting*, is seen when both men and women have the same alleles, but it is expressed in one sex but not the other. Sex imprinting is thought to be responsible for shutting off the expression of the alleles on one of the X chromosome in women so that they express the same levels of proteins as men who only have one X chromosome. Some diseases are affected by sex imprinting, like diabetes, psoriasis, and some types of mental retardation.

Polygenic Traits

Some phenotypic traits are the result of alleles at more than one locus. These traits are known as polygenic traits. Human polygenic traits include height, skin pigmentation, and eye color. Human height is the result of multiple alleles at many loci acting in an additive manner. The inheritance of eye color was thought to be a simple example of dominant and recessive alleles. Early models suggested that there were two alleles for eye color: a dominant brown and a recessive blue. People with two brown alleles, or a brown and a blue allele, would always have brown eyes. Only a person with two blue alleles would have blue eyes. This simple model did not explain the presence of green, hazel, and other eye colors. It is apparent that there are at least three genes with dominant and recessive alleles that affect eye color. In some polygenic traits, it is common to have continuous variation, where the trait varies between wide ranges of phenotypes.

Incomplete Dominance

In some traits, a dominant allele does not completely suppress the recessive allele. The resulting phenotype lies somewhere between the two alleles. In this case, the dominant allele expresses *incomplete dominance*. Human traits that express incomplete dominance include skin pigmentation and the pitch of human male voices. Careful examination of male singers with high and low pitch has suggested that they are homozygous for dominant and recessive alleles (*PP* and *pp*). Male singles with intermediate pitch are heterozygotic (*Pp*). Incomplete dominance is also seen in several genetic diseases. Tay Sachs disease is characterized by incomplete dominance. Heterozygotic carriers of the Tay Sachs allele only express 50% of the normal levels of hexosaminidase activity, which is absent in the Tay Sachs disease. Sickle cell anemia is also characterized by incomplete dominance. Heterozygotic individuals have half of their red blood cells filled with normal hemoglobin and half filled with abnormal hemoglobin. This condition is known as the *sickle-cell trait*, and a person with this condition usually exhibits no adverse traits.

Codominance

In some traits, both alleles are expressed equally (with equal dominance) and are said to be *codominant*. A classic example of codominance is the human ABO blood groups (see chapter 18). A type A homozygous father (*AA*) and a type B homozygous mother (*BB*) genotype would produce heterozygous offspring (*AB*) with a phenotype distinct from either parent.

Multiple Allele Series

Some traits are controlled by multiple alleles at a single locus. The human ABO blood group is also an example of a multiple allele series, there are three alleles (A, B, and O) but children only inherit two, one from each parent. The major histocompatibility complex (MHC) is another example of a multiple gene series. There are an estimated 25 million to 35 million different genotypes and individuals only inherit one allele from each parent. Multiple allele series are more common than the simple two-allele model.

Modifying and Regulating Alleles

The product of one allele can modify the expression of other alleles at different loci. These modifying and regulating alleles can alter the phenotype produced by different alleles. A good example is provided by human cataract development (see chapter 15). Cataracts are commonly the result of disease (like diabetes) or environmental factors (like excessive exposure to UV light)

Clinical-In Context: Inheritance of Eye Color

Scientists once thought that a single gene pair, in a dominant/recessive inheritance pattern, controlled human eye color (Figure 11.25). The allele for brown eyes was considered dominant over the allele for blue eyes. However, the genetic basis for eye color is actually far more complex. Presently, three gene pairs are known to control human eye color: Two of the gene pairs occur on chromosome 15 and one occurs on chromosome 19. The *Bey-2* gene, on chromosome 15, has a brown allele and a blue allele. A second gene, located on chromosome 19 (the *Gey* gene), has a blue allele and a green allele. A third gene, *Bey-1*, located on chromosome 15, has two brown alleles.

The brown allele is always dominant over the blue allele so even if a person is heterozygous (one brown allele and one blue allele) for the *Bey-2* gene on chromosome 15, the brown allele will be expressed. The *Gey gene* also has two alleles, one green, and one blue. The green allele is dominant to the blue allele on either chromosome but is recessive to the brown allele on chromosome 15. This means that there is a dominance order among the two gene pairs. If a person has a brown allele on chromosome 15 and all other alleles are blue or green, the person will have brown eyes. If there is a green allele on chromosome 19 and the rest of the alleles are blue, eye color will be green. Blue eyes will occur only if all four alleles are for blue eyes. This model explains the inheritance of blue, brown, and green eyes but cannot account for gray, hazel, or multiple shades of brown, blue, green, and gray eyes. It cannot explain how two blue-eyed parents can produce a brown-eyed child or how eye color can change over time. This suggests that there are other genes, yet to be discovered, that determine eye color or that modify the expression of the known eye color genes.

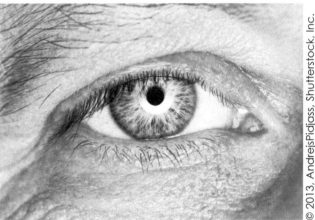
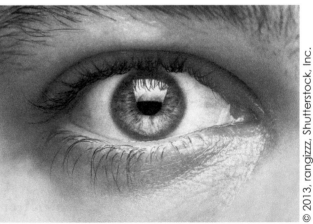

FIGURE 11.25

Differences in human eye color

but the level of vision impairment is influenced by a dominant modifying allele. The presence of this modifying allele can increase the development of cataracts.

Human development is under the regulation of a wide range of regulating alleles. Regulating genes (also known as Homeobox genes) work as *master switches* that turn on or off other alleles during development. Regulating genes also appear to influence cellular changes that occur during aging.

Incomplete Penetrance

In some traits, the phenotype is only seen when trigger by an outside influence. This incomplete penetrance is suggested to occur in several disease states. The development of adult onset diabetes (see chapter xx) has a clear genetic component in some families. Offspring of parents who develop adult onset diabetes are more likely to develop the disease themselves. The development of diabetes in these offspring is sometimes triggered by an outside influence, like obesity, viral infection, or stress. Knowledge of these factors makes it possible for a person to limit the development of certain diseases or to minimize their impact. It is thought that in some cases, multiple sclerosis (see chapter 9) can be triggered after an infection by the Epstein Barr virus, which stimulates the expression of the altered allele that is responsible for the disease.

Late Onset

Some traits are not expressed until late in life. In our previous discussion about human aging, the antagonistic pleiotropy theory proposed that late-acting deleterious genes are favored by selection and actively accumulated if they have any beneficial effects early in life. In this example of late onset, parents can pass on mutant alleles to their offspring without even knowing. Scientists think that Alzheimer disease, Parkinson disease, and Huntington's chorea are all example of late-onset human traits. The age of onset of these diseases varies from the late teens to the late sixties, but the average onset is around 40 years of age. In contrast with incomplete penetrance traits, 100% of the people with late onset traits will develop disease and eventually die. Examination of the altered alleles cannot indicate when the effects of the diseases will be manifested.

Epigenetics

In epigenetics, a genetic locus on one chromosome will affect the expression of alleles at a second locus. This is similar to sexual imprinting. Epigenetics is suggested to affect expression of cancer oncogenes and be responsible for several disorders like, fragile X syndrome, and Rett syndrome. Rett syndrome a neurologic disorder thought to be related to X-chromosome inactivation. Epigenetics is usually the result of the chemical modification of one locus by an enzyme produced by a different locus, thereby affecting its expression.

Pleiotropy

A single gene may be responsible for a wide range of phenotypic traits. This is known as **pleiotropy** (G. *pleion*, more). Examples of pleiotropy are found in the sickle cell disorder, cystic fibrosis, and the Y chromosome SRY gene. During the sickle cell disorder, the presence of altered hemoglobin forms affects the structure of red blood cells. The altered (sickle-cell shaped) red blood cells do not pass through small blood vessels, blocking them, causing pain and damage. The damage can alter physical development and directly damage other organ systems that will also affect the expression of other traits. The SRY gene codes for proteins that regulates testis formation in a male fetus. The presence of testes results in the production of testosterone, which changes the expression of hundreds of other genes.

Environmental Influences

The phenotype of a child is due to both the alleles inherited from each parent but also in a complex way with the environment that the child is reared in. An extreme example is given by teratogens. If a mother is exposed to teratogens when she is pregnant, the baby will be born with developmental defects because of changes in gene expression. Fetal alcohol syndrome is caused by maternal alcohol ingestion during pregnancy resulting in congenital abnormalities. Poor nutrition, excessive stress, drug exposure (crack babies), oxygen deprivation, hormonal imbalances, accidents can all influence gene expression and subsequent phenotypes. Environmental influences can also be seen in identical twins reared in different environments.

Scientists have discovered and characterized well more than 13,000 genetically inherited human traits, with at least 5,000 of them associated with a disorder or disease. The great majority of these traits are not the simple dominant-recessive model proposed by Mendel in his experiments. In fact, simple dominant-recessive traits are quite rare. There are at thirteen different modes of inheritance discussed here, and there will undoubtedly be new modes of inheritance discovered in the future.

Congenital and Genetic Disorders

Genetic traits associated with diseases and disorders are the result of alleles sequence changes forming aberrant forms of normal proteins or preventing their synthesis. Genetic disorders can affect individual genes, chromosomal sections, or entire chromosomes. Genetic disorders can be present at birth, not be evident until exposure to an environment factor, or only appear after a person has reached 40 years of age (late onset). Genetic disorders differ from congenital disorders that are caused by errors in fetal development (like after exposure to teratogens). Congenital disorders always lead to birth defects but only 15% have a known genetic component. Approximately 15% are due to known environmental (teratogens) causes or a combination of genetic and environmental causes (like incomplete penetrance). The remainder has no known cause and is idiopathic (without any known cause) congenital defects.

Our knowledge of teratogens has grown enormously since 1990 and we now know that thalidomide, alcohol, coumarin anticoagulants, phenytoin, and valproic acid are all teratogens, to name a few. Thalidomide produces limb malformation, and valproic acid produces neural tube defects. To act as teratogens, these chemicals must cross the placenta into the fetal blood supply. Teratogens produce specific abnormalities at certain times during gestation.

Genetic disorders are always caused by an alteration in the normal sequence of alleles. These alterations, or **mutations**, change protein structure and protein function. Mutations occur by chance during DNA replication, or in the presence of certain chemicals, radiation, viruses, or other factors. Mutations can range from lethal to silent. Lethal mutations alter the structure of a critical protein and the fetus dies during development. Silent mutations alter protein structure but not its function. Between these two extremes, mutations can alter the sequence and structure of proteins, either impairing their activity or altering it. Mutation is a means by which new proteins are formed and acted on by evolutionary forces. Once a nonlethal mutation has occurred, the abnormal trait can be passed from one generation to the next. Factors that induce mutations are known as mutagens.

Examples of genetic disorders mapped to specific chromosomes include tremors, phenylketonuria, breast cancer, and amyotrophic lateral sclerosis (ALS, Lou Gerhig's disease). Neurologic disorders like Parkinson disease or strokes commonly induce tremors, or uncontrolled shaking. Many people have tremors without any known cause (idiopathic). Idiopathic tremors are inherited as an autosomal dominant trait and mapped to loci on chromosome 2 (ETM2) and chromosome 3 (ETM1). This is an example of a polygenic trait that requires mutations in more than one gene to cause a phenotype (tremor).

Phenylketonuria is caused by a deficiency in phenylalanine hydroxylase activity and causes mental retardation, organ failure, and other physiologic problems Phenylketonuria is an inherited trait that has been mapped to a locus on chromosome 12. People heterozygous for this trait produce only 50% of the normal phenylalanine hydroxylase activity (an example of codominance) but do not suffer any ill effects.

Breast cancer is likely caused by environmental factors but clearly has a genetic component. In 1994, two genes that increase the risk for breast cancer were mapped to chromosome 17 (BRCA1) and chromosome 13 (BRCA2). Both of these genes produce proteins that repair radiation-induced breaks in double-stranded DNA. Mutations in both genes increase the risk of breast cancer but environmental factors may be required for breast cancer to occur.

Amyotrophic lateral sclerosis is a neurologic disorder caused by the degeneration of brain and spinal cord motor neurons. Eventually the disorder results in paralysis and death. This trait has been localized to chromosome 21 (SOD1). The SOD1 gene encodes for super oxide dismutase, an enzyme that removes oxygen free radicals from cells. Oxygen free radicals are normally produce during cellular respiration and can cause extensive cell damage if not removed. Amyotrophic lateral sclerosis has been associated with multiple mutations in the SOD1 gene.

Treatment of genetic disorders usually involves treatment of symptoms but does not result in a cure because the treatment does not correct the genetic mistake. Current research hopes to change this by allowing doctors to directly alter somatic cell DNA to correct for genetic mistakes.

Genetic Counseling

The rapid explosion in genetic information since 1980 and the realization that there are at least 5,000 genetic traits associated with disorders and diseases, has led physicians and scientists to urge for the development of specialists, genetic counselors, with the knowledge and expertise to discuss these matters with the general public.

More than 100 years ago, many people realized that some diseases "ran in families." Medical geneticists who worked in the early "hereditary counseling clinics" met with families, examined affected individuals, and drew **pedigrees** in an attempt to help clarify the genetic component of diseases and birth defects. A pedigree is a family history of a genetic disorder. They worked primarily with relatively common diseases and conditions such as achondroplasia, and cleft palate. The practice of genetic counseling took a giant leap forward in 1967 with the first amniocentesis. Physicians and geneticists soon realized that communicating genetic information required specialized training. Genetic counseling is a communication process that deals with the human problems associated with the occurrence, or the risk of an occurrence, of a genetic disorder in a family. Genetic counselors help individuals and families to: (1) comprehend the medical facts, including the diagnosis, probable course of the disorder, and the available management; (2) appreciate how heredity contributes to the disorder, and the risk of recurrence in specified relatives; (3) understand the alternatives for dealing with the risk of occurrence; (4) choose the course of action that seems appropriate in view of their risk, their family goals, and their ethical and religious standards, and to act in accordance with that decision; and (5) make the best possible adjustment to the disorder in an affected family member and/or the risk of recurrence of that disorder.

Clinical–In Context: Advances in Molecular Medicine

The recent completion of the human genome project has energized a new frontier in medicine, molecular medicine. The identification of all of the genes in the 23 pairs of chromosomes opens up new possibilities for the treatment of human diseases. Patients in the future will likely be treated in a completely different manner. Shortly after a person is born, his or her genotype might be recorded at a physician's office and the information transmitted to a secure database. Assisted by a decision support system, physicians may prescribe a personalized immunization and screening schedule, and recommend specific measures to prevent disease development. At any point, screening may lead to recommendations for lifestyle changes and nutrition, or to detection of early stages of a disease. It is hoped that this personalized therapy will be supported by an expanded spectrum of drugs developed to target particular disease subtypes for a specific genetic background. This approach has the potential to dramatically alter the treatment of genetically based diseases, but there are some controversial aspects.

Many people fear that they may be discriminated against based on their genetic makeup and potential for disease. Employers might be reluctant to hire someone with a genetic predisposition to a particular disorder or disease, especially if it will involve high healthcare costs in the future. Even with these problems, the coming decades will likely bring a revolution in how many diseases are treated. The causes of cleft lip/palate are not well understood. Studies suggest that a number of genes, as well as environmental factors, such as drugs (including several different antiseizure drugs), infections, maternal illnesses, maternal smoking, and alcohol use and, possibly, deficiency the vitamin folic acid may be involved. Cleft lip/palate may occur alone or with other abnormalities that may be hidden or obvious. Some cases involve genetic syndromes that may pose specific problems for the baby and may have a high risk of affecting others in the family.

In Review...

28.30 What is the study of genetics?

28.31 What is a chromosome? What is DNA? What is a karyotype?

28.32 How many chromosomes does a somatic cell normally have?

28.33 What is a gene? An allele?

28.34 What is a dominant gene? A recessive gene?

28.35 What is a sex-linked trait?

28.36 What are genetic defects? Congenital defects?

28.37 What is genetic counseling?

Clinical Perspective

Pregnancy Problems

Ectopic Pregnancy

Ectopic (L., out-of-place) pregnancy is the development of the embryo or fetus outside of the uterine cavity. An ectopic pregnancy usually occurs when the movement of the fertilized ovum is impaired through the uterine tube. Scarring can block the uterine tube caused by a prior tubal infection, decreased smooth muscle motility, or abnormal uterine tube morphology. The most common sites of ectopic pregnancies are in the ampullar and infundibular portions of the uterine tube, but they can also occur in the ovary, abdominal cavity, or uterine cervix. Mothers who smoke are twice as likely to have an ectopic pregnancy as nonsmokers. The nicotine in cigarette smoke paralyzes the uterine tube cilia as it does in the respiratory airways. Scars from previous uterine surgery, pelvic inflammatory disease, and previous ectopic pregnancies may also hinder movement of the fertilized oocyte.

The ectopic pregnancy can lead to acute pelvic and abdominal pain. Unless removed, the developing embryo can rupture the uterine, causing the mother to die of internal bleeding.

Placenta Previa

In some cases, the placenta, or part of it, may become implanted in the inferior uterus, near or covering the internal os of the cervix. This condition is known as *placenta previa* (L., before or in front of). Placenta previa may lead to spontaneous abortion, and it occurs in approximately 0.4% of live births. Placenta previa causes premature births and hypoxia because maternal bleeding. Maternal mortality is increased by hemorrhage and infection. The most common symptom of placenta previa is the sudden appearance of painless, bright-red vaginal bleeding in the third trimester. Cesarean section is the preferred method for the delivery in placenta previa.

Preeclampsia

About 10% to 15% of all pregnant women in the United States experience preeclampsia, or *pregnancy-induced hypertension*. The major cause of preeclampsia is an abnormal pregnancy-related condition characterized by sudden hypertension, large amounts of protein in the urine, and generalized edema that appears after 20 weeks of fetal development. In addition, pregnant women experience blurred vision, headaches, and swollen ankles. Preeclampsia may be related to an autoimmune or allergic reaction to the presence of the fetus. When the condition is associated with convulsions and coma, it is known as eclampsia.

Dystocia and Cesarean Section

Dystocia (L., painful or difficult birth), or *difficult labor*, can result either from an abnormal fetal position or from an inadequate vaginal canal. In a breech presentation, which often occurs during premature births, the fetal buttocks or lower limbs, rather than the head, enter the birth canal first. If fetal or maternal distress makes a vaginal birth difficult or dangerous, the baby can be delivered surgically through an abdominal incision. A low, horizontal cut is made through the abdominal wall and lower uterus through which the baby and placenta are removed. This type of birth was first described in Roman Law, *lex cesaria*, about 600 years before the birth of Julius Caesar, hence the term *caesarean section*. Improvements in the cesarean section make it possible for women to have subsequent vaginal deliveries.

Spontaneous Abortions

Spontaneous abortions, or *miscarriages*, occur because of fetal developmental or physiologic problems that cause the pregnancy to be terminated. Spontaneous abortions result from chromosomal abnormalities, hormonal problems, or placental problems. Common hormonal problems include inadequate maternal LH production, reduced corpus luteum sensitivity to LH, inadequate uterine sensitivity to progesterone, or failure of the placenta to produce hCG. Spontaneous abortions occur in 15% of all pregnancies.

Therapeutic abortions are performed when the continuing pregnancy represents a threat to the life of the mother. Induced abortions are performed at the mother's request. The ratios of induced abortions to live births for married women are 1 to 10, but it is 2 to 1 in unmarried women. In most states, induced abortions are permissible only in the first trimester.

Premature Infants

Delivery of a physiologically immature baby carries substantial risks to the baby. A premature infant is a baby who weighs less than 2,500 g at birth. Poor prenatal care, drug abuse, smoking, a history of premature births, and age (younger than 16 or older than 35) all increase the risk for premature birth. The fetus may not be able to breathe on its own because of a lack of lung surfactant production, leading to respiratory distress syndrome. Respiratory distress syndrome is treated by administration of artificial surfactant and ventilation to deliver oxygen until the lungs are capable of independent operation.

Clinical Terms

amniocentesis	A common prenatal test in which a small sample of the amniotic fluid surrounding the fetus is removed and examined.
chorionic villus sampling	Collection of chorionic villi cells to diagnose birth defects.
cleft palate	A separation of the tissues in the lip or mouth.
congenital birth defects	Mistakes in development leading to congenital (present at birth) defects.
conjoined twins	Twins that develop physically attached to one another.
ectopic pregnancies	Pregnancies that occur outside the uterus.
in vitro fertilization	Fertilization of an egg in a Petri dish and then implantation of the fertilized egg into the uterus.
IVF-ET	Fertilization of an oocyte outside of the uterine tube and then the transfer of an embryo into the uterus.
multiple births	Giving birth to more than one baby. The most common multiple births are twins. Humans have given birth to triplets, quadruplets, and quintuplets.
teratogens	A number of chemicals, radiation, or other factors that induce congenital birth defects.

Chapter Review

Factual
(to define, list, state and name)

1. A gene is...
    a. found within DNA.
    b. a part of a chromosome.
    c. the functional unit of heredity.
    d. instructions to produce a protein.
    e. All of the above.

2. Major organ development occurs during...
   a. fetal development.
   b. the germinal phase.
   c. the embryonic period.
   d. development of the blastomere.

3. The placenta...
   a. invades the fetal lacunae.
   b. develops from the trophoblast layer.
   c. allows maternal blood to mix with fetal blood.
   d. All of the above.

4. The...causes blood to bypass the fetal...
   a. ductus arteriosus...lung.
   b. foramen ovale...liver.
   c. fetal umbilicus...left atria.
   d. ductus venous...left atria.
   e. fossa ovalis...lung.

5. Capacitation...
   a. is the initial step in ova fertilization.
   b. is the process by which sperm are killed before they have a chance to fertilize the ova.
   c. is the process by which polyspermy is prevented.
   d. is the process by which the ovulated egg is swept into the uterine tube by fimbrae.
   e. is the alteration of the sperm into its final form ready for fertilization.

6. Labor is induced by...
   a. increased uterine stretch.
   b. increased pituitary oxytocin secretion.
   c. increased placental glucocorticoid secretion.
   d. increased placental estrogen and progesterone secretion.
   e. All of the above.

7. Which of the following is ovulated?
   a. Primordial follicle
   b. Primary oocyte
   c. Secondary oocyte
   d. Mature ovum

8. In a normal pregnancy, implantation occurs...
   a. during the luteal phase of the ovarian cycle.
   b. during the proliferative phase of the uterine cycle.
   c. in the Fallopian tubes.
   d. in the uterine myometrium.
   e. prior to the acrosome reaction.

9. About seven days following fertilization...
   a. FSH and LH levels will rise.
   b. ovulation will occur.
   c. the blastocyst will implant in the endometrium.
   d. the morula will implant in the Fallopian tube.
   e. the placenta will form.

10. Which hormone causes the differentiation of sex organs in the developing male fetus?
    a. FSH and LH
    b. LH and testosterone
    c. Estrogen and testosterone
    d. Estrogen and progesterone
    e. Testosterone and dihydrotestosterone

11. Following parturition...
    a. stomach pH decreases.
    b. the ductus arteriosus remains open.
    c. the fossa ovalis becomes the foramen ovale.
    d. blood flow through the pulmonary arteries increases.
    e. All of the above.

12. Which one of the following is not a physiological role of the uterus?
    a. It helps to provide mechanical protection of the fetus.
    b. Its rich vascularization provides nutrients for the developing fetus.
    c. The endometrium feeds the developing fetus until the placenta is developed.
    d. The smooth muscle (myometrium) provides contraction to promote parturition.
    e. It is the primary source of estrogen and progesterone during the last 7 months of pregnancy.

13. An infection that scarred the epididymus causes infertility...
    a. because now the seminiferous tubule is incapable of producing spermatocytes.
    b. because of a decrease in the blood levels of testosterone.
    c. because the fluids produced by accessory tissues would not enter the ejaculatory duct.
    d. because sperm could not move from the epididymus into the ductus deferens.
    e. All of the above.

14. Damage to...prevents ova transport from the ovary to the uterus.
    a. the uterine tube
    b. the ductus deferens
    c. the cervix
    d. the vagina
    e. the efferent ductule

15. The *Morning After Pill* is an effective contraceptive if given within a day or so after fertilization. Which of the following describes its most probable mechanism of action?
    a. The drug likely prevents the blastocyst from implanting itself into the uterine endometrium.
    b. The drug likely prevents the release of hCG.
    c. The drug likely blocks the release of LH and FSH from the pituitary.
    d. The drug likely directly blocks LH receptors on thecal cells.
    e. The drug likely stimulates estrogen and progesterone synthesis by the corpus luteum.

16. Which life stage is correctly matched with the time that the stage occurs?
    a. Puberty, 10 to 12 years
    b. Child, 6 months to 5 years
    c. Middle age, 20 to 40 years
    d. Infant, 1 month to 6 months
    e. Neonate, birth, to 1 month after birth

17. During the last week of pregnancy, this change is an important factor that initiates the onset of parturition.
    a. The dilation of the cervix
    b. High levels of blood progesterone
    c. Release of hCG by the corpus luteum
    d. Increased numbers of oxytocin receptors in the uterus
    e. Decreased spontaneous uterine contractions (Braxton-Hicks)

18. Which of the following statements is false?
    a. Women produce about 50 times as much estrogen as testosterone.
    b. Both men and women produce estrogen.
    c. Testosterone is at much higher concentrations inside the blood than in seminiferous tubule.
    d. Testosterone induces male secondary sex characteristics.
    e. Men are more aggressive than women, in part, because of their higher testosterone levels.

19. During the last week of pregnancy,…initiates parturition.
    a. the dilation of the cervix
    b. decreased spontaneous uterine contractions (Braxton-Hicks)
    c. release of hCG by the corpus luteum
    d. high levels of blood progesterone
    e. decreased progesterone blood levels

20. Which of the following terms is correctly linked with its definition?
    a. Autosomal, an X or Y chromosome
    b. Phenotype, the genetic makeup of an individual
    c. Dominant, a trait that is hidden by a superior allele
    d. Heterozygous, having two identical genes at a locus
    e. Allele, genes occupying the same locus on homologous chromosomes

21. The rate of live births in a small town near an industrial complex is 50% less than in a nearby cleaner area. Many of the newborns also have a higher than normal number of congenital birth defects. What is the likely problem?

22. Is it possible to make an XY fetus a biologically functional female by preventing the production of testosterone during fetal development?

23. Nursing mothers complain that their breasts leak milk when they are in public places and see a young baby. Why is this the case?

24. A woman who has a history of a severe blood disorder in her family comes to you and asks if it is safe to marry a distant second cousin. There are no tests to determine the genetic nature of this trait. What advice would you give her?

25. A woman with a fertilized egg in her body won't go through an additional menstrual cycle because...
    a. trophoblast cells secrete hCG to maintain high blood levels of estrogen and progesterone.
    b. zygotes secrete inhibin into the uterine fluid that in turn causes the release of hCG.
    c. once the fertilized egg reaches the blastocyst stage it secretes a powerful inhibitor of ovulation.
    d. trophoblast cells secrete hCG that directly inhibits the release of GnRH from the hypothalamus.
    e. high blood levels of estrogen prevent any increase in GnRH.

26. Which of the following statements is false?
    a. Women are likely born with all of the ova that they will ever have.
    b. Men continuously produce sperm throughout their lives if they continue to produce testosterone.
    c. Primordial oocytes are arrested during Meiosis I.
    d. After ovulation, oocytes are arrested during Meiosis II.
    e. A polar body can be fertilized.

Answers to end-of-chapter questions

1. a	6. e	11. e	16. a	25. a
2. a	7. c	12. e	17. d	26. e
3. b	8. a	13. d	18. c	
4. a	9. c	14. a	19. e	
5. e	10. b	15. a	20. e	

21. Mothers are likely being exposed to teratogens during pregnancy that is causing severe birth defects within fetuses and the termination of many pregnancies.

22. If you blocked the expression of the SRY gene then the fetus would develop as a normal female, even though it was XY and not XX.

23. Seeing a baby in public induces the release of oxytocin in nursing mothers and milk letdown.

24. The only advice you could give her is a determination of the genetic relatedness of her to her second cousin, and the likelihood that he might have the trait and the chances that her children could have the trait.

Thought Questions

1. Polyspermy is incompatible to life.
2. It means you can follow maternal inheritance through history.
3. The ovum is actually much larger than a normal cell and the dividing cells have limited access to nutrients within the uterine tube. Becoming smaller with each division ensures that there is enough cytoplasm for each cell and lets each cell become normal size.
4. These cells are derived from the mother.
5. The CNS must be present to direct cells to the targets that need to be innervated.
6. The mother filters the fetus's blood indirectly though placental exchange.
7. It is a small fraction of the adult CO.
8. A sensory cell, an interneuron, and an effector cell.
9. The fetus is undergoing rapid growth and cell division that gives off a great deal of heat.
10. Yes, because the fetal tissues are recognized as nonself by the maternal immune system.
11. Yes, if you prevent expression of the SRY gene on the Y chromosome.
12. This would allow a mother to carry larger babies, which is common in subsequent births.
13. That means that systemic and pulmonary blood are mixing in the heart, decreasing the efficiency of respiration.
14. Placental blood is full of nutrients obtained from maternal blood; the fetal liver must metabolize and interchange those nutrients as the adult liver does after a meal.
15. This implies a CNS developmental abnormality preventing the correct control of skeletal muscle.
16. Babies with Down syndrome have an extra copy of the 21st chromosome, so it is not necessarily lethal, but having 45 chromosomes is usually lethal.

Index

A

"ABCS" of heart disease, 53
abdominal obesity, 110
abdominal wall, 124
abortions, 406
ABP. *See* androgen-binding protein
absorption methods, 224
abuse, heroin, 268
accessory glands, 306–307
acetylcholine, 29
acquired immunodeficiency syndrome (AIDS), 223
acrosome, 313
ACSM. *See* American College of Sports Medicine
ACTH. *See* adrenocorticotropin hormone
action, SCM, 12–14
activity, cardiovascular exercise, 118
activity nervosa, 192
acupressure, 91
acupuncture, 89, 90
acute muscle soreness, 127
Adderall, 277
adolescence, smoking and, 256
adrenaline, 28, 32
adrenarche, 310, 325
adrenocorticotropin hormone (ACTH) secretion, 310
adult stem cells, 359
aerobic exercise, 110, 114–115
    cardiovascular fitness, 112–114
    common types of, 117
aerobic metabolism, 166
aerobic training, 120
    changes in body composition, 123
age, CVD, 63
aging, 42–46
    creativity and, 47
    and diabetes, 70
    and disease, 41–74

AHA. *See* American Heart Association
AIDS. *See* acquired immunodeficiency syndrome
air pollution, 137
Al-Anon and Alateen organizations, 244
alarm stage, stressor, 26
Alcock's canal, 305
alcohol, 32, 236
    absorption and elimination, 239
    BAC, 237–238
    digestion and metabolism, 238–239
    intoxication, 239
    laws relating to, 247
    neuroscience, 237
    societal problems, 244–247
    tolerance, 240–244
    types of drinks, 237
alcohol cirrhosis, 243
alcohol dehydrogenase, 238
alcohol dependence, 241–242
alcoholic hepatitis, 243
Alcoholics Anonymous (AA), 243
alcoholism, 241–242
alcohol pellagra, 243
alcohol poisoning, 246–247
allantois, formation of, 375
allele, 396
allergens, 137
α-fetoprotein test, 367
alternative healthcare systems, 87, 89–91
alternative medicine, 84
amenorrhea, 74, 334
American College of Sports Medicine (ACSM), 74, 107
American Heart Association (AHA), 107
    blood pressure, 61
American Medical Association, 92

amniocentesis, fetal development, 367–368
amniotic fluid, fetal development, 366–367
amniotic sac, formation of, 374
amphetamines, 270–271
ampulla, 320
amyotrophic lateral sclerosis, 404
anabolic steroids, 273–274
anabolism, 165
anaerobic exercise, 114–115
anaerobic metabolism, 166, 167
andrenocorticotropic hormone (ACTH), 25
androgen-binding protein (ABP), 313
aneuploidy, 397
aneurysm, 54
angel dust, 272–273
anger, stress and, 26–27
angina pectoris, 113
animal-assisted therapy, 99–100
Annals of Internal Medicine, 91
anorexia nervosa, 190–191
anorexics, 189
antagonistic pleiotropy theory, 393
anterior muscles in human body, 125
anterograde amnesia, 272
antidepressant bupropion, 258
antioxidants, 202–203
anxiety, 33
aortic arches, 362
Apgar scores, postnatal life, 389–390
appetite suppressants, 196–197
applied psycho-neuro-immunology, 96
arteriosclerosis, 53–55
artificial insemination, 318
atherosclerosis, 54–55, 57, 113, 248, 394
athletic shoes, 133

413

atrial septal defects, 371
autoimmune disorder, 318
autosomal chromosomes, 395–396
axon, 227
Ayurveda, 89

B
BAC. *See* Blood Alcohol Concentration
back muscles, 124
bacterial diseases, 337–339
ballistic stretching, 130–131
bariatrics, 181
basal metabolic rate (BMR), 169
BBB. *See* blood-brain barrier
BDNF. *See* brain-derived neurotrophic factor
beer, 237
behavioral methods, reproductive system, 340–342
behavioral treatments, for tobacco addiction, 258–259
behavior change and goal setting, 12–15
benzodiazepines, 274
beriberi, 159
Bey-1 gene, 401
Bey-2 gene, 401
bicycle seat neuropathy, 305
bilirubin, 389
binge drinking, 246
binge-eating disorder, 191
bioenergy practitioners, 100
biological age, 42
biological-based therapies, 93–96
birth control methods, failure rates of, 340
birthmarks, 389
blackout, 242
blastocele, 356
blastocyst formation, 356
blastomere, 355, 356
Blood Alcohol Concentration (BAC), 237–238
blood-brain barrier (BBB), 229
blood, maternal changes, 382
blood pressure (BP), 60, 61
blood testosterone, 315
blood vessels, formation of, 379
BMI. *See* body mass index
BMR. *See* basal metabolic rate
body composition, 182–183
body image, 189–190
    activity nervosa, 192
    anorexia nervosa, 190–191
    binge-eating disorder, 191
    bulimia nervosa, 191
    eating disorders, causes of, 193
    women at risk, 192–193

body mass index (BMI), 109, 169, 182, 183
    vs. percentage of fat content, 170
body temperature, 168–169
body weight, regulation of, 169
bone loss, during spaceflight, 73
bone marrow stem cells, 359
bone mineral density, 123
Borg RPE scale, 117
bouncing, 130
brain attack, 60
brain-derived neurotrophic factor (BDNF), 45
brain plasticity, 45
brainstem, 225
brain wave patterns, 370
Braxton-Hicks contractions, 384–385
BRCA gene mutation, 398
breast cancer, 335–336, 404
    physical activity effect, 68
breast milk, 390
breast-milk jaundice, 389
breathing, initiation of, 387
broad ligaments, 321
bulbourethral glands, 299, 307
bulimia nervosa, 191
bulimics, 189
bulk minerals, 160
burn-out, training program, 127
B vitamins, 31

C
cadmium, 263
caesarean section, 406
caffeine, 30, 32, 269–270
caloric balance, 179
caloric expenditure and death, 48, 49
caloric needs, determination of, 185
calories, 169, 205
CAM. *See* complementary and alternative medicine
cancer, 68–69, 243
Candida albicans, 339
cannabinoids, 265–268
Cannabis (marijuana), 230
carbohydrate metabolism, 166–167
carbohydrates, 157
carbon monoxide, 248–249
cardiac output (Q), 112, 121
cardiogenic region, 378
cardiotocograph, 368
cardiovascular disease (CVD), 50–51
    childhood obesity, 64–65

    contributing factors, 63
    controllable factors, 62
    obesity, 63–64
        causes, 65–68
        physiological response to, 68
    prevention, 52–53
    risk for, 51–52
    uncontrollable factors, 63
cardiovascular endurance, 110
cardiovascular fitness, 110–112
    aerobic exercise, 112–114
    evaluating, 120–121
        muscular fitness, 122–123
    heart rate response, 116
cardiovascular system
    defects, 371–372
    formation of, 378–379
    maternal changes, 382–383
catabolism, 165
cell senescence, development of, 393
cellular respiration, 165
central nervous system, 225–229
cerebellum, 225
cerebral cortex, 226
cervical cancer, 322
cervical mucous glands, 322
cervix, 321
chancre, 338
chancroid, 338
Chantix (varenicline tartrate), 258
chemical castration, 317
chewing tobacco, 262, 263
childhood obesity, 48–49, 64–65
chiropractic, 91
Chlamydia trachomatis, 337
cholesterol, 57, 62
ChooseMyPlate, 205–208
chorion, 357
chorionic villi, 357
    formation, 359
chorionic villus sampling, 367
chromosomes, 395–396
chronic stress, 31
    health disorders associated with, 25
cigarette smoking, 62, 247, 249, 251–253
    vs. water pipe smoking session, 262
circulatory system, 113
circumcision, 306
cleft lips, 377
Clery Act, 11
clinical perineum, 324
clinical program, weight loss, 198
clove cigarettes, 249
club drugs, 271–272
cocaine, 270

codeine, 277
codominance, 400
cognitive distraction, 8
collateral arteries, 55
collateral circulation, 112
colon cancer, physical activity effect, 68
colostrum, 390
comfort foods, 164
ComfortMax, 134
community, aging, 43–46
complementary and alternative medicine (CAM), 84
   alternative healthcare systems, 87, 89–91
   biological-based therapies, 93–96
   common therapies among adults 2007, 87
   diseases/conditions for, 91
   energy therapies, 100
   ethnicity among adults 2007, 86
   manipulative and body-based therapies in, 91–93
   mind-body medicine, 96–100
   modalities, 87
   U.S. adults and children, 85, 86
complete proteins, 157
congenital birth defects
   cardiovascular system, 371–372
   gastrointestinal system, 373
   musculature system, 372
   nervous system, 372–373
   respiratory system, 374
   trinucleotide repeat diseases, 374
   urogenital system, 373–374
congenital disorders, 403–404
congenital limb reduction, 372
conjoined twins, 354
contemplation, SCM, 12, 14
contraceptive methods, reproductive system, 340
Controlled Substance Act, The, 236
conventional medicine, 85
cooldown and stretch, 118–119
core musculature, importance of, 123–124
corona radiata, 329
coronary artery disease, 54, 57
coronary collateral circulation, 59
corpora cavernosa, 306
corpus albicans, 330
corpus callosum, agenesis of, 372–373
corpus luteum, 329–330
corpus spongiosum, 306

cortisol, 28
crack cocaine, 270
craniosacral therapy, 92
creativity and aging, 47
   credible medical information, Internet for, 100–102
creeping obesity, 185
cremaster muscle contraction, 300
Crime Awareness and Campus Security Act of 1990, 11
cross dependence, 230
cross-tolerance, 230
cross training, 119
cryptorchidism, 309, 374
CVD. See cardiovascular disease
Cyclic GMP, 315
cytochrome P-450 enzymes, 231
cytotrophoblast cells, 357

D

Daily Values (DVs), 208–209
dairy products, 163
"date rape" drug, 272
deaths
   and caloric expenditure, 49
   causes of, 6
decidua, 374
deep breathing, stress technique, 32
deferens, 299
dehydration
   adverse effects of, 136
   avoiding, 135
delayed menarche, 334
delayed-onset muscle soreness, 127
deleterious mutation theory, 393
dendrites, 227
deoxyribonucleic acid (DNA), 395
Depo Provera, 317
depressants, 274–277
Destructive Emotions (2003) (Goleman), 96
DEXA. See dual energy X-ray absorptiometry
dextrocardia, 372
DHT, 314
diabetes, 62
   aging and, 70
   walking, 72
diarrhea, 162
diastolic blood pressure, 112
diastolic reading, 55
diencephalon, 226
diet, 85
dietary cholesterol, 62
dietary supplements, 195–196
digestion
   alcohol, 238–239
   stress and, 31

digestive system, movement of fluid in, 161
digestive tract, maternal changes, 380–381
diploid cells, 395
dissociative drugs, 272–273
distilled spirits, 237
distracted driving, 8–9
distraction, types of, 8
DNA. See deoxyribonucleic acid
dominant genes, 397–398
dopamine, 29, 227
Doppler, 368
dowager's hump, 72
Down syndrome, 372
drinking and driving, 244–245
drinks, types of, 237
driving under the influence (DUI), 247
driving while intoxicated (DWI), 247
drowsy driving, 9–10
drug, 32
   absorption methods, 224
   BBB, 229
   central nervous system, 225–229
   defined, 222
   drug dependence, 229–230
   duration of action, 225
   inhalation, 222–223
   injection, 223
   method of administration, 224
   oral ingestion, 222
   psycho-active drug, 224–225
   suppositories, 224
Drug Induced Rape Prevention and Punishment Act, 272
dual energy X-ray absorptiometry (DEXA), 183
Duchenne muscular dystrophy, 399
duct system, 301–302
ductus, 299
ductus arteriosus, 388
ductus deferens, 304–305
DUI. See driving under the influence
duration of action, 225
DVs. See Daily Values
DWI. See driving while intoxicated
dysmenorrhea, 334
dystocia, 406

E

eating disorders
   body image, 193
   guidelines for, 194
   medical treatment for, 195

obesity, 188–189
  risk factors for, 189
ecstasy, 271
ectoderm cells, 359, 360
ectopic pregnancies, 357, 405–406
effective training, 124, 127
  muscle soreness, 127
  weight training myths, 127, 129
ejaculation, male sexual response, 316
ejaculatory ducts, 299, 305
electronic fetal monitoring, 368
eliability theory, 393
embryonic blood vessels, 379
embryonic development
  of male reproductive system, 308–309
  stages in, 362
embryonic stem cells, 359
emission, male sexual response, 315
emotional wellness, 3, 5
endoderm cells, 359, 360
endometrial hyperplasia, 333
endometriosis, 337
endometrium, 322
endorphins (opioids), 28, 29
  functions of, 31
energy drinks, 196
energy therapies, 100
environment, 85
environmental influences, 403
environmental tobacco smoke (ETS), 263–264
environmental wellness, 4, 5
epidermis, 223
epididymis, 299, 302–304
epigenetics, 44–45, 402
  maternal, 369
epinephrine, 28
episiotomy, 324
erectile dysfunction, 316–318
erection, male sexual response, 315
essential amino acids, 157
essential oils, 88
estrogen, beneficial effect of, 327
ethyl alcohol, 236
ETS. *See* environmental tobacco smoke
eugenol, 249
Ewing, John, 242
exercise, 85, 171, 172
  benefits of, 53, 54
  and cancer, 69
  components of, 118–119
  and diabetes, 70

environmental conditions, 134–135
in heat, guidelines, 134
hormone, 46
and hypertension, 59
hyponatremia, 135–136
illness, 136–137
injuries, 131–132
intensity, measuring, 117
and low back pain, 71
and osteoporosis, 74
outside activity, sense concerns for, 137
and pregnancy, 383–384
proper footwear, 132–134
risk, 138
workout considerations, 137–138
exhaustion stage, stressor, 26
external genitalia, female reproductive system, 323–324
external urethral orifice, 306

F
face, formation of, 376
facial primordia, 376
fad diets, 199, 200
fatness. *See* fitness
fats, 32
fat-soluble vitamins, 158–159
fatty liver, 243
FDA. *See* Food and Drug Administration
Federal Trade Commission (FTC), 93
Feldenkrais method, 99
Female Athlete Triad, 192
female condoms, 341
female infertility, causes of, 336–337
female reproductive cycle, hormone interactions and, 333
female reproductive system, 298
  anatomy of, 319
  development of, 325–327
  disorders of, 334–335
  external genitalia, 323–324
  mammary glands, 324
  ovaries, 319
  physiology of, 325
  uterine tubes, 319–321
  uterus, 321–322
  vagina, 322–323
female sex hormones and reproductive cycle, 330
female sexual response, 333–334

breast cancer, 335–336
female infertility, causes of, 336–337
female reproductive system, disorders of, 334–335
fermentation, 236
fertility, 318
fertilization, 351–352
fetal abnormalities, detection of, 368
fetal alcohol syndrome, 243, 373
fetal blood sampling, 368
fetal development, 350
  amniocentesis and fetal monitoring, 367–368
  amniotic fluid, 366–367
  embryonic development, 350–351, 359–365
  germinal development, stages of, 351–359
  specific body regions formation, 374–380
  stages of, 366, 368–371
fetal distress, 367–368
fetal monitoring, 367–368
fetal surgery, 384
fetus, 366
fimbriae, 320
financial wellness, 4–6
fire safety 101, 10–11
first year of life
  aging, 392–395
  growth and development in, 391–392
  life stages, 392
fitness, 110
  training, principles of, 119–120
FITT, 115
  frequency, 115
  intensity, 115, 117
  time, 117
  type, 117–118
FiveFingers shoe, 134
flexibility, 111, 129–131
flexibility training, 120
flunitrazepam, 272
follicles, 319
follicular maturation, 327–329
follicular phase, 330–332
Food and Drug Administration (FDA), 316
food groups, 162
food label, 208–210
food pyramid, 162–164
food shopping, healthy habits, 201
food *vs.* nutrients, 162

INDEX

carbohydrate metabolism, 166–167
food groups, 162
food pyramid, 162–164
lipid metabolism, 167
metabolic reactions, types of, 165–166
metabolism, 165
forebrain, 362
formaldehyde, 263
formation, 72
formyl-Met-Leu-Phe (fMLP), 351
fornix, 322
fraternal twins, 354
frequency, 124
FITT, 115
FTC. *See* Federal Trade Commission
functional foods, 95, 203
functional movement, 123
core musculature, importance of, 123–124
functional strength, 130

G
GABA. *See* gamma amino butyric acid
gametes, 307, 311
gamma amino butyric acid (GABA), 29
gamma hydroxybutyrate (GHB), 272
Gardnerella vaginalis, 339
gastric production, during pregnancy, 381
gastrointestinal system defects, 373
gastrulation, 359
gender, CVD, 63
genes, 396–397
dominant and recessive genes, 397–398
linkage analysis, 397
sex-linked traits, 399
genetic counseling, 404
genetic disorders, 403–404
genetic traits, 403
genital herpes, 339
genome imprinting, 400
genotype, 398
Gey gene, 401
GFR. *See* glomerular filtration rate
GHB. *See* gamma hydroxybutyrate
glans penis, 306
glomerular filtration rate (GFR), 381
gluconeogenesis, 166
glucose, 166
metabolism, 167

glucosuria, during pregnancy, 381–382
glutamate, 228
glutes, 124
glycogen, 114, 122
glycolysis, 166
GnRH. *See* gonadotropin-releasing hormone
Go4life, 46
gonadarche, 310
gonadotropin-releasing hormone (GnRH), 309
gonorrhea, 339
good mental health, 85
gout, 400
G-protein coupled receptor (GPCR), 313
grains, 163
granulosa cells, 327, 329
guided imagery, 99
gut and body cavities, formation of, 375–376
gynecomastia, 324

H
Haemophilus ducreyi, 338
hallucinogens, 273
'halo' effect, 181
haploid cells, 395
hardiness, 27
hard liquor, 237
hashish, 265
hash oil, 265
hCG. *See* human chorionic gonadotropin
hCS, *See* human chorionic somatomammaotropin
HD. *See* Huntington disease
health
defined, 2
disorders, associated with chronic stress, 25
risk, measuring, 109–110
and wellness, factors influence, 6
health-related fitness, 110
five components of, 111
health span, 42
healthy body weight, 182–183
healthy eater, tips for, 184
healthy habits
antioxidants, 202–203
building healthy plate, 203–208
fast foods/eating out, 201–202
fitness/fatness, 202
food shopping, 201
functional foods, 203
Nutrition Facts Label, 208–210
organic foods, 203

vegetarianism, 211
weight gain, 200–201
Healthy People 2020, 15–16
healthy plate, 203–208
heart attack, 58–61
heart chambers, 362
heart disease
"ABCS" of, 53
risk of, 57
heart rate, 112
cardiovascular exercise, 116
heart, stress and, 113
heat cramps, 134
heat exhaustion, 134
heat stroke, 134
herbals and dietary supplements, 93
heredity, 395
CVD, 63
heroin, 268
herpes simplex, 339
heterozygous alleles, 396
high blood pressure. *See* hypertension
hindbrain, 362
hip muscles, 124
holistic self-care, 85
homeopathy, 89
homeostasis, 172
homozygous alleles, 396
hormonal imbalances, 317
hormonal regulation of spermatogenesis, 312
hormonal therapy, 336
hormone, 28
hormone interactions and female reproductive cycle, 333
horseshoe kidney, 373
HPV. *See* human papillomavirus
human body
anterior muscles in, 125
posterior muscles in, 126
six substances, 157
human chorionic gonadotropin (hCG), 308
secretion, 356–357
human chorionic somatomammaotropin (hCS), 375
human fetal kidney, formation of, 378
human genetics, 395
chromosomes, 395–396
congenital and genetic disorders, 403–404
exceptions to inheritance, 399–403
genes, 396–399

human height, 400
human menopausal gonadotropin, 358
human papillomavirus (HPV), 339
Huntington disease (HD), 374
hydrocodone, 277
hypertension, 55, 58, 62
hypokinetic conditions, 48
   prevention, 46–47
   types
      arteriosclerosis, 53–55
      CVD, 50–53
      heart attack, 58–59
      hypertension, 55, 58
      peripheral vascular disease, 55
hyponatremia, 135–136
hypoplastic left heart syndrome, 372
hypothalamus, 226
hypoxia, 368

I

IBS. *See* irritable bowel syndrome
identical triplets, 355
identical twins, 354
idiopathic tremors, 403–404
IgG antibodies, 389
illicit prescription use, ethical considerations of, 282–283
illness, 136–137
immune system, stress and, 25
implantation, 356–357
inactivity, heart disease, 112
   physical, 62
incomplete dominance, 400
incomplete penetrance, 402
incomplete proteins, 157
Indinavir, 93
individual differences, principle of, 119–120
induced abortions, 406
infundibulum, 320
inguinal hernias, 301
inhalants, 274
inhalation of drugs, 222–223
inheritance
   exceptions to, 399–403
   of eye color, 401
initiation, male sexual response, 314–315
injection, drug, 223
injuries, 131–132
injury facts, 6–7
   motor vehicle safety, 8–10
insulin, 170
   purpose of, 163
intellectual wellness, 3, 5

intensity, 127
   FITT, 115, 117
interval training, 114–115
intestinal malrotation, 373
intoxication, 239
intoxication assault, 247
intoxication manslaughter, 247
intradermal injection, 223
intramuscular injection, 223
intranasal use, 224
intrathecal injection, 223
intrauterine device (IUD), 340, 341
intravenous injection, 223
in vitro fertilization and embryo transfer (IVF-ET), 358
IUD. *See* intrauterine device

J

"Just Say No" slogan, 236

K

karyotype, 395, 396
ketamine hydrochloride, 272
ketone bodies, 167
K-hole, 272
kidneys, maternal changes, 381–382
Korsakoff's syndrome, 243
Krebs cycle, 167
kyphosis, 72

L

lactation, 390
lactovegetarians, 211
lacunae-like finger, 357
late-onset human traits, 402
lat muscles, 124
LDL-C. *See* low-density lipoprotein cholesterol
leptin hormone, 188
lesser vestibular glands, 323
Letterman, David, 63
leukoplakia, 263
Leuprolide (Lupron), 358
lex cesaria, 406
Leydig cells, 301
life expectancy of smokers, 249
lifespan factors, 46
lifestyle activity, 47–50
lightning, 137
limb buds, 362
limbic system, 225–226
linkage analysis, 397
linked alleles, 397
lipid metabolism, 167
lipids, 157–158
lipogenesis, 167

lipolysis, 167
liver disease, 243
load, effective training, 124
locus, 396
longevity, wellness and, 1–16
long-term abuse, 228–229
low back pain, 71
low-carbohydrate diets, 198
low-carbohydrate high-protein diets, 198
low-density lipoprotein cholesterol (LDL-C), 52
LSD. *See* lysergic acid diethylamide
luteal cells, 329
luteal phase, 332
lysergic acid diethylamide (LSD), 273

M

macrobiotics, 93
mainlining, 223
maintenance, SCM, 13, 14
major histocompatibility complex (MHC), 401
male, 309–310
male baby hypogonadism, 374
male condoms, 341
male infertility, causes of, 318
male reproductive system, 298
   accessory glands, 306–307
   anatomy of, 298–299
   disorders of, 316–318
   penis, 306
   physiologic functions
      embryonic development of, 308–309
      puberty, 309–311
   scrotum, 299–301
   testes, 301–306
male sexual response, 314–316
male urethra, 305
malnutrition, 243
mammary glands, 319, 324
mammary ligaments, 324
manipulative and body-based therapies in CAM, 91–93
mantle zone, spinal cord, 363
manual distraction, 8
MAO. *See* monoamine oxidase
marginal zone, spinal cord, 363
marijuana, 230
   addictive, 267–268
   behavioral effects on, 265
   description, 265
   effect in brain, 265–266
   health effects of, 266–267
   usage, 265